LONDON MATHEMATICAL SOCIETY LECTURE NOTE SERIES

Managing Editor: Professor J.W.S. Cassels, Department of Pure Mathematics and Mathematical Statistics, University of Cambridge, 16 Mill Lane, Cambridge CB2 1SB, England

The books in the series listed below are available from booksellers, or, in case of difficulty, from Cambridge University Press.

34 Representation theory of Lie groups, M.F. ATIYAH *et al*
36 Homological group theory, C.T.C. WALL (ed)
39 Affine sets and affine groups, D.G. NORTHCOTT
46 p-adic analysis: a short course on recent work, N. KOBLITZ
49 Finite geometries and designs, P. CAMERON, J.W.P. HIRSCHFELD & D.R. HUGHES (eds)
50 Commutator calculus and groups of homotopy classes, H.J. BAUES
57 Techniques of geometric topology, R.A. FENN
59 Applicable differential geometry, M. CRAMPIN & F.A.E. PIRANI
66 Several complex variables and complex manifolds II, M.J. FIELD
69 Representation theory, I.M. GELFAND *et al*
74 Symmetric designs: an algebraic approach, E.S. LANDER
76 Spectral theory of linear differential operators and comparison algebras, H.O. CORDES
77 Isolated singular points on complete intersections, E.J.N. LOOIJENGA
79 Probability, statistics and analysis, J.F.C. KINGMAN & G.E.H. REUTER (eds)
80 Introduction to the representation theory of compact and locally compact groups, A. ROBERT
81 Skew fields, P.K. DRAXL
82 Surveys in combinatorics, E.K. LLOYD (ed)
83 Homogeneous structures on Riemannian manifolds, F. TRICERRI & L. VANHECKE
86 Topological topics, I.M. JAMES (ed)
87 Surveys in set theory, A.R.D. MATHIAS (ed)
88 FPF ring theory, C. FAITH & S. PAGE
89 An F-space sampler, N.J. KALTON, N.T. PECK & J.W. ROBERTS
90 Polytopes and symmetry, S.A. ROBERTSON
91 Classgroups of group rings, M.J. TAYLOR
92 Representation of rings over skew fields, A.H. SCHOFIELD
93 Aspects of topology, I.M. JAMES & E.H. KRONHEIMER (eds)
94 Representations of general linear groups, G.D. JAMES
95 Low-dimensional topology 1982, R.A. FENN (ed)
96 Diophantine equations over function fields, R.C. MASON
97 Varieties of constructive mathematics, D.S. BRIDGES & F. RICHMAN
98 Localization in Noetherian rings, A.V. JATEGAONKAR
99 Methods of differential geometry in algebraic topology, M. KAROUBI & C. LERUSTE
100 Stopping time techniques for analysts and probabilists, L. EGGHE
101 Groups and geometry, ROGER C. LYNDON
103 Surveys in combinatorics 1985, I. ANDERSON (ed)
104 Elliptic structures on 3-manifolds, C.B. THOMAS
105 A local spectral theory for closed operators, I. ERDELYI & WANG SHENGWANG
106 Syzygies, E.G. EVANS & P. GRIFFITH
107 Compactification of Siegel moduli schemes, C-L. CHAI
108 Some topics in graph theory, H.P. YAP
109 Diophantine analysis, J. LOXTON & A. VAN DER POORTEN (eds)
110 An introduction to surreal numbers, H. GONSHOR
111 Analytical and geometric aspects of hyperbolic space, D.B.A. EPSTEIN (ed)
113 Lectures on the asymptotic theory of ideals, D. REES
114 Lectures on Bochner-Riesz means, K.M. DAVIS & Y-C. CHANG
115 An introduction to independence for analysts, H.G. DALES & W.H. WOODIN
116 Representations of algebras, P.J. WEBB (ed)
117 Homotopy theory, E. REES & J.D.S. JONES (eds)
118 Skew linear groups, M. SHIRVANI & B. WEHRFRITZ
119 Triangulated categories in the representation theory of finite-dimensional algebras, D. HAPPEL
121 Proceedings of *Groups - St Andrews 1985*, E. ROBERTSON & C. CAMPBELL (eds)
122 Non-classical continuum mechanics, R.J. KNOPS & A.A. LACEY (eds)
124 Lie groupoids and Lie algebroids in differential geometry, K. MACKENZIE
125 Commutator theory for congruence modular varieties, R. FREESE & R. MCKENZIE
126 Van der Corput's method of exponential sums, S.W. GRAHAM & G. KOLESNIK
127 New directions in dynamical systems, T.J. BEDFORD & J.W. SWIFT (eds)
128 Descriptive set theory and the structure of sets of uniqueness, A.S. KECHRIS & A. LOUVEAU
129 The subgroup structure of the finite classical groups, P.B. KLEIDMAN & M.W.LIEBECK
130 Model theory and modules, M. PREST

131 Algebraic, extremal & metric combinatorics, M-M. DEZA, P. FRANKL & I.G. ROSENBERG (eds)
132 Whitehead groups of finite groups, ROBERT OLIVER
133 Linear algebraic monoids, MOHAN S. PUTCHA
134 Number theory and dynamical systems, M. DODSON & J. VICKERS (eds)
135 Operator algebras and applications, 1, D. EVANS & M. TAKESAKI (eds)
136 Operator algebras and applications, 2, D. EVANS & M. TAKESAKI (eds)
137 Analysis at Urbana, I, E. BERKSON, T. PECK, & J. UHL (eds)
138 Analysis at Urbana, II, E. BERKSON, T. PECK, & J. UHL (eds)
139 Advances in homotopy theory, S. SALAMON, B. STEER & W. SUTHERLAND (eds)
140 Geometric aspects of Banach spaces, E.M. PEINADOR and A. RODES (eds)
141 Surveys in combinatorics 1989, J. SIEMONS (ed)
142 The geometry of jet bundles, D.J. SAUNDERS
143 The ergodic theory of discrete groups, PETER J. NICHOLLS
144 Introduction to uniform spaces, I.M. JAMES
145 Homological questions in local algebra, JAN R. STROOKER
146 Cohen-Macaulay modules over Cohen-Macaulay rings, Y. YOSHINO
147 Continuous and discrete modules, S.H. MOHAMED & B.J. MÜLLER
148 Helices and vector bundles, A.N. RUDAKOV *et al*
149 Solitons nonlinear evolution equations & inverse scattering, M. ABLOWITZ & P. CLARKSON
150 Geometry of low-dimensional manifolds 1, S. DONALDSON & C.B. THOMAS (eds)
151 Geometry of low-dimensional manifolds 2, S. DONALDSON & C.B. THOMAS (eds)
152 Oligomorphic permutation groups, P. CAMERON
153 L-functions and arithmetic, J. COATES & M.J. TAYLOR (eds)
154 Number theory and cryptography, J. LOXTON (ed)
155 Classification theories of polarized varieties, TAKAO FUJITA
156 Twistors in mathematics and physics, T.N. BAILEY & R.J. BASTON (eds)
157 Analytic pro-*p* groups, J.D. DIXON, M.P.F. DU SAUTOY, A. MANN & D. SEGAL
158 Geometry of Banach spaces, P.F.X. MÜLLER & W. SCHACHERMAYER (eds)
159 Groups St Andrews 1989 volume 1, C.M. CAMPBELL & E.F. ROBERTSON (eds)
160 Groups St Andrews 1989 volume 2, C.M. CAMPBELL & E.F. ROBERTSON (eds)
161 Lectures on block theory, BURKHARD KÜLSHAMMER
162 Harmonic analysis and representation theory for groups acting on homogeneous trees, A. FIGA-TALAMANCA & C. NEBBIA
163 Topics in varieties of group representations, S.M. VOVSI
164 Quasi-symmetric designs, M.S. SHRIKANDE & S.S. SANE
165 Groups, combinatorics & geometry, M.W. LIEBECK & J. SAXL (eds)
166 Surveys in combinatorics, 1991, A.D. KEEDWELL (ed)
167 Stochastic analysis, M.T. BARLOW & N.H. BINGHAM (eds)
168 Representations of algebras, H. TACHIKAWA & S. BRENNER (eds)
169 Boolean function complexity, M.S. PATERSON (ed)
170 Manifolds with singularities and the Adams-Novikov spectral sequence, B. BOTVINNIK
171 Squares, A.R. RAJWADE
172 Algebraic varieties, GEORGE R. KEMPF
173 Discrete groups and geometry, W.J. HARVEY & C. MACLACHLAN (eds)
174 Lectures on mechanics, J.E. MARSDEN
175 Adams memorial symposium on algebraic topology 1, N. RAY & G. WALKER (eds)
176 Adams memorial symposium on algebraic topology 2, N. RAY & G. WALKER (eds)
177 Applications of categories in computer science, M.P. FOURMAN, P.T. JOHNSTONE, & A.M. PITTS (eds)
178 Lower K- and L-theory, A. RANICKI
179 Complex projective geometry, G. ELLINGSRUD, C. PESKINE, G. SACCHIERO & S.A. STRØMME (eds)
180 Lectures on ergodic theory and Pesin theory on compact manifolds, M. POLLICOTT
181 Geometric group theory I, G.A. NIBLO & M.A. ROLLER (eds)
182 Geometric group theory II, G.A. NIBLO & M.A. ROLLER (eds)
183 Shintani zeta functions, A. YUKIE
184 Arithmetical functions, W. SCHWARZ & J. SPILKER
185 Representations of solvable groups, O. MANZ & T.R. WOLF
186 Complexity: knots, colourings and counting, D.J.A. WELSH
187 Surveys in combinatorics, 1993, K. WALKER (ed)
189 Locally presentable and accessible categories, J. ADAMEK & J. ROSICKY
190 Polynomial invariants of finite groups, D.J. BENSON
191 Finite geometry and combinatorics, F. DE CLERCK *et al*
192 Symplectic geometry, D. SALAMON (ed)
197 Two-dimensional homotopy and combinatorial group theory, C. HOG-ANGELONI W. METZLER & A.J. SIERADSKI (eds)
198 The algebraic characterization of geometric 4-manifolds, J.A. HILLMAN

London Mathematical Society Lecture Note Series. 184

Arithmetical Functions

An Introduction to Elementary and Analytic Properties of Arithmetic Functions and to some of their Almost-Periodic Properties

Wolfgang Schwarz
Johann Wolfgang Goethe-Universität, Frankfurt am Main

Jürgen Spilker
Albert-Ludwigs-Universität, Freiburg im Breisgau

CAMBRIDGE
UNIVERSITY PRESS

CAMBRIDGE UNIVERSITY PRESS
Cambridge, New York, Melbourne, Madrid, Cape Town, Singapore, São Paulo

Cambridge University Press
The Edinburgh Building, Cambridge CB2 2RU, UK

Published in the United States of America by Cambridge University Press, New York

www.cambridge.org
Information on this title: www.cambridge.org/9780521427258

First published 1994

A catalogue record for this publication is available from the British Library

ISBN-13 978-0-521-42725-8 paperback
ISBN-10 0-521-42725-8 paperback

Transferred to digital printing 2006

TO OUR WIVES DORIS and HELGA

Contents

Preface

This book is an attempt to provide an introduction to some parts, more or less important, of a subfield of elementary and analytic number theory, namely the field of arithmetical functions. There have been countless contributions to this field, but a general theory of arithmetical functions does not exist, as yet. Interesting questions which may be asked for arithmetical functions or "sequences" are, for example,

(1) the size of such functions,
(2) the behaviour in the mean,
(3) the local behaviour,
(4) algebraic properties of spaces of arithmetical functions,
(5) the approximability of arithmetical functions by "simpler" ones.

In this book, we are mainly concerned with questions (2), (4) and (5). In particular, we aim to present elementary and analytic results on mean-values of arithmetical functions, and to provide some insight into the connections between arithmetical functions, elements of functional analysis, and the theory of almost-periodic functions.

Of course, standard methods of number theory, such as the use of convolution arguments, TAUBERIAN Theorems, or detailed, skilful estimates of sums over arithmetical functions are used and given in our book. But we also concentrate on some of the methods which are not so common in analytic number theory, and which, perhaps for precisely this reason, have not been refined as have the above. In respect of applications and connections with functional analysis, our book may be considered, in part, as providing special, detailed examples of well-developed theories.

We do not presuppose much background in these theories; in fact, only the rudiments of functional analysis are required, and we are ever hopeful that mathematicians better acquainted with this theory may provide yet further applications. In the interest of speedy reference, some of the material is gathered in an appendix to the book.

Our book is not intended to be a textbook. In spite of this, some of the chapters could be used in courses on analytic number theory. Both authors quite independently, have led courses on arithmetical functions, and the present text is - in part - an extended version of these courses, in particular of lectures on arithmetical functions given in Frankfurt am Main and in Freiburg im Breisgau in the 1992 summer term to third- and fourth-year students.

Our book presupposes some knowledge of the theory of complex functions, some fundamental ideas and basic theorems of functional analysis and - on two or three occasions - a little knowledge of the theory of integration. Some acquaintance with elementary number theory would be helpful, and [sometimes] a good deal of patience in performing long and troublesome calculations is demanded.

An attentive reader will notice that certain techniques are used again and again, and this may be interpreted as a hint to develop these techniques independently into a universally applicable scheme. We have attempted to do this for one particular case in Chapter III, where a general theorem on "related" arithmetical functions is presented with some applications. The underlying idea is to replace multiplicative arithmetical functions by "related", simpler ones. Thus, it is often possible to reduce proofs of complicated theorems to simpler special cases.

The main topics of the book are the following:

- a study of elementary properties of arithmetical functions centered on the concept of convolution of arithmetical functions;
- a study of mean-values of arithmetical functions, in part by simple, in part by more complicated, elementary methods, and by analytic methods;
- the study of spaces of arithmetical functions defined as the completion of the spaces of even, respectively periodic, functions;
- the characterization of arithmetically interesting functions (in particular multiplicative functions) in these spaces: we discuss important theorems by P. D. T. A. ELLIOTT, H. DELANGE and H. DABOUSSI. The more general theorems of K.-H. INDLEKOFER [1980] will not be proved in this book, and INDLEKOFER's "New method in Probabilistic Number Theory" (1993) will not be dealt with.

The idea of presenting a book on arithmetical functions grew out of a series of papers presented by the authors, beginning in 1971. Our aim was to replace some number-theoretical techniques, as far as possible [for us] by "soft" techniques that are more common in mathematics. The papers mentioned and this book itself are an attempt to draw together number theory and some aspects of main-stream mathematics.

We have tried to write the book for third- and fourth-year mathematics students rather than for specialists in number theory, and we have tried to produce a book which is more or less self-contained. Exercises of varying degrees of difficulty are given at the end of most of the chapters. These are intended to provide material leading to greater insight into some of the methods used in number theory by applying these to more or less special problems.

"Pictures" of arithmetical functions give some impression of the behaviour of [well-known] arithmetical functions. Hopefully, visualization of arithmetical functions will be helpful for some readers; mathematics is abstract, but concrete, two-dimensional geometry can illustrate abstract ideas of arithmetical functions. Of course, those diagrams that illustrate inequalities are not intended to be proofs for these inequalities; proofs could be provided by any first-year student, by means of the TAYLOR formula, for example, or using similar techniques. However, in the authors' opinion, a diagram is both striking and convincing, while an exact proof is often tedious.

The relevant literature on the topics treated in the book is enormous, and we thus had to omit many important and interesting results from the bibliography. However, an extensive list of references is given, for example, in ELLIOTT's books.

There are many books which deal with arithmetical functions, some of which we list below, although we feel that there are distinct differences between these and our own book.

K. CHANDRASEKHARAN [1970]; his *Arithmetical Functions* deal with analytic aspects of prime number theory, making use of the properties of the RIEMANN zeta-function and of estimates of exponential sums,

P. J. McCARTHY's *Arithmetical Functions* [1986], and

R. SIVARAMAKRISHNAN, *Classical Theory of Arithmetical Functions,* [1989].

Texts covering topics similar to ours seem to be those by P. D. T. A. ELLIOTT [1979, 1980a], J. KUBILIUS [1964], and J. KNOPFMACHER [1975]. Many interesting aspects of a theory of arithmetical functions may be found in the books by G. H. HARDY & E. M. WRIGHT [1956], L. K. HUA [1982], and T. APOSTOL [1976].

Acknowledgements

The authors are solely responsible for any errors still remaining. However, they are grateful to Rainer TSCHIERSCH for generous assistance with some proof-reading.

The manuscript was written on an ATARI 1040 ST Computer, using the word processing system SIGNUM2 designed by F. SCHMERBECK, Application Systems, Heidelberg, which in the authors' opinion seems to be suitable for the preparation of mathematical texts.

The diagrams, intended to give some indication of the behaviour of arithmetical functions, were produced by the first author, using the PASCAL-SC system (A PASCAL Extension for Scientific Computation) created by U. KULISCH and his group at the University of Karlsruhe (version for the ATARI ST, A. TEUBNER Verlag); this said author alone is responsible for programming errors or inaccuracies.

The cartoons at the beginning of each chapter were designed by the artist ULRIKE DÜKER from Stegen, and we are grateful for her kind assistance.

For help with photographs and permission for publication we are grateful to many mathematicians and to some institutions (for example Miss VORHAUER (Ulm), The Mathematisches Forschungsinstitut Oberwolfach, The Librarian of the Trinity College, Cambridge and many others). Their help is acknowledged on page 367.

Finally we wish to thank the staff of Cambridge University Press, in particular DAVID TRANAH and ROGER ASTLEY, and an unknown lector for their help and patience during the preparation of this book.

Wolfgang Schwarz & Jürgen Spilker, August 1993

Notation

a) Standard Notation for Some Sets

$\mathbb{N} = \left\{ 1, 2, \dots \right\}$, the set of *positive integers*,

$\mathbb{N}_0 = \mathbb{N} \cup \{0\}$,

$\mathbb{Z} = \{ \dots, -2, -1, 0, 1, 2, \dots \}$, the set of *integers*,

$\mathbb{Q} = \left\{ \frac{a}{b};\ a, b \in \mathbb{Z},\ b \neq 0 \right\}$, the set of *rational numbers*,

$\mathbb{R}$ the set of *real numbers*,

$\mathbb{C}$ the set of *complex numbers*,

Re(z), Im(z) real [imaginary] part of $z \in \mathbb{C}$,

$B(a, r) = \{ z \in \mathbb{C};\ |z - a| < r \}$,

$\mathcal{P}$ the set of *prime numbers*,

[the letter p [in general] denotes a prime]

$\#(\mathcal{A})$ is the *number of elements* of the [finite] set $\mathcal{A}$,

$\mathbb{Z}/m\mathbb{Z}$ is the [additive] group of integers mod m,

$(\mathbb{Z}/m\mathbb{Z})^{\times}$ is the [multiplicative] group of *residue-classes* mod m, prime to m.

b) Divisibility, Factorization

gcd(a,b): *greatest common divisor* of [the integers] a and b; often also written as (a,b);

lcm[a.b]: *lowest common multiple* of [the integers] a and b,

$d|n$: d is a divisor of n,

$d \nmid n$: d does not divide n,

$p^k \| n$: p^k is the exact power of the prime p, dividing n: $p^k | n$, but $p^{k+1} \nmid n$,

$n = \prod_{p|n} p^{\nu_p(n)}$ gives the prime factor decomposition of n according to the fundamental theorem of elementary number theory,

P(n) denotes [sometimes] the maximal prime divisor of n.

c) Some Notation for Intervals and Functions on $\mathbb{R}$

$[\beta]$ denotes the greatest integer $\leq \beta$ (where β is real),

$\{\beta\} = \beta - [\beta]$ is the *fractional part* of the real number β,

$B_o(\beta) = \beta - [\beta] - \frac{1}{2}$ is the first BERNOULLI-polynomial [sometimes also denoted by $\psi(\beta)$ - we avoid this notation],

$[\alpha,\beta]$ closed interval $\{x \in \mathbb{R};\ \alpha \le x \le \beta\}$,

$]\alpha,\beta[$ open interval $\{x \in \mathbb{R};\ \alpha < x < \beta\}$,

$\mathrm{li}\ x = \mathrm{li}(2) + \int_2^x \{\log u\}^{-1}\, du$ is the *integral logarithm*,

$\mathscr{C} = 0.577\ 2\ldots$ EULER's constant,

$e_\alpha : n \mapsto \exp(2\pi i \cdot \alpha \cdot n)$,

$\Gamma(x) = \int_0^\infty t^{x-1} \cdot e^{-t}\, dt$, the *Gamma-Function*,

exp, log: *exponential* function and *logarithm* function,

$\mathcal{O}(\ \ldots\ )$, $o(\ \ldots\ )$ are LANDAU's symbols; $f = \mathcal{O}(g)$ is sometimes also written as $f \ll g$,

$f \sim g$ means $\lim_{x \to \infty} f(x)/g(x) = 1$.

d) Arithmetical Functions

$\Lambda(n) = \log p$, if $n = p^k$ is a power of a prime, $= 0$ otherwise,

$\pi(x) = \sum_{p \le x} 1$, the *number of primes* less or equal to x,

$\vartheta(x) = \sum_{p \le x} \log p$,

$\psi(x) = \sum_{n \le x} \Lambda(n)$,

$f * g$ *convolution* of f and g (see I.(2.1)),

$f^{-1(*)}$ convolution *inverse* of f,

$f' = \mu * f$,

$f_{(a+)}$, f_a, resp. $f_{(b\cdot)}$, f_b : *shifted functions* $n \mapsto f(a+n)$ resp. $f(b \cdot n)$,

μ, φ, τ, τ_k, σ, ω, Ω are standard notations for arithmetical functions (MÖBIUS, EULER, divisor function, higher divisor functions, sum of divisors, number of distinct prime divisors, number of prime divisors,

χ: in general a DIRICHLET character,

μ_r: characteristic function of the set of r-free numbers,

$c_r : n \mapsto \sum_{d \mid \gcd(n,r)} d \cdot \mu(r/d)$ denotes the RAMANUJAN sum,

$\varphi_f(p,s) = 1 + p^{-s} \cdot f(p) + p^{-2s} \cdot f(p^2) + \ldots$, p^{th} factor of an EULER product.

e) Some Dirichlet Series

$\zeta(s) = \sum_{n=1}^\infty n^{-s}$: RIEMANN Zeta-Function,

$L(s,\chi)$ DIRICHLET L-functions,

$\mathcal{D}(f,s) = \sum_{n=1}^\infty f(n) \cdot n^{-s}$ DIRICHLET series associated with $f: \mathbb{N} \to \mathbb{C}$,

f) Mean-Values, Densities, Fourier-Coefficients

$M(f) = \lim_{x\to\infty} x^{-1}\cdot\sum_{n\le x} f(n)$ denotes the *mean-value* of f if this limit exists,

$\delta(\mathcal{A}) = \lim_{x\to\infty} x^{-1}\cdot\sum_{n\le x, x\in\mathcal{A}} 1$ is the *density* of the set $\mathcal{A} \subset \mathbb{N}$,

$\delta^{-}(\mathcal{A})$, resp. $\delta_{-}(\mathcal{A})$, denotes the *upper*, resp. *lower*, *density* of the set $\mathcal{A}$

[$\lim_{x\to\infty}$ has to be replaced by $\limsup_{x\to\infty}$, resp. $\liminf_{x\to\infty}$],

$f^\wedge(\alpha) = M(f\cdot e_{-\alpha})$ is the α-th *Fourier-coefficient* of f (if it exists), also written $\hat{f}(\alpha)$,

$a_r(f) = \dfrac{M(f\cdot c_r)}{\varphi(r)}$ denotes the r-th RAMANUJAN-(FOURIER-)coefficient.

g) Norms, Spaces of Arithmetical Functions

$\|f\|_u = \sup_{n\in\mathbb{N}} |f(n)|$, supremum-norm,

$\|f\|_q = \left\{ \limsup_{x\to\infty} x^{-1}\cdot\sum_{n\le x} |f(n)|^q \right\}^{1/q}$,

$\mathcal{B}$, $\mathcal{D}$, $\mathcal{A}$: *Vector-spaces* of linear combinations of RAMANUJAN sums, exponential functions $e_{a/r}$, resp. e_α, $\alpha \in \mathbb{R}/\mathbb{Z}$,

$\mathcal{B}^u$, $\mathcal{D}^u$, $\mathcal{A}^u$: closures of $\mathcal{B}$, $\mathcal{D}$, $\mathcal{A}$ with respect to $\|.\|_u$,

$\mathcal{B}^q$, $\mathcal{D}^q$, $\mathcal{A}^q$: closures of $\mathcal{B}$, $\mathcal{D}$, $\mathcal{A}$ with respect to $\|.\|_q$,

$\mathcal{N}(\mathcal{L})$: null-space of $\mathcal{L}$,

$\Delta_{\mathcal{B}}$, $\Delta_{\mathcal{D}}$, $\Delta_{\mathcal{A}}$ are the GELFAND *maximal ideal spaces* of $\mathcal{B}^u$, $\mathcal{D}^u$, $\mathcal{A}^u$ (see IV),

$\langle f, g \rangle$: inner product in $\mathcal{A}^2$,

$\mathcal{C}(X)$ is the vector-space of *continuous complex-valued functions* defined on the [topological] space X,

$f^{\#}$: continuous image of $f \in \mathcal{B}^u$, $\mathcal{D}^u$, $\mathcal{A}^u$ in $\mathcal{C}(\Delta_{\mathcal{B}})$, $\mathcal{C}(\Delta_{\mathcal{D}})$, $\mathcal{C}(\Delta_{\mathcal{A}})$ under the GELFAND transform.

h) Some Special Series

$S_1(f) = \sum_p p^{-1}\cdot(f(p)-1)$, $\quad S_2(f) = \sum_p p^{-1}\cdot|f(p)-1|^2$,

$S_2'(f) = \sum_{p, |f(p)|\le 5/4} p^{-1}\cdot|f(p)-1|^2$, $S_{2,q}''(f) = \sum_{p,|f(p)|>5/4} p^{-1}\cdot|f(p)|^q$,

$S_{3,q}(f) = \sum_p \sum_{k\ge 2} p^{-k}\cdot|f(p^k)|^q$,

$\mathcal{E}_q = \{f$ multiplicative, $S_1(f)$, $S_2'(f)$, $S_{2,q}''(f)$, and $S_{3,q}(f)$ are convergent$\}$.

Chapter I

Tools from Number Theory

Abstract. *This preparatory chapter forms the basis of our presentation of arithmetical functions. Such techniques as* EULER*'s summation formula and partial summation are introduced, as is the notion of convolution. Examples of standard arithmetical functions are provided; some properties of* RAMANUJAN *sums are introduced, and* MÖBIUS *inversion formulae are proved. The* TURÁN-KUBILIUS *inequality is discussed, prior to its application in Chapter II, VI, and IX. Finally many results from prime number theory (including some results on characters and the prime number theorem in arithmetic progressions) are presented without proofs.*

I.1. PARTIAL SUMMATION

Assume for some given complex-valued function a: $n \mapsto a(n)$, defined on the set $\mathbb{N}_0$ of non-negative integers, that some knowledge concerning the sum $\sum_{n\le x} a(n)$ is available; then the problem of obtaining information about the sum

$$\sum_{n\le x} a(n)\cdot g(n),$$

where $g\colon [0,\infty[\to \mathbb{C}$ is a sufficiently smooth function (think of $g(n) = n^\alpha$ or $g(n) = \log n$, for example) can often easily be solved using partial summation. The following version of this technique is taken from PRACHAR [1957].

Theorem 1.1 (Partial summation). *Assume that a sequence a_n of complex numbers, and a sequence λ_n of real numbers, satisfying $\lambda_1 < \lambda_2 < \dots$, $\lambda_n \to \infty$, are given; then for any continuous, piecewise continuously differentiable function*

$$g : [\lambda_1, x] \to \mathbb{C},$$

the formula

$$\sum_{\lambda_n \le x} a_n \cdot g(\lambda_n) = g(x)\cdot \sum_{\lambda_n \le x} a_n - \int_{\lambda_1}^{x} \Big(\sum_{\lambda_n < u} a_n\Big)\cdot g'(u)\,du \tag{1.1}$$

is true.

Corollary. *If g satisfies the assumptions of Theorem 1.1 in $[\lambda_1,\infty[$, then*

$$\sum_{1\le n<\infty} a_n \cdot g(\lambda_n) = \lim_{x\to\infty}\Big(g(x)\cdot \sum_{\lambda_n\le x} a_n \Big) - \int_{\lambda_1}^{\infty}\Big(\sum_{\lambda_n<u} a_n\Big)\cdot g'(u)\,du,$$

if (at least two of) the limits exist.

The proof of Theorem 1.1 is nothing more than an application of the formula for partial integration for STIELTJES-integrals. An important special case is obtained, if all the reals λ_n are equal to n and if the a_n are all equal to 1: The heuristic expectation that the sum $\sum_{a<n\le b} g(n)$ is nearly equal to the corresponding integral $\int_a^b g(u)du$ is true, as is shown by the following theorem.

Theorem 1.2 (EULER's summation formula). *Assume that the complex-valued function* g *defined on the interval* [a,x] *is continuous and piecewise continuously differentiable. Define the* BERNOULLI *polynomial* $B_0(x)$ *by*[1)]

$$B_0(x) = x - [x] - \tfrac{1}{2}. \tag{1.2}$$

Then

$$\left\{ \begin{aligned} \sum_{a<n\le x} g(n) &= \int_a^x g(u)\,du + \int_a^x B_0(u)\cdot g'(u)\,du \\ &\quad - g(x)\cdot B_0(x) + g(a)\cdot B_0(a). \end{aligned} \right. \tag{1.3}$$

The proof begins with

$$\sum_{n\le x} g(n) = g(x)\cdot[x] + \int_1^x \left\{ B_0(u)\cdot g'(u) - u\cdot g'(u) + \tfrac{1}{2}g'(u) \right\} du,$$

and an easy, somewhat lengthy, computation gives (1.3).

Abbreviating the constant $\frac{1}{2} - \int_1^\infty B_0(u)\cdot u^{-2}\,du = 0.577\ 215 \ldots$ (EULER's constant) by $\mathscr{C}$, Theorem 1.2 implies

$$\Big| \sum_{n\le x} n^{-1} - (\log x + \mathscr{C}) \Big| \le x^{-1}, \text{ if } x \ge 1. \tag{1.4}$$

The proof of (1.4) is achieved by a direct application of EULER's summation formula. One additional idea is necessary: replace

$$\int_1^x B_0(u)\cdot g'(u)du \quad \text{by} \quad \left\{ \int_1^\infty - \int_x^\infty \right\} B_0(u)\cdot u^{-2}du$$

and estimate the second integral by $\frac{1}{2}\cdot x^{-1}$; the first integral from 1 to ∞ is convergent.

Similarly, replace $\int_1^n B_0(u)\cdot u^{-1}\,du$ (via partial integration) by

$$\int_1^u B_0(w)dw\cdot u^{-1}\Big|_1^n + \int_1^n u^{-2}\left(\int_1^u B_0(w)dw \right) du,$$

---

1) The common notation $\psi(x) = x - [x] - \frac{1}{2}$ will not be used in this book in order to avoid confusion with the function ψ from prime-number theory. [x] denotes for the largest integer less than or equal to x.

and estimate $\left|\int_1^u B_0(w)dw\right|$ by $^1/_8$. Then Theorem 1.2 gives

$$(1.5)\qquad |\log(n!) - n\cdot\log n + n - \tfrac{1}{2}\log n - D| \le \frac{1}{8}\cdot\frac{1}{n},$$

where D is some constant. Note that the same technique leads to summation formulae of higher order (see exercise 1).

A discrete version of partial summation, which can be proved by inverting the order of summation on the right-hand side, is provided by the following result.

Theorem 1.3 (ABEL's summation formula). *For given sequences* a_n, b_n *of complex numbers and integers* M, N *define*

$$A_n = \sum_{M<\nu\le n} a_\nu\ ,\quad M < n \le N.$$

Then

$$(1.6)\qquad \sum_{M<n\le N} a_n\cdot b_n = A_N b_N - \sum_{M<n\le N-1} A_n\cdot(b_{n+1} - b_n).$$

I.2. ARITHMETICAL FUNCTIONS, CONVOLUTION, MÖBIUS INVERSION FORMULA

In this section, simple, but useful, notations and results concerning the notion of *convolution* of arithmetical functions are given, together with a number of examples of standard arithmetical functions.

An *"arithmetical function"* is a map $f\colon \mathbb{N} \to \mathbb{C}$, defined on the set $\mathbb{N}$ of positive integers. The set $\mathbb{C}^{\mathbb{N}}$ of arithmetical functions becomes a $\mathbb{C}$-vector-space

$$(\ \mathbb{C}^{\mathbb{N}},\ +,\ \cdot\ )$$

by defining addition and scalar multiplication as follows:

$$(f+g)\colon n \mapsto f(n) + g(n),\quad \lambda\cdot f : n \mapsto \lambda\cdot f(n).$$

This vector-space is [obviously] not finite-dimensional. A pointwise definition of multiplication is near at hand. However, the existence of divisors of zero suggests the following definition of a different kind of multiplication, the *convolution* of two arithmetical functions f, g.

Definition. For arithmetical functions $f, g \in \mathbb{C}^{\mathbb{N}}$ the **convolution** $f*g$ is the map

$$f*g : n \mapsto \sum_{d|n} f(d)\cdot g\left(\frac{n}{d}\right), \tag{2.1}$$

where the summation is extended over all the [positive] divisors d of n.

Theorem 2.1. *Convolution is commutative, associative and distributive with respect to addition. Moreover, there is a unit element* ε, *defined by*

$$\varepsilon(n) = 1, \text{ if } n = 1, \quad \varepsilon(n) = 0 \text{ otherwise.} \tag{2.2}$$

Finally, $f * g = 0$ *implies* $f = 0$ *or* $g = 0$.

Proof. For the moment, denote the set of divisors of n by $\mathcal{T}(n)$. The map $\iota: d \mapsto {}^{n}/_{d}$ from $\mathcal{T}(n)$ onto $\mathcal{T}(n)$, associating with a divisor d the "complementary divisor" n/d, is bijective. Therefore,

$$(f*g)(n) = \sum_{d\in\mathcal{T}(n)} f(d)\cdot g\left(\frac{n}{d}\right) = \sum_{d\in\mathcal{T}(n)} f(\iota d)\cdot g\left(\frac{n}{\iota d}\right)$$

$$= \sum_{d\in\mathcal{T}(n)} f\left(\frac{n}{d}\right)\cdot g(d) = (g*f)(n).$$

Associativity of convolution is proved by careful handling of [finite] double sums:

$$\big(f*(g*h)\big)(n) = \sum_{d|n} f(d \cdot \sum_{t|(n/d)} g(t)\cdot h({}^{n}/_{dt}) = \Sigma_1, \tag{2.3'}$$

say, and

$$\big((f*g)*h\big)(n) = \sum_{D|n} \Big(\sum_{T|D} f(T)\cdot g({}^{D}/_{T}) \Big)\cdot h({}^{n}/_{D}) = \Sigma_2. \tag{2.3''}$$

But, putting $d\cdot t = D$ in (2.3'), one obtains

$$\Sigma_1 = \sum_{d\cdot t|n}\sum f(d)\cdot g(t)\cdot h({}^{n}/_{dt}) = \sum_{D|n} h({}^{n}/_{D}) \cdot \sum_{d|D} f(d)\cdot g({}^{D}/_{d}) = \Sigma_2 .$$

The remaining assertions are easily proved. □

Definition. An arithmetical function f is ***multiplicative*** if $f \neq 0$ and if for all pairs m, n of positive integers the condition $\gcd(m,n) = 1$ implies

$$f(m \cdot n) = f(m) \cdot f(n). \tag{2.4}$$

An arithmetical function g is ***additive*** if

$$g(m \cdot n) = g(m) + g(n) \tag{2.5}$$

whenever the greatest common divisor of m and n is 1.

The following **remarks** are trivial, but useful. Every multiplicative function satisfies $f(1) = 1$, since $f(n \cdot 1) = f(n) \cdot f(1)$, and an integer n may be chosen for which $f(n) \neq 0$. If g is additive, then $g(1) = 0$. If f_1 and f_2 are multiplicative, then the pointwise product $f_1 \cdot f_2$ is also multiplicative; if f is multiplicative and $f(n) \neq 0$ for every n, then $n \mapsto 1/f(n)$ is multiplicative. *A multiplicative [resp. additive] function* f *[resp.* g *] is determined by its values at the prime-powers:*

$$f\Big(\prod_p p^{\nu_p(n)} \Big) = \prod_p f\big(p^{\nu_p(n)} \big). \tag{2.6'}$$

$$g\Big(\prod_p p^{\nu_p(n)} \Big) = \sum_p g\big(p^{\nu_p(n)} \big). \tag{2.6''}$$

In these formulae, according to the fundamental theorem of arithmetic, an integer n is written uniquely as

$$n = \prod_p p^{\nu_p(n)}.$$

The set of all additive functions on $\mathbb{N}$ is a subspace of the vector-space $\mathbb{C}^{\mathbb{N}}$. Of course, this is not true for the set of multiplicative functions: a multiplicatively defined concept is seldom compatible with addition.

Examples of additive functions are $\omega(n) = \sum_{p|n} 1$ [the number of distinct prime factors of n], and $\Omega(n) = \sum_{p|n} \nu_p(n)$, if $n = \prod_p p^{\nu_p(n)}$ [the total number of prime factors of n], and the logarithm function $n \mapsto \log n$.

If f and g are multiplicative, then the pointwise product $f \cdot g$ is multiplicative, as mentioned above. The same is true for the convolution-product, as is shown by the following theorem.

Theorem 2.2. *Let* f, g *be arithmetical functions.*

(1) *If* f, g *are both multiplicative, then the convolution* $f*g$ *is multiplicative.*

(2) *If* f *satisfies* $f(1) \neq 0$, *then there exists a uniquely defined* *inverse* $f^{-1(*)}$ *with respect to convolution.*

(3) *If* f *is multiplicative, then* $f^{-1(*)}$ *is multiplicative.*

Proof. (1) Write $n = n_1 \cdot n_2$, where $\gcd(n_1,n_2) = 1$, and let d be a divisor d of n. Then $d = d_1 \cdot d_2$, where $d_\varkappa = \gcd(n_\varkappa, d)$; the greatest common divisors (d_1,d_2) and $(n_1/d_1,\ n_2/d_2)$ are equal to 1, and the map

$$\iota\colon \mathcal{T}(n) \to \mathcal{T}(n_1) \times \mathcal{T}(n_2),\ d \mapsto (d_1,d_2)$$

is bijective. Thus

$$(f*g)(n_1 \cdot n_2) = \sum_{d \mid n_1 \cdot n_2} f(d) \cdot g\left(\frac{n_1 \cdot n_2}{d}\right)$$

$$= \sum_{d_1 \mid n_1} \sum_{d_2 \mid n_2} f(d_1 \cdot d_2) \cdot g\left(\frac{n_1}{d_1} \cdot \frac{n_2}{d_2}\right) = (f*g)(n_1) \cdot (f*g)(n_2).$$

(2) The equation $X*f = \varepsilon$ (with an unknown arithmetical function X) is equivalent to the system of infinitely many linear equations

$$\sum_{d \mid n} f\left(\frac{n}{d}\right) \cdot X(d) = \begin{cases} 1, & \text{if } n = 1, \\ 0, & \text{if } n = 2, 3, \ldots, \end{cases}$$

in the unknowns X(d), d = 1, 2, ... The coefficient matrix is triangular:

$$\begin{bmatrix} f(1) & 0 & 0 & 0 & \cdots \\ f(2) & f(1) & 0 & 0 & \cdots \\ f(3) & 0 & f(1) & 0 & \cdots \\ f(4) & f(2) & 0 & f(1) & 0 \ . \\ \cdots & \cdots & \cdots & \cdots & \cdots \end{bmatrix},$$

and so the system of equations is recursively solvable, and the solution is unique.

(3) If f is multiplicative, f(1) = 1, then we proceed by induction. Let $n > 1$, $n = n_1 \cdot n_2$, $\gcd(n_1, n_2) = 1$. Assume that the relation

$$f^{-1(*)}(m_1 \cdot m_2) = f^{-1(*)}(m_1) \cdot f^{-1(*)}(m_2)$$

is true for all pairs (m_1, m_2), satisfying $m_1 \cdot m_2 < n$, $\gcd(m_1, m_2) = 1$. Then

$$f^{-1(*)}(n_1\cdot n_2) = -\sum_{\substack{d_1|n_1}} \sum_{\substack{d_2|n_2 \\ d_1d_2 \neq n_1n_2}} f^{-1(*)}(d_1\cdot d_2)\cdot f\Big(\frac{n_1}{d_1}\cdot\frac{n_2}{d_2}\Big)$$

$$= -\sum_{\substack{d_1|n_1 \\ d_1 \neq n_1}} \sum_{\substack{d_2|n_2 \\ d_2 \neq n_2}} \dots \; -\sum_{\substack{d_1|n_1 \\ d_1 = n_1}} \sum_{\substack{d_2|n_2 \\ d_2 \neq n_2}} \dots \; -\sum_{\substack{d_1|n_1 \\ d_1 \neq n_1}} \sum_{\substack{d_2|n_2 \\ d_2 = n_2}} \dots$$

$$= -f^{-1(*)}(n_1)\cdot f^{-1(*)}(n_2) + f^{-1(*)}(n_1)\cdot f^{-1(*)}(n_2) + f^{-1(*)}(n_1)\cdot f^{-1(*)}(n_2)$$

$$= f^{-1(*)}(n_1)\cdot f^{-1(*)}(n_2).$$ □

Remark 1. A simpler proof for (3) is possible: by (2), the convolution inverse $f^{-1(*)}$ exists. Define the multiplicative function g by

$$g(n) = \prod_{p^k \| n} f^{-1(*)}(p^k).$$

Then, for prime-powers p^k, g and $f^{-1(*)}$ have identical values, and so

$$(f*g)(p^k) = (f*f^{-1(*)})(p^k) = 0 = \varepsilon(p^k).$$

$f*g$ and ε are multiplicative, therefore $f^{-1(*)}$ equals g and is multiplicative.

Remark 2. If f is an arithmetical function satisfying $f(1) = 1$, then for primes p the following values are obtained:

$$(2.7)\quad \begin{cases} f^{-1(*)}(p) = -f(p), \\ f^{-1(*)}(p^2) = -f(p^2) + f^2(p), \\ f^{-1(*)}(p^3) = -f(p^3) + 2f(p^2)\cdot f(p) - f^3(p), \\ \dots\ \dots, \end{cases}$$

and so the values of the convolution-inverse of a multiplicative function for prime-powers (and, by (2.6'), for any integer) are recursively computable.

Denote the characteristic function of a subset $\mathcal{A}$ of the set $\mathbb{N}$ of positive integers by $1_{\mathcal{A}}$. Then, in the special case $\mathcal{A} = \mathbb{N}$, the function $1 = 1_{\mathbb{N}}$ is [trivially] multiplicative, and so its convolution inverse, the *MÖBIUS function*

$$(2.8)\qquad \mu = 1_{\mathbb{N}}^{-1(*)},$$

is multiplicative. The identical map $\mathbb{N} \to \mathbb{N}$, $n \mapsto n$, is denoted by $id_{\mathbb{N}}$ or id, and we define *EULER's function* φ by $\varphi = \mu * id_{\mathbb{N}}$, the *divisor function* by $\tau = 1 * 1$, and higher divisor functions by $\tau_m = 1 * 1 * \ldots * 1$ (m times). Some properties of these functions are listed below:

(a) By definition $1 * \mu = \varepsilon$, and so

$$(2.9) \qquad \sum_{d|n} \mu(d) = \begin{cases} 1, & \text{if } n = 1, \\ 0 & \text{otherwise.} \end{cases}$$

(b) The functions μ, φ, τ, τ_k are multiplicative.

(c) $\mu(p) = -1$ for every prime p, $\mu(p^k) = 0$, if $k \geq 2$.[2)]

(d) $\tau(p^k) = k+1$, $\quad \tau_m(p^k) = \binom{k+m-1}{m-1}$.

(e) $\varphi(p^k) = p^{k-1} \cdot (p-1) = p^k \cdot (1 - p^{-1})$,

$\varphi(n) = n \cdot \prod_{p|n} \left(1 - p^{-1}\right)$.

(f) $\varphi(n)$ is the number of residue-classes mod n, which are coprime with n.

Proof: Denoting this number by $\Phi(n)$ for the moment[3)], $\Phi(n/d) = \#(\mathscr{C}_d)$, where

$$\mathscr{C}_d = \{\nu \leq n, \ \gcd(\nu,n) = d\} = \{\nu \leq {}^n/_d, \ \gcd(\nu, {}^n/_d) = 1\}.$$

The union $\bigcup_{d|n} \mathscr{C}_d$ consists of all integers in $[1,\ldots,n]$, and is disjoint, and so $n = \sum_{d|n} \#(\mathscr{C}_d) = \sum_{d|n} \Phi(n/d)$. Therefore $id_{\mathbb{N}} = 1 * \Phi$; hence $\Phi = \mu * id_{\mathbb{N}} = \varphi$. □

The next theorem, known as the **MÖBIUS Inversion Formula**, does not express anything other than the fact that the functions $1_{\mathbb{N}}$ and μ are inverse to each other with respect to convolution.

Theorem 2.3. *Given two arithmetical functions* f *and* g, *the relations*

---

2) by induction from (a) and $1 + \mu(p) + \mu(p^2) + \ldots + \mu(p^k) = 0$.

3) For a [finite] set $\mathscr{D}$ we denote by $\#(\mathscr{D})$ the number of elements in $\mathscr{D}$.

(2.10') $$g(n) = \sum_{d|n} f(d) \text{ for every } n \in \mathbb{N},$$

and

(2.10") $$f(n) = \sum_{d|n} \mu(d) \cdot g\left(\frac{n}{d}\right) \text{ for every } n \in \mathbb{N}$$

are equivalent.

Alternatively, these relations may be expressed in the following way.

Theorem 2.3'. *The map* $T: \mathbb{C}^{\mathbb{N}} \to \mathbb{C}^{\mathbb{N}}$, *defined by*

(2.11') $$Tf : n \mapsto \sum_{d|n} f(d),$$

is linear and bijective, with the inverse map T^{-1},

(2.11") $$T^{-1}f : n \mapsto \sum_{d|n} \mu(d) \cdot f\left(\frac{n}{d}\right).$$

This follows from $Tf = 1 * f$ and $T^{-1}f = \mu * f$ by the associativity of convolution. □

The results just given may be generalized: let $h: \mathbb{N} \to \mathbb{C}$ be a *completely multiplicative* function $h: \mathbb{N} \to \mathbb{C}$ [this means, that $h(1) = 1$ and $h(m \cdot n) = h(m) \cdot h(n)$, thus h is a non-trivial homomorphism from the semi-group $(\mathbb{N}, \cdot)$ into the semi-group $(\mathbb{C}, \cdot)$]. *Then the map* $T_h: \mathbb{C}^{\mathbb{N}} \to \mathbb{C}^{\mathbb{N}}$, *defined by*

(2.12') $$T_h f : n \mapsto \sum_{d|n} h(d) \cdot f(n/d),$$

is linear and bijective, with an inverse map T_h^{-1} *given by*

(2.12") $$T_h^{-1} g : n \mapsto \sum_{d|n} \mu(d) \cdot h(d) \cdot g(n/d),$$

which follows from $h^{-1(*)} = \mu \cdot h$ (this is the pointwise product!), $T_h f = h * f$, and $T_h^{-1} g = (\mu \cdot h) * g$. □

A second kind of MÖBIUS inversion formula is now presented as follows.

Theorem 2.4. *Let* $h: \mathbb{N} \to \mathbb{C}$ *be completely multiplicative and consider the vector-space*

$$\mathbb{C}^{[1,\infty[} = \{ F: [1, \infty[\rightarrow \mathbb{C} \}.$$

Then the map $\mathcal{T}_h : \mathbb{C}^{[1,\infty[} \rightarrow \mathbb{C}^{[1,\infty[}$, *defined by*

(2.13') $$\mathcal{T}_h F : x \mapsto \sum_{n \le x} h(n) \cdot F(x/n), \ (x \ge 1),$$

is linear and bijective, its inverse being

(2.13") $$\mathcal{T}_h^{-1} F : x \mapsto \sum_{n \le x} \mu(n) \cdot h(n) \cdot F(x/n), \ (x \ge 1).$$

Proof. Obviously, $\mathcal{T}_h$ is linear. If $\mathcal{T}_h F = 0$, then for any $x \ge 1$,

$$0 = \sum_{n \le x} \mu(n) \cdot h(n) \cdot \mathcal{T}_h F(x/n) = \sum_{n \le x} \mu(n) \cdot h(n) \cdot \sum_{k \le (x/n)} h(k) \cdot F\left(x/nk\right),$$

and, putting $t = k \cdot n$, this double sum

$$= \sum_{t \le x} F(x/t) \cdot h(t) \cdot \sum_{n | t} \mu(n) = F(x) \cdot h(1),$$

and so $F = 0$. Hence $\mathcal{T}_h$ is injective. The surjectivity of $\mathcal{T}_h$ is proved similarly. □

An application of this result is given next.[4)]

Corollary 2.5. *If* $x \ge 1$, *then*

(2.14) $$\left| \sum_{n \le x} n^{-1} \cdot \mu(n) \right| \le 1 + x^{-1}.$$

Proof. Choose $h = 1$, $F = 1$ in Theorem 2.4. Then $\mathcal{T}_h F(x) = [x]$ and

$$1 = \sum_{n \le x} \mu(n) \cdot \left[x/n \right] = x \cdot \sum_{n \le x} n^{-1} \cdot \mu(n) + \sum_{n \le x} \mu(n) \cdot \vartheta_n,$$

where $|\vartheta_n| \le 1$. The obvious observation $\left| \sum_{n \le x} \mu(n) \cdot \vartheta_n \right| \le x$ gives the result. □

To obtain an impression of the erratic behaviour of the μ-function, see Figure I.1, which gives values of the MÖBIUS μ-function in the range $1 \le n \le 298$.

---

4) We remark that inequality (2.14), obtained elementarily, is rather weak. From the prime number theorem (see I.6) it follows that $\sum n^{-1} \cdot \mu(n)$ is convergent to zero (see also Figure 2, next page).

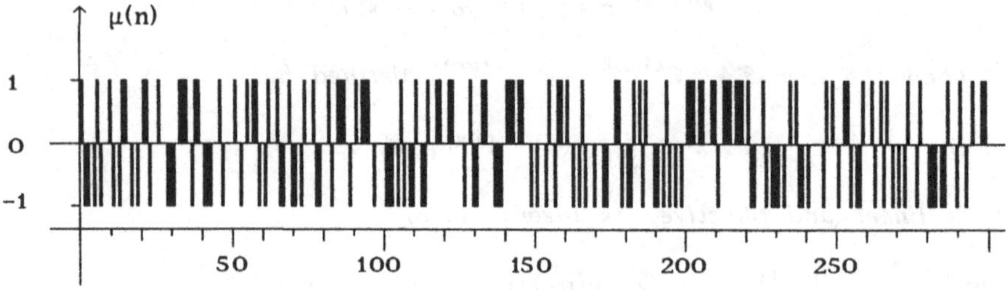

Figure I.1, Values of the Möbius Function

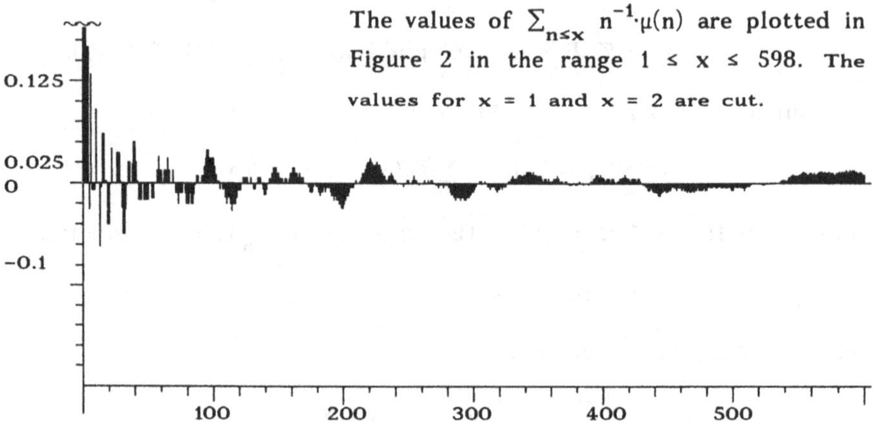

The values of $\sum_{n\le x} n^{-1}\cdot\mu(n)$ are plotted in Figure 2 in the range $1 \le x \le 598$. The values for $x = 1$ and $x = 2$ are cut.

Figure I.2, Values of $\sum_{n\le N} n^{-1}\cdot\mu(n)$

Finally, we give some results on the divisor function τ. For any $\varepsilon > 0$ there is a constant $C(\varepsilon)$, for which the estimate

$$\tau(n) \le C(\varepsilon) \cdot n^{\varepsilon}$$

holds; this is proved, in HARDY-WRIGHT [1956], Theorem 315, for example. A more general result is given in the same work, § 18.1 (see also SCHWARZ [1987a]), as follows

Theorem 2.6. *If* f *is multiplicative and satisfies* $\lim_{p^k \to \infty} f(p^k) = 0$, *then*

$$\lim_{n \to \infty} f(n) = 0.$$

Proof. Given $\varepsilon > 0$, there is a constant $N(\varepsilon)$ such that $|f(p^k)| < \varepsilon$, if $p^k \ge N(\varepsilon)$. In particular, $|f(p^k)| < 1$ if $p^k \ge N(1)$. Therefore there is some constant γ, independent of ε, p, k, for which $|f(p^k)| \le \gamma$.

The number of integers, composed entirely from prime-powers $p^\ell \le N(\varepsilon)$, is finite, and so any of these numbers is less than some $N^*(\varepsilon)$. If $n > N^*(\varepsilon)$, then there is some prime-power $p^\ell > N(\varepsilon)$, which divides n. Denote by NPP(ε) the number of prime-powers below N(ε). The function f being multiplicative, we obtain

$$|f(n)| \le \gamma^{\mathrm{NPP}(1)} \cdot 1 \cdot \varepsilon, \text{ if } n > N^*(\varepsilon). \qquad \square$$

To obtain an impression of the behaviour of the divisor-function, this function is plotted in the range $1 \le n \le 298$ (see Figure I.3, with the mean - value plotted inversely), and in the range $10001 \le n \le 10598$ (see Figure I.4).

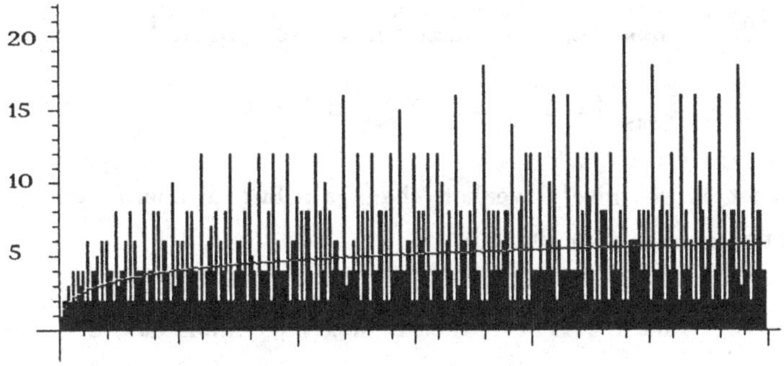

Figure I.3. The Divisor Function in the range $1 \le n \le 298$.

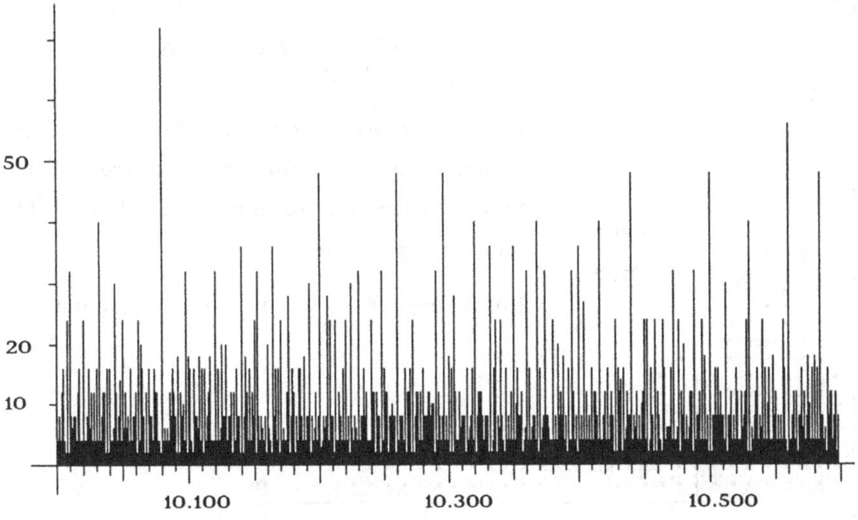

Figure I.4. The Divisor Function in the range $10001 \le n \le 10598$

Theorem 2.7. *The following asymptotic formulae are true for the divisor-function* τ:

(a) $\sum_{n\le x} \tau(n) = x\cdot \log x + (2\mathcal{C}-1)\cdot x + \mathcal{O}(\sqrt{x})$,

(b) $\sum_{n\le x} n^{-1}\cdot\tau(n) = \frac{1}{2}\log^2 x + 2\mathcal{C}\cdot\log x + K + \mathcal{O}\left(1/\sqrt{x}\right)$,

with some constant K,

(c) $\sum_{n\le x} \tau^{\ell}(n) \le C(\ell)\cdot x\cdot(1+\log x)^{2^{\ell}-1}$ *for* $\ell = 1, 2, \ldots$.

Proof. (a) The simple attempt of interchanging the order of summation,

$$\sum_{n\le x}\tau(n) = \sum_{n\le x}\sum_{d|n} 1 = \sum_{d\le x}\sum_{n\le x,\ n\equiv 0 \bmod d} 1$$

$$= \sum_{d\le x}[x/d] = x\cdot\sum_{d\le x} d^{-1} + \mathcal{O}(x),$$

gives a result that is definitely weaker than (a). But a useful trick, due to DIRICHLET, proves formula (a):

$$\sum_{n\le x}\tau(n) = \sum_{d\cdot m\le x} 1 = \sum_{m\le B}\ \sum_{d\le x/m} 1 + \sum_{d\le x/B}\ \sum_{m\le x/d} 1 - \sum_{m\le B}\ \sum_{d\le x/B} 1;$$

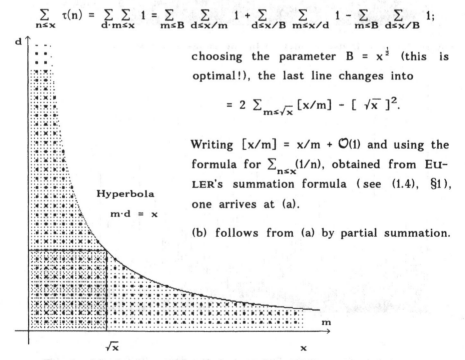

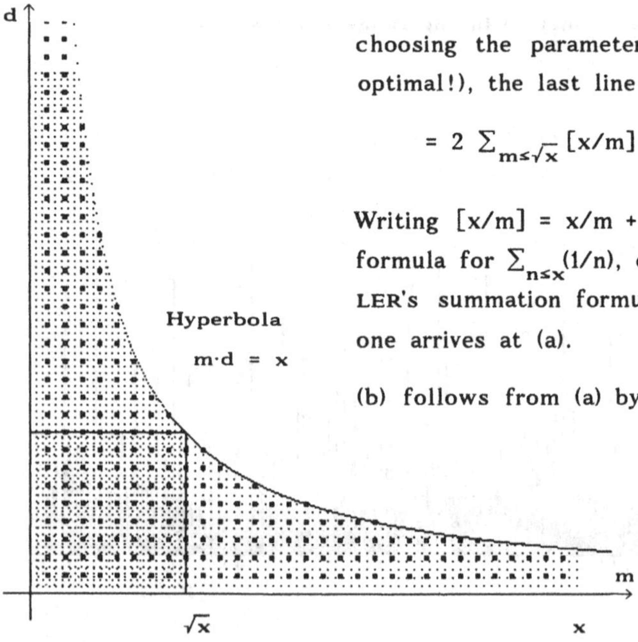

Figure I.5. Lattice points below m·d = x

choosing the parameter $B = x^{\frac{1}{2}}$ (this is optimal!), the last line changes into

$$= 2\sum_{m\le\sqrt{x}}[x/m] - [\sqrt{x}\,]^2.$$

Writing $[x/m] = x/m + \mathcal{O}(1)$ and using the formula for $\sum_{n\le x}(1/n)$, obtained from EULER's summation formula (see (1.4), §1), one arrives at (a).

(b) follows from (a) by partial summation.

(c) is proved by induction; the assertion is true for $\ell = 1$. Assume that (c) is true for ℓ. Then, by partial summation, for $x \geq 1$,

$$\sum_{n \leq x} n^{-1} \cdot \tau^{\ell}(n) \leq C(\ell) \cdot (1 + \log x)^{2^{\ell}}. \tag{2.15}$$

Using multiplicativity and $\tau(p^k) = k+1$, we see that for every pair (m,n) of positive integers the inequality

$$\tau(m \cdot n) \leq \tau(m) \cdot \tau(n)$$

holds [so τ is *"sub-multiplicative"*]. Therefore, writing $n = m \cdot d$, we obtain

$$\sum_{n \leq x} \tau^{\ell+1}(n) = \sum_{n \leq x} \tau^{\ell}(n) \sum_{d|n} 1 = \sum_{d \leq x} \sum_{m \leq x/d} \tau^{\ell}(md)$$

$$\leq \sum_{d \leq x} \tau^{\ell}(d) \cdot \sum_{m \leq x/d} \tau^{\ell}(m).$$

Using the induction hypothesis for the sum over m and then (2.15), a short calculation gives the assertion. □

I.3. PERIODIC FUNCTIONS, EVEN FUNCTIONS, RAMANUJAN SUMS

Definition. Let r be a positive integer and p a prime. An arithmetical function f is called

r-*periodic*, if $f(n+r) = f(n)$ for every positive integer n,

r-*even*, if $f(n) = f(\gcd(n,r))$ for every positive integer n,

p-*fibre-constant*, if $f(n) = f(p^{\nu_p})$,

where the exponents ν_p are taken from the prime factor decomposition

$$n = \prod_{q^{\nu_q} \| n} q^{\nu_q}.$$

f is termed *periodic* [resp. *even*], if there is some r for which f is r-periodic [resp. r-even].

Obviously, an r-even function is r-periodic. Standard examples of r-periodic functions are the exponential functions $e(\alpha)$, where $\alpha = \frac{a}{r}$, $a \in \mathbb{Z}$, $r \in \mathbb{N}$, and where

(3.1) $$e(\alpha) : n \mapsto \exp(2\pi i \cdot \alpha \cdot n).$$

These exponential functions satisfy the following **orthogonality relations**:

Let $d|r$, $t|r$, *and* $\gcd(a,d) = \gcd(b,t) = 1$. *Then*

$$\frac{1}{r}\sum_{m=1}^{r} e\left(\frac{a}{d}\cdot m\right)\cdot e\left(\frac{b}{t}\cdot m\right) = \begin{cases} 0, & \text{if } d \neq t \quad \textit{or} \quad a+b \not\equiv 0 \bmod d \\ 1, & \text{if } d = t \quad \textit{and} \ a+b \equiv 0 \bmod d. \end{cases}$$

The sum on the left-hand side is

$$r^{-1}\cdot\sum_{1\le m\le r} e\left(m \cdot \frac{at+bd}{dt} \right) = \begin{cases} 0, & \text{if } d\cdot t \nmid (at+bd), \\ 1 & \text{otherwise.} \end{cases}$$

The **RAMANUJAN sum** c_r is a special exponential sum:

(3.2) $$c_r(n) = \sum_{1\le a\le r,\ \gcd(a,r)=1} \exp\left(2\pi i \cdot \frac{a}{r}\cdot n\right).$$

Important properties of the RAMANUJAN sums are given in the following theorem.

Theorem 3.1. *RAMANUJAN sums have the following properties.*

(a) *The RAMANUJAN sum* c_r *is r-periodic.*

(b) $c_r(n) = \sum_{d|\gcd(r,n)} d \cdot \mu\left(\frac{r}{d}\right)$.

(c) *The RAMANUJAN sum* c_r *is r-even.*

(d) *For any fixed* n *the map* $r \mapsto c_r(n)$ *is multiplicative.*

(e) *The RAMANUJAN sums satisfy the following* ***orthogonality relations:***

(3.3) $$\begin{cases} \textit{If } t|r \textit{ and } d|r, \textit{ then} \\ \frac{1}{r}\sum_{m=1}^{r} c_d(m)\cdot c_t(m) = \begin{cases} 0, & \textit{if } d \neq t, \\ \varphi(d), & \textit{if } d = t. \end{cases} \end{cases}$$

Proof. (a) is obvious.
(b) Using $1 * \mu = \varepsilon$, the value $c_r(n)$ is

$$c_r(n) = \sum_{1\le a\le r} e\left(\frac{a}{r}\cdot n\right)\cdot \sum_{d|a,\ d|r} \mu(d)$$

$$= \sum_{d|r} \mu(d) \cdot \sum_{1\le a\le r, a\equiv 0 \bmod d} e\left(\frac{a}{r}\cdot n\right);$$

the latter part of the equation above is equal to $\sum_{1\le b\le r/d} e\left(\frac{b}{r/d}\cdot n\right)$, and this expression is 0 if $(r/d) \nmid n$, and is equal to r/d otherwise. Therefore,

$$c_r(n) = \sum_{d|r,\ (r/d)|n} \mu(d)\cdot (r/d) = \sum_{t|r,\ t|n} \mu(r/t)\cdot t.$$

(c) The r-evenness of the map $n \mapsto c_r(n)$ is obvious from (b).
(d) In (b) $c_r(n) = (\mu * F_n)(r)$ was obtained, where

$$F_n(t) = \begin{cases} t, & \text{if } t|n, \\ 0 & \text{otherwise.} \end{cases}$$

The functions F_n and μ are multiplicative, therefore the same is true for the convolution $\mu * F_n$ (see Theorem 2.2).
(e) By the definition (3.2) of the RAMANUJAN sum the proof of the orthogonality relations (3.3) is reduced to an application of the corresponding relations for the exponential functions. More explicitely:

$$\frac{1}{r}\sum_{m=1}^{r} c_d(m)\cdot c_t(m) = \frac{1}{r}\cdot\sum_{m\le r}\ \sum_{\substack{a\le d\\ \gcd(a,d)=1}} e\left(\frac{a}{d}m\right)\cdot\sum_{\substack{b\le t\\ \gcd(b,t)=1}} e\left(\frac{b}{t}m\right)$$

$$= \sum_{\substack{a\le d\\ \gcd(a,d)=1}}\ \sum_{\substack{b\le t\\ \gcd(b,t)=1}} \frac{1}{r}\cdot\sum_{m\le r} e\left(\frac{a}{d}m\right)\cdot e\left(\frac{b}{t}m\right)$$

$$\begin{cases} = 0, \text{ if } d \ne t, \text{ or } a+b \not\equiv 0 \bmod d, \\ = \sum_{\substack{a\le d\\ \gcd(a,d)=1}}\ \sum_{\substack{b\le d\\ \gcd(b,d)=1\\ a+b\equiv 0 \bmod d}} 1 = \varphi(d) \text{ otherwise.} \end{cases}$$

The reason for the last equality-sign is that for every a there is exactly one b, satisfying $a + b \equiv 0 \bmod d$. □

In Chapter IV we shall need some special values of $c_r(n)$. If the index r is a prime power p^k, then, as is easily verified,

$$(3.4) \qquad c_{p^k}(n) = \begin{cases} p^k - p^{k-1}, & \text{if } p^k | n, \\ -p^{k-1}, & \text{if } p^{k-1} \| n, \\ 0, & \text{if } p^{k-1} \nmid n. \end{cases}$$

Figures I.6 and I.7 illustrate the periodic behaviour of RAMANUJAN sums rather instructively. The functions c_r with index $r = 30$ and $r = 210$, resprectively, are plotted in the range $1 \le n \le 299$.

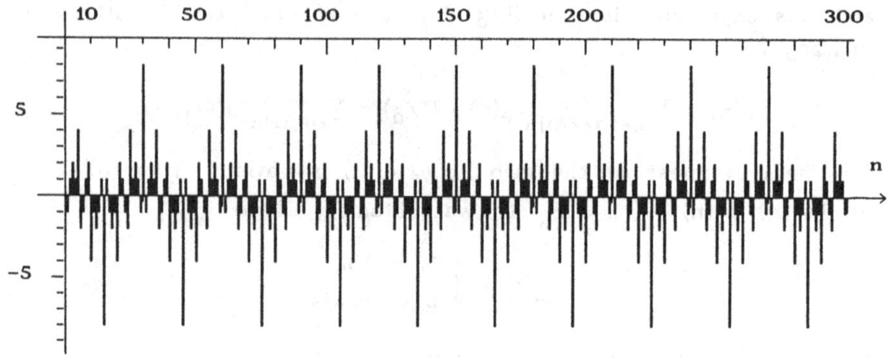

Figure I.6: RAMANUJAN sum c_{30} in the range $1 \le n \le 299$

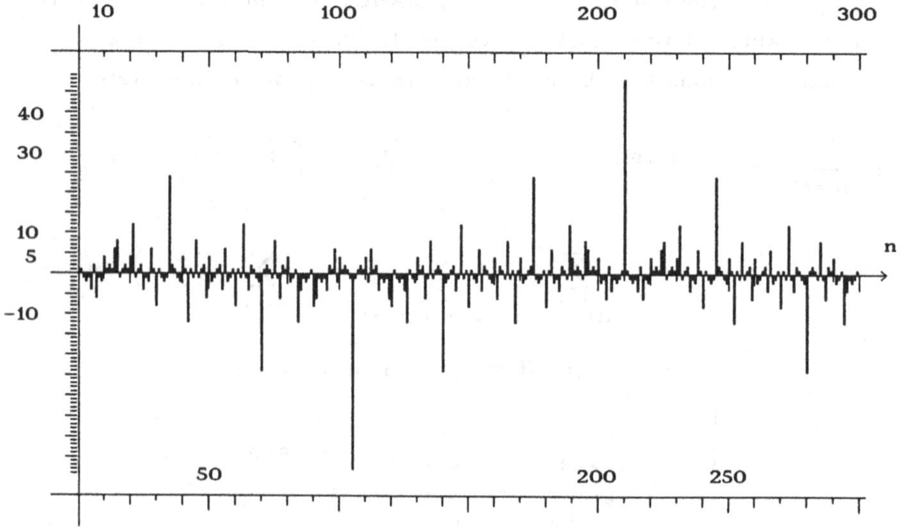

Figure I.7

RAMANUJAN sum c_{210} in the range $1 \le n \le 299$

Other **examples** of r-even functions are

$$g_d : n \mapsto \begin{cases} 1, & \text{if } \gcd(n,r) = d, \\ 0, & \text{otherwise.} \end{cases}$$

The functions g_d, where $d|r$, as well as the RAMANUJAN sums c_d, where $d|r$, form a basis of the $\mathbb{C}$-vector-space of r-even functions (this space is of dimension $\tau(r)$). This is obvious for the functions g_d, and for the RAMANUJAN sums the assertion easily follows from the orthogonality relations.

The KRONECKER-LEGENDRE symbol $\left(\frac{a}{p}\right)$ is equal to zero if $p|a$; otherwise, if $p\nmid a$, it is equal to 1 or -1 if a is a quadratic residue [resp. non-residue] modulo the prime p.

This function $a \mapsto \left(\frac{a}{p}\right)$ is a completely multiplicative, p-periodic function (considered as a function of the "nominator" a). For a thorough investigation of the LEGENDRE symbol as a function of its "denominator" p, see, for example, H. HASSE [1964].

Generally, given a character χ of the group $(\mathbb{Z}/m\mathbb{Z})^{\times}$ of residue-classes prime to m, in other words, given a group-homomorphism

$$\chi : (\mathbb{Z}/m\mathbb{Z})^{\times} \to (\mathbb{C}, \cdot),\ |\chi(n)| = 1,$$

we obtain a completely multiplicative, m-periodic function

$$\chi : \mathbb{N} \to \{ z \in \mathbb{C},\ |z| = 1 \text{ or } z = 0 \},$$

defined by $\chi(n) = \chi(n \bmod m)$ if $\gcd(n,m) = 1$, and $\chi(n) = 0$ otherwise.

I.4. THE TURÁN-KUBILIUS INEQUALITY

An additive function $w: \mathbb{N} \to \mathbb{C}$ is called *strongly additive* if the values of w at prime-powers are restricted by the condition

$$w(p^k) = w(p), \text{ if } k = 1, 2, \dots .$$

In 1934, Paul TURÁN [1934] discovered the following inequality for the strongly additive function $n \mapsto \omega(n)$, the number of prime divisors of n:

$$\sum_{n\le x} (\omega(n) - \log\log x)^2 \le c \cdot x \cdot \log\log x \tag{4.1}$$

with some constant c. P. TURÁN used this result to reprove HARDY and RAMANUJAN's theorem [1917] that $\omega(n)$ has normal order loglog n. Inequality (4.1) was generalized by J. KUBILIUS [1964] to additive functions, and later "dualized" by P. D. T. A. ELLIOTT [1979]. If w is strongly additive, then

$$\sum_{n\le x} w(n) = \sum_{n\le x} \sum_{p|n} w(p) = \sum_{p\le x} w(p) \cdot [x/p],$$

and so w(n) is, on average, heuristically approximate to $\sum_{p\le x} p^{-1}\cdot w(p)$. The so-called TURÁN-KUBILIUS inequality gives an estimate for the difference of the values of the function minus the "expectation":

$$\left| w(n) - \sum_{p\le x} p^{-1}\cdot w(p) \right|$$

in mean square.

In its general form the TURÁN-KUBILIUS inequality has often been applied to the study of arithmetical functions. We use this inequality in Chapter VII in order to approximate functions in $\mathcal{B}^1$ by even functions and to outline criteria for additive and multiplicative functions to belong to $\mathcal{B}^q$.

For an arithmetical function w (and $x > 0$) we define the expressions

(4.2) $$A(x) = \sum_{p^k\le x} p^{-k}\cdot w(p^k),$$

(4.3) $$E(x) = \sum_{p^k\le x} p^{-k}\cdot w(p^k)\cdot\left(1-\frac{1}{p}\right)$$

and

(4.4) $$D^2(x) = \sum_{p^k\le x} p^{-k}\cdot |w(p^k)|^2.$$

Theorem 4.1 (Turán–Kubilius inequality).

There exist constants C_1, C_2 *with the property that for every* $x \ge 2$ *and for any additive function* w *the inequalities*

(4.5) $$x^{-1}\sum_{n\le x} \left|w(n) - A(x)\right|^2 \le C_1\cdot D^2(x)$$

and

(4.6) $$x^{-1}\sum_{n\le x} \left|w(n) - E(x)\right|^2 \le C_2\cdot D^2(x)$$

are true. In fact, it is possible to have $C_1 = 30$, $C_2 = 20$.

Remark. If w is *strongly* additive, then the CAUCHY–SCHWARZ inequality gives

$$| \, A(x) - \sum_{p\le x} p^{-1}\cdot w(p) \, | \le \sum_{p^k\le x, k\ge 2} p^{-k}\cdot|w(p^k)| \le 2 \sum_{p\le x} p^{-2}\cdot|w(p)|$$

$$\le \Big(\sum_{p\le x} p^{-1}\cdot|w(p)|^2\Big)^{\frac12}\cdot 2\Big(\sum_{p\le x} p^{-3}\Big)^{\frac12} \le \Big(\sum_{p\le x} p^{-1}\cdot|w(p)|^2\Big)^{\frac12},$$

and

$$D^2(x) \le 2 \sum_{p\le x} p^{-1}\cdot|w(p)|^2.$$

Therefore, from (4.5) we deduce

$$(4.7) \qquad x^{-1}\cdot \sum_{n\le x} \Big| \, w(n) - \sum_{p\le x} p^{-1}\cdot w(p) \, \Big|^2$$

$$\le 2\, x^{-1} \sum_{n\le x} \big|w(n) - A(x)\big|^2 + 2\, \Big|A(x) - \sum_{p\le x} p^{-1}\cdot w(p)\Big|^2$$

$$\le (\, 4C_1 + 2\,)\cdot \sum_{p\le x} p^{-1}\cdot|w(p)|^2$$

for every strongly additive function w. Note that the constants are far from being the best possible.

Proof of Theorem 4.1. Inequality (4.5) is a consequence of (4.6). By appropriate application of the CAUCHY–SCHWARZ inequality we obtain

$$\Big(\sum_{n\le x} |w(n) - E(x)|\Big)^2 \le x\cdot\sum_{n\le x} |w(n) - E(x)|^2 \le C_2\cdot x^2\cdot D^2(x),$$

and

$$x^{-1}\cdot\sum_{n\le x} |w(n)-A(x)|^2$$

$$\le x^{-1}\cdot\sum_{n\le x} |w(n)-E(x)|^2 + 2x^{-1}\sum_{n\le x} |w(n)-E(x)| \cdot\sum_{p^k\le x} p^{-k-1}\cdot|w(p^k)|$$

$$+ \Big(\sum_{p^k\le x} p^{-k-1}\cdot|w(p^k)|\Big)^2 \le (\,C_2 + 2\,C_2^{\frac12} + 1\,)\cdot D^2(x).$$

We follow the **proof** given in ELLIOTT [1979], p. 148. There is another proof, due to ELLIOTT [1970], which uses the "large sieve". First, the assertion for complex-valued functions is reduced to the corresponding assertion for real-valued functions in an obvious manner, and then the assertion for these functions is reduced to a problem concerning non-negative functions.

(i) Assume initially that w is real-valued and non-negative. Then

$$S = \sum_{n\le N} \big(w(n)-E(N)\big)^2 = \sum_{n\le N} w^2(n) - 2\,E(N)\cdot\sum_{n\le N} w(n) + N\cdot E^2(N)$$

$$= S_1 - 2\,S_2 + N \cdot E^2(N)\ , \text{ say.}$$

First,

$$S_1 = \sum_{n\le N} w^2(n) = \sum_{n\le N} \sum_{p^k\| n} w(p^k) \cdot \sum_{q^\ell \| n} w(q^\ell)$$

$$= \sum_{p^k \le N} w^2(p^k) \cdot \sum_{\substack{n\le N\\ p^k \| n}} 1 + \sum_{\substack{p^k q^\ell \le N\\ p\ne q}} w(p^k) \cdot w(q^\ell) \cdot \#(N),$$

where, for distinct primes p and q,

$$\#(N) = \sum_{n\le N,\ p^k\| n,\ q^\ell \| n} 1$$

counts the number of integers, which are exactly divisible by the prime powers p^k, q^ℓ. Then

$$\text{(4.8)}\quad \begin{cases} \#(N) = \left[N/p^k\cdot q^\ell\right] - \left[N/p^{k+1}\cdot q^\ell\right] - \left[N/p^k\cdot q^{\ell+1}\right] \\ \qquad\qquad + \left[N/p^{k+1}\cdot q^{\ell+1}\right] \\ \quad = \left(N/p^k\cdot q^\ell\right)\cdot\left(1-\frac{1}{p}\right)\cdot\left(1-\frac{1}{q}\right) + 2\Theta, \end{cases}$$

where $|\Theta| \le 1$, and therefore

$$S_1 \le N \cdot D^2(N) + N \cdot E^2(N) + 2 \sum_{p^k q^\ell \le N,\ p\ne q} w(p^k) \cdot w(q^\ell).$$

Second,

$$\sum_{n\le N} \sum_{p^k\| n} w(p^k) = \sum_{p^k\le N} w(p^k) \cdot \left(\left[p^{-k}\cdot N\right] - \left[p^{-k-1}\cdot N\right]\right)$$

$$\ge \sum_{p^k\le N} p^{-k}\cdot w(p^k)\cdot(1-p^{-1}) \cdot N - \sum_{p^k\le N} w(p^k).$$

Putting these estimates together, the term $E^2(N)$ cancels. Application of the CAUCHY-SCHWARZ inequality gives

$$N^{-1}\cdot S \le D^2(N) + 2\cdot N^{-1}\cdot \sum_{p^k q^\ell\le N,\ p\ne q} w(p^k)\cdot w(q^\ell) + 2\cdot N^{-1}\cdot E(N) \cdot \sum_{p^k\le N} w(p^k)$$

$$\le \left\{1 + 2\cdot N^{-1}\cdot\Big(\sum_{p^k q^\ell\le N,\ p\ne q} p^k\, q^\ell\Big)^{\frac12} + 2\cdot N^{-1}\cdot\Big(\sum_{p^k\le N} p^k \cdot \sum_{p^k\le N} p^{-k}\Big)^{\frac12}\right\}\cdot D^2(N).$$

Some standard estimates complete the first part of the proof:

$$2\cdot N^{-1}\cdot \Big(\sum_{p^k q^\ell\le N,\ p\ne q} p^k\, q^\ell\Big)^{\frac12} \le 2\cdot N^{-1}\cdot\Big(2 \sum_{n\le N} n\Big)^{\frac12} \le 2\cdot\sqrt{2}\ ,$$

$$\sum_{p^k \le N} p^{-k} \le \sum_{2 \le n \le N} n^{-1} \le \log N,$$

$$\sum_{p^k \le N} p^k \le N \cdot \sum_{p^k \le N} 1 \le 8\, N^2 \cdot \log^{-1}(N),\ N \ge 2.$$

The last estimate uses $\sum_{p \le N} \log p \le 2 \cdot \log 2 \cdot N$ (for $N \ge 2$). This implies

$$\sum_{p \le N} 1 + \sum_{p^k \le N, k \ge 2} 1 \le$$

$$\le \sum_{p \le \sqrt{N}} 1 + \sum_{p \le N} (\log N^{\frac{1}{2}})^{-1} \cdot \log p + \sum_{2 \le k \le \log N/\log 2} \sum_{p \le N^{1/k}} 1$$

$$\le N^{\frac{1}{2}} + (2/\log N) \cdot 2 \log 2 \cdot N + (\log N/\log 2) \cdot N^{\frac{1}{2}} \le 8\, N \cdot \log^{-1}(N).$$

Thus the method gives the constant $1 + 2\sqrt{2} + 2 \cdot \sqrt{8} \le 10$. Due to part (ii) this implies $C_2 = 20$.

(ii) If w is real-valued, define additive functions f^+, f^-, where $f^+(p^k) = \max(0, w(p^k))$, $f^-(p^k) = -\min(0, w(p^k))$. Define $E^+(N)$, ..., $D^-(N)$ in the same manner as $E(N)$ and $D(N)$, but now using the functions f^+ and f^-. Then

$$| w(n) - E(N) |^2 \le 2 \left(|f^+(N) - E^+(N)|^2 + |f^-(N) - E^-(N)|^2 \right),$$

and utilizing the relation $D^+(N)^2 + D^-(N)^2 = D^2(N)$, we obtain the result for real-valued functions, unfortunately with an additional factor of 2.

(iii) The case of complex-valued functions is reduced to the real-valued case in the usual way by decomposing f into its real and imaginary part. □

Next, we are going to "dualize" the TURÁN-KUBILIUS inequalities. Consider the complex vector-space $\mathbb{C}^M$ of all vectors $\mathfrak{z} = (z_1, z_2, \ldots, z_M)$ with Euclidean norm

$$\| \mathfrak{z} \| = \Big(\sum_{m \le M} |z_m|^2 \Big)^{\frac{1}{2}} \tag{4.9}$$

and the usual inner product $\langle\ .\ ,\ .\ \rangle$. For linear maps $L\colon \mathbb{C}^M \to \mathbb{C}^N$ the "operator-norm"

$$\| L \| = \sup_{\|\mathfrak{z}\| = 1} \| L(\mathfrak{z}) \| \tag{4.10}$$

is used. If, with respect to the canonical basis, the matrix $\mathfrak{C} = (c_{m,n})$,

$1 \le m \le M$, $1 \le n \le N$, is associated with L, then the adjoint operator L^* (defined by $\langle Lx, y \rangle = \langle x, L^* y \rangle$) is connected to the matrix $\bar{C}^t$. Because of $\| L \| = \| L^* \|$ we obtain the following result.

Theorem 4.2 *(ELLIOTT's Dualization Principle). Let* $c_{m,n}$, $m = 1, \dots, M$, $n = 1, \dots, N$, *be complex numbers, and let* $c > 0$ *be given. Then the inequality*

$$(4.11) \qquad \sum_{n\le N} \Big| \sum_{m\le M} c_{m,n} z_m \Big|^2 \le c \cdot \sum_{m\le M} |z_m|^2$$

is valid for all $z \in \mathbb{C}^M$ *if and only if the "dual inequality"*

$$(4.12) \qquad \sum_{m\le M} \Big| \sum_{n\le N} c_{m,n} w_n \Big|^2 \le c \cdot \sum_{n\le N} | w_n |^2$$

is true for every $w \in \mathbb{C}^N$.

Applying this principle to Theorem 4.1 yields the following theorem.

Theorem 4.3. *For all* $x \ge 2$ *and for all complex sequences* (w_n) *the following inequalities are valid:*

$$(4.5') \quad \sum_{p^k\le x} p^{-k} \cdot \Big| \frac{p^k}{x} \sum_{n\le x, p^k\|n} w_n - \frac{1}{x} \sum_{n\le x} w_n \Big|^2 \le C_1 \cdot \frac{1}{x} \cdot \sum_{n\le x} |w_n|^2,$$

$$(4.6') \quad \sum_{p^k\le x} p^{-k} \cdot \Big| \frac{p^k}{x} \cdot \sum_{n\le x, p^k\|n} w_n - \frac{1}{x} \cdot (1-p^{-1}) \cdot \sum_{n\le x} w_n \Big|^2 \le C_2 \cdot \frac{1}{x} \cdot \sum_{n\le x} |w_n|^2.$$

Further examples of applications of the dualization principle are given in the exercises, p.41. Finally, there is the following generalization of Theorem 4.1 to higher powers:

Theorem 4.4 *(TURÁN-KUBILIUS-ELLIOTT Inequality). Given* $q \ge 0$, *there is a constant* $c > 0$, *so that the inequalities*

$$x^{-1} \cdot \sum_{n\le x} | w(n) - A(x) |^q \le c \cdot \Big(D^q(x) + \sum_{p^k\le x} p^{-k} \cdot | w(p^k) |^q \Big), \text{ if } q > 2,$$

$$x^{-1} \cdot \sum_{n\le x} | w(n) - A(x) |^q \le c \cdot D^q(x), \text{ if } 0 \le q \le 2,$$

are valid for every additive function w *and every* $x \ge 2$.

The special case where $q = 2$ is Theorem 4.1 (only the numerical value of the constant c is not specified). We do not use this generalization, and so we do not prove it, but, rather refer the reader to P. D. T. A. ELLIOTT [1980c].

I.5. GENERATING FUNCTIONS, DIRICHLET SERIES

The study of meromorphic functions near their singularities leads to arithmetical insight. In order to obtain meromorphic functions associated with arithmetical functions, different kinds of *generating functions* are used which are often treated purely formal; among the best known are LAMBERT series, generating power series and DIRICHLET series.

(a) LAMBERT series: associate with a given arithmetical function $f\colon \mathbb{N} \to \mathbb{C}$ the infinite series

$$L(f,z) = \sum_{n\geq 1} f(n) \cdot z^n \cdot (1 - z^n)^{-1}.$$

Then, assuming absolute convergence in the [open] unit disc

$$B(0,1) = \{z \in \mathbb{C};\ |z| < 1\},$$

the series $L(f,z)$ can be transformed into

$$L(f,z) = \sum_{n\geq 1} f(n) \cdot \sum_{k\geq 0} z^{n(1+k)} = \sum_{r\geq 1} z^r \cdot \sum_{d|r} f(d)$$
$$= \sum_{r\geq 1} z^r \cdot (1 * f)(r).$$

Examples. In $|z| < 1$ the LAMBERT series $L(1,z) = \sum_{n\geq 1} \tau(n) \cdot z^n$, since $1 * 1 = \tau$, and $L(\lambda,z) = \sum_{n=1}^{\infty} z^{n^2}$, where λ is the completely multiplicative function taking the value -1 at every prime p. It is easily checked that $1 * \lambda = 1_{sq}$, the characteristic function of the set *sq* of squares.

If some suitable condition restricts the growth of the arithmetical function f, then the LAMBERT series $L(f,z)$ is holomorphic in $B(0,1)$. In general, there will be a singularity of $L(f,z)$ at $z = 1$. This is true when the convolution $1 * f$ is non-negative and infinitely many of the values $(1 * f)(r)$ are non-zero, for example.

Conclusions about the behaviour of the coefficients are often possible with the aid of *Tauberian Theorems*; some of these theorems, important in number theory, are summarized in the Appendix (A.4).

(b) Generating power series.

We associate with the function $f:\mathbb{N}_0 \to \mathbb{C}$ the power series

(5.1) $$\mathcal{P}(f,z) = \sum_{n\geq 0} f(n) \cdot z^n.$$

If the function f is not too large, the power series (5.1) will converge in the complex unit circle $B(0,1) = \{ z \in \mathbb{C}, |z| < 1 \}$. In order to obtain arithmetical conclusions, the most interesting singularity is generally the point $z = 1$ (if this point is a singularity. This is certainly true, if f is non-negative and if infinitely many values of f are non-zero). However, in the case of the *partition function* $n \mapsto p(n)$, for example, with generating power series

(5.2) $$\sum_{n=0}^{\infty} p(n) \cdot z^n = \prod_{k=1}^{\infty} (1 - z^k)^{-1},$$

there are many other singularitites which have to be investigated if better estimates of the remainder term are desired. The method to be used is the analytic HARDY-LITTLEWOOD-RAMANUJAN circle method; the coefficients p(n) of the power series (5.2) are expressed through a contour-integral, the main contribution to this integral being from small arcs of the integration path near the singularities of the function on the right-hand-side of (5.2).

A useful device is outlined in HALL-TENENBAUM [1988]. If $f \geq 0$, and $\mathcal{P}(f,z)$ converges in some interval I of the real axis including the point 1, then, obviously, for any $N > 0$,

(5.3') $$\sum_{n\leq N} f(n) \leq \inf_{0<x\leq 1} x^{-N} \cdot \mathcal{P}(f,x),$$

(5.3") $$\sum_{n\geq N} f(n) \leq \inf_{x\geq 1, x\in I} x^{-N} \cdot \mathcal{P}(f,x),$$

and

(5.4) $$f(N) \leq \inf_{x>0, x\in I} x^{-N} \cdot \mathcal{P}(f,x).$$

For example, using $f(n) = (n!)^{-1}$, the last equation gives $(N!)^{-1} \leq (e/N)^N$, a rather good lower estimate of (N factorial).

Another application of this principle is also taken from HALL-TENENBAUM [1988], section 0.5. If E is a set of primes with least element $p_0(E)$, $E(x) = \sum_{p\leq x, p\in E} p^{-1}$, and $\Omega(n,E)$ the total number of prime-divisors of n [counted with multiplicity] which lie inside E, then, for $0 < y < p_0(E)$, the following inequality is a consequence of II. Theorem 3.2:

$$\sum_{n\le x} y^{\Omega(n,E)} \ll \left(x/\log x \right)\cdot \prod_{\substack{p\le x\\ p\in E}} \left(1-\frac{y}{p}\right)^{-1}\cdot \prod_{\substack{p\le x\\ p\notin E}} \left(1-p^{-1}\right)^{-1} \ll x\, e^{(y-1)E(x)}.$$

Uniformly in $0 \le k \le (p_0(E)-\varepsilon)\cdot E(x)$, taking $y = k/E(x)$, (5.4) yields

$$\#\{\, n \le x,\ \Omega(n,E)= k \,\} \le \inf_{y>0}\, y^{-k}\cdot \sum_{n\le x} y^{\Omega(n,E)} \ll x\cdot e^{-E(x)}\cdot e^{k}\cdot (E(x)/k)^{k}.$$

$$\ll \quad x\cdot e^{-E(x)}\cdot \frac{E^{k}(x)}{k!}\cdot \sqrt{k}\,.$$

As an example of the use of power series in additive number theory, we mention, for a given sequence $\mathcal{A} = \{a_\nu\}$ of integers $0 \le a_1 \le a_2 \le \ldots$ (where at most $\mathcal{O}(m)$ repetitions for a_m are allowed) the power series

$$g(z) = \sum_{n=1}^{\infty} z^{a_n}\ ;$$

this series forms the basis of the proof of the famous ERDÖS-FUCHS Theorem. Abbreviating the number of representations of n in the form $n = a_\nu + a_\mu$ to R_n, this theorem states that, as $N \to \infty$, the relation

$$\sum_{n=0}^{N} R_n = c\cdot N + o(\, N^{\frac14}\cdot \log^{-\frac12} N\,)$$

cannot hold. The proof exploits the generating power series

$$\sum_{N=0}^{\infty} \left(\sum_{n=0}^{N} R_n \right)\cdot z^N = (\,1-z\,)^{-1}\cdot g^2(z)\ .$$

For details see H. HALBERSTAM and K. F. ROTH, *Sequences* [1966], II.§4.

(c) DIRICHLET series.

The most suitable of the generating functions for the investigation of *multiplicative* arithmetical functions are generating DIRICHLET series,

$$\mathcal{D}(f,s) = \sum_{n\ge1} f(n)\cdot n^{-s}, \tag{5.5}$$

which are defined and holomorphic in some half-plane $\sigma = \operatorname{Re} s > \sigma_0$, if f does not increase too quickly. Multiplicativity of f implies a product representation for $\mathcal{D}(f,s)$ by the following lemma.

Lemma 5.1. *If f is a multiplicative function for which $\sum_{n\ge1} f(n)$ is absolutely convergent, then there is a product representation*

$$(5.6) \qquad \sum_{n=1}^{\infty} f(n) = \prod_{p} \left(1 + f(p) + f(p^2) + \dots \right),$$

where all the series $1 + f(p) + f(p^2) + \dots$ *and the product itself are absolutely convergent.*

Lemma 5.1 is well known (see, for example, HARDY-WRIGHT [1956], Theorem 286); its proof is left as Exercise 19. Applying the lemma to the DIRICHLET series $\mathcal{D}(f,s)$, we obtain the following corollary.

Corollary 5.2. *If* f *is multiplicative, then in the region of absolute convergence of the DIRICHLET series* $\mathcal{D}(f,s)$, *the product representation*

$$(5.7) \qquad \sum_{n\geq 1} f(n)\cdot n^{-s} = \prod_{p} \left(1 + f(p)\cdot p^{-s} + f(p^2)\cdot p^{-2s} + \dots \right)$$

is valid.

One of the simplest DIRICHLET series is the RIEMANN zeta-function

$$(5.8) \qquad \zeta(s) = \mathcal{D}(1,s) = \sum_{n\geq 1} n^{-s} = \prod_{p} \left(1 + p^{-s} + p^{-2s} + \dots \right) = \prod_{p} \left(1 - p^{-s} \right)^{-1},$$

which is absolutely convergent in Re s > 1. This product representation indicates some connection with the theory of prime numbers. EULER's summation formula (see Theorem 1.2), applied to $\sum_{n\geq 1} n^{-s}$, gives

$$(5.9) \qquad \zeta(s) = (s-1)^{-1} + \tfrac{1}{2} - s \cdot \int_1^{\infty} B_0(u)\cdot u^{-(s+1)}\, du .$$

The integral defines a holomorphic function in Re s > 0, and so formula (5.9) provides an analytic continuation of $\zeta(s)$ into the half-plane Re s > 0, showing that $\zeta(s)$ has a simple pole at s = 1 with residue 1.

Further integrations by parts of the integral occurring in (5.9) give the analytic continuation of $\zeta(s)$ into the whole complex plane; which can be achieved in one stroke by the *functional equation*

$$(5.10) \qquad \zeta(s) = 2^s \pi^{s-1} \sin\left(\tfrac{1}{2}\pi s \right) \Gamma(1-s) \cdot \zeta(1-s)$$

of the RIEMANN zeta-function.

The pointwise product of two DIRICHLET series is (in the region of absolute convergence of both DIRICHLET series) given by

$$\mathcal{D}(f,s) \cdot \mathcal{D}(g,s) = \sum_{n\geq 1} \sum_{d|n} f(d) \cdot g(n/d) \cdot n^{-s} = \mathcal{D}(f*g,s),$$

and so the pointwise product of the DIRICHLET series corresponds to the convolution product of arithmetical functions. Thus, consideration of DIRICHLET series is useful for multiplicative functions and also in connection with convolutions of arithmetical functions.

Noting this remark and the convolution identities $\tau = 1*1$, $1*\mu = \varepsilon$, $\varphi = \mathrm{id}*\mu$, one obtains formulae for some generating DIRICHLET series:

$$(5.11) \quad \begin{cases} \sum_{n=1}^{\infty} \tau(n) \cdot n^{-s} = \zeta^2(s), \ \mathrm{Re}\ s > 1, \\ \sum_{n=1}^{\infty} \mu(n) \cdot n^{-s} = \zeta^{-1}(s), \ \mathrm{Re}\ s > 1, \\ \sum_{n=1}^{\infty} \varphi(n) \cdot n^{-s} = \zeta(s-1)/\zeta(s), \ \mathrm{Re}\ s > 2. \end{cases}$$

Many other formulae of this type are given in HARDY-WRIGHT [1956], and a general theory of *"Zeta-Formulae"* is developed in J. KNOPFMACHER's book [1975] on abstract analytic number theory.

The connection of the RIEMANN zeta-function with the theory of prime numbers arises from the generating DIRICHLET series

$$(5.12) \quad \zeta'(s)/\zeta(s) = - \sum_{n=1}^{\infty} \Lambda(n) \cdot n^{-s}, \ \mathrm{Re}\ s > 1,$$

where the VON MANGOLDT function Λ is given by

$$(5.13) \quad \Lambda(n) = \begin{cases} \log p, & \text{if } n \text{ is a power } p^k \text{ of the prime } p, \\ 0 & \text{otherwise.} \end{cases}$$

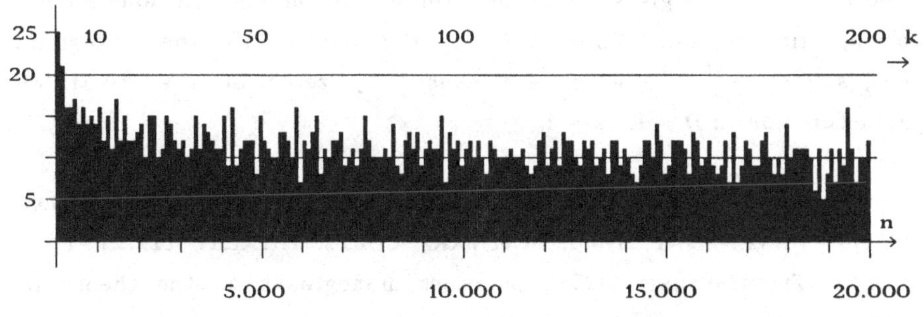

F i g u r e I. 8. Primes in intervals of length 100

The number $\pi(x)$ of primes in the interval $[1,x]$ behaves rather erratic locally. This is illustrated in Figure I.8 on the foregoing page, giving the number of primes in intervals of length one hundred, from $k\cdot 100+1$ to $(k+1)\cdot 100$, where $0 \le k \le 199$. The first interval contains twenty-five primes, the next one twenty-one, etc., but there is also an interval containing only five primes.

The problem of obtaining an asymptotic formula for the number $\pi(x)$ of primes up to x,

$$(5.14)\qquad \pi(x) = \sum_{p \le x} 1 ,$$

is equivalent (via partial summation) to a suitable approximation to

$$(5.15)\qquad \vartheta(x) = \sum_{p \le x} \log p$$

or to

$$(5.16)\qquad \psi(x) = \sum_{n \le x} \Lambda(n),$$

via the easily verified relation [use the fact that higher powers are rare]

$$\vartheta(x) = \sum_{n \le x} \Lambda(n) + \mathcal{O}\left(x^{\frac{1}{2}} \cdot (\log x)^2 \right).$$

The function $\psi(x) = \sum_{n \le x} \Lambda(n)$ has an integral representation (by a complex inversion formula),

$$(5.17)\qquad \psi(x) = (2\pi i)^{-1} \cdot \int_{c-i\infty}^{c+i\infty} \left(-\zeta'(s)/\zeta(s) \right) \cdot s^{-1} x^s \, ds,$$

where $c > 1$. The *"Method of Complex Integration"* allows approximation of the integral in (5.17) by shifting the path of integration to the left. The pole at $s = 1$ gives the main term x. Further contributions to the asymptotic formula follow from the poles of the integrand $(-\zeta'(s)/\zeta(s))\cdot s^{-1} x^s$, which are caused by zeros of the RIEMANN zeta-function in $0 < \operatorname{Re} s < 1$.

For the [lengthy] details of this method see, for example, PRACHAR [1957], DAVENPORT [1967], SCHWARZ [1969], HUXLEY [1972], IVIĆ [1985], TITCHMARSH [1951], or other monographs on the theory of primes.

I.6. SOME RESULTS ON PRIME NUMBERS

It will frequently be necessary to use asymptotic formulae or estimates for sums or products running over primes. We cannot prove these [standard] results, but quote some of them for easy reference. The method described at the end of the preceding section, in combination with some deeper knowledge on the distribution of the zeros of the zeta-function in the critical strip $\frac{1}{2} < \mathrm{Re}\ s < 1$, gives the **Prime Number Theorem**:

Theorem 6.1. *For* $x \to \infty$, *with some positive constant* γ, *the following asymptotic formulae hold:*

$$\psi(x) = x + \mathcal{O}\left(x \cdot \exp(- \gamma (\log x)^{\frac{1}{2}}) \right), \tag{6.1}$$

$$\vartheta(x) = x + \mathcal{O}\left(x \cdot \exp(- \gamma (\log x)^{\frac{1}{2}}) \right), \tag{6.2}$$

$$\pi(x) = \mathrm{li}\ x + \mathcal{O}\left(x \cdot \exp(- \gamma (\log x)^{\frac{1}{2}}) \right), \tag{6.3}$$

where li x *is an abbreviation for the so-called* *integral-logarithm*

$$\mathrm{li}\ x = \mathrm{li}\ e + \int_e^x (\log u)^{-1}\, du, \quad \mathrm{li}\ e = 1.895\ 117\ 8\ldots\ . \tag{6.4}$$

Some [rounded] values of $\pi(x)$, li x, and $x/\log x$ are given in Table I.1.

x	100	1000	10^4	10^5	10^6
$\pi(x)$	25	168	1229	9592	78498
li x	30	178	1246	9630	78628
$x/\log x$	21.7	144.8	1086	8686	72382

Table I.1

The function li(x) is connected with the Exponential-Integral Function Ei(x) by the formula li(x) = Ei(log x), and may be calculated from the series development

$$\mathrm{li}\ x = \mathscr{C} + \log\log x + \sum_{1 \le n < \infty} (\log x)^n \Big/ (n \cdot n!), \text{ for } x > 1.$$

$\mathscr{C}$ = 0.577 215 664 90. ... is EULER's constant. Roughly speaking, the integral-logarithm behaves as x/log x, so that $\lim_{x \to \infty} \operatorname{li} x \big/ (x/\log x) = 1$. It is possible to deduce from (6.4) an asymptotic development of li x by partial integrations, for example (with three main terms on the right-hand side):

$$\operatorname{li} x = x \cdot \log^{-1} x + x \cdot \log^{-2} x + 2 \cdot x \cdot \log^{-3} x + \mathcal{O}(x \cdot \log^{-4} x).$$

A. SELBERG, P. ERDÖS showed in 1948 (independently) that the prime number theorem may also be obtained by "*elementary methods*", We present one of these elementary proofs, due to H. DABOUSSI, in II.§9. Rather simple, elementary methods lead to the estimates given below, which are frequently required. For proofs, see, for example, PRACHAR [1957] or SCHWARZ [1969].

Theorem 6.2. *There are constants* $0 < c_1 < 1 < c_2$ *such that*

$$(6.5)\qquad \begin{cases} c_1 \cdot (x/\log x) < \pi(x) < c_2 \cdot (x/\log x), \\ c_1 \cdot x < \vartheta(x) < c_2 \cdot x, \\ c_1 \cdot x < \psi(x) < c_2 \cdot x. \end{cases}$$

Furthermore, as $x \to \infty$,

$$(6.6)\qquad \sum_{p \le x} p^{-1} \cdot \log p = \log x + \mathcal{O}(1),$$

$$(6.7)\qquad \sum_{p \le x} p^{-1} = \log\log x + \gamma_2 + o(1),$$

and

$$(6.8)\qquad \prod_{p \le x} (1 - p^{-1}) = e^{-\mathscr{C}} \cdot (\log x)^{-1} \cdot (1 + o(1)).$$

The remainder terms in these formulae may be improved by using the prime number theorem. For example, with the "standard remainder term" $\mathcal{O}\left(x \cdot \exp(-\gamma\sqrt{\log x})\right)$ of the prime number theorem, we obtain

$$(6.9)\qquad \sum_{p \le x} p^{-1} \cdot \log p = \log x + \gamma_1 + \mathcal{O}\left(\exp(-\gamma\sqrt{\log x})\right),$$

$$(6.10)\qquad \sum_{p \le x} p^{-1} = \log\log x + \gamma_2 + \mathcal{O}\left(\exp(-\gamma\sqrt{\log x})\right),$$

and

$$(6.11)\quad \prod_{p \le x} \left(1 - p^{-1}\right) = e^{-\mathscr{C}} \cdot (\log x)^{-1} \cdot \left(1 + \mathcal{O}\left\{ x \cdot \exp\{-\gamma \sqrt{\log x}\} \right\}\right),$$

where $\gamma_2 = \mathscr{C} - \sum_p \sum_{k \ge 2} (kp^k)^{-1}$ and $\mathscr{C} = 0.577\ 215\ 664\ 901\ldots$ is EULER's constant (see, for example, PRACHAR [1957]), and x tends to infinity. Many estimates and inequalities of this nature, with explicit constants and often very deep, are given in ROSSER & SCHOENFELD [1962].

The "mean-value" $M(\mu) = \lim_{x \to \infty} x^{-1} \cdot \sum_{n \le x} \mu(n)$ of the MÖBIUS-function is zero. More exactly,

$$(6.12)\qquad \sum_{n \le N} \mu(n) = \mathcal{O}\left(N \cdot \exp\{-\gamma \sqrt{\log N}\}\right).$$

The function $N \mapsto N^{-\frac{1}{2}} \cdot \sum_{n \le N} \mu(n)$ is plotted below for N = 2, 4, ...

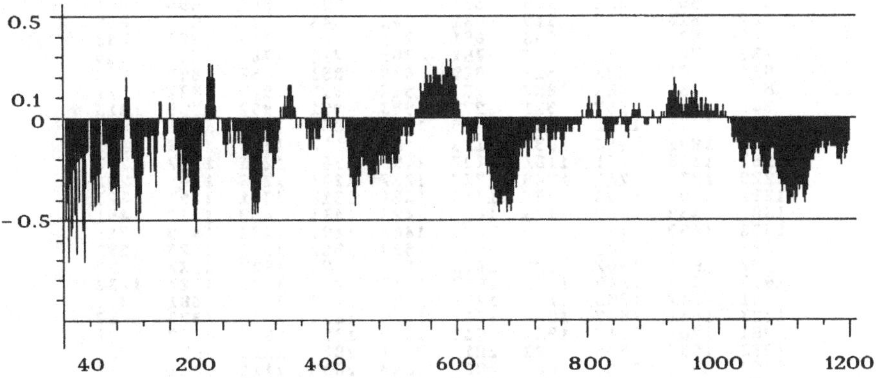

Figure I.9

Sum over the Möbius function

VON STERNECK's conjecture, supported by Figure I.9, states that

$$|N^{-\frac{1}{2}} \cdot \sum_{n \le N} \mu(n)| \le \tfrac{1}{2}, \text{ if } N > 200.$$

This conjecture is not true (G. NEUBAUER [1963]), and the weaker MERTENS conjecture, where $\frac{1}{2}$ is replaced with 1, is also not true (A. M. ODLYZKO & H. J. J. TE RIELE [1985]. See also TE RIELE [1985]. JURKAT & PEYERIMHOFF proved a weaker result in 1976).

The 560 primes in the interval [2, 4057] are given in Table I.2.

Figure I.10 represents the primes between 1 and 10.000. Some explanation is necessary. The small rectangles mark the integers, beginning with 1 in the bottom line up to 100, from 101 to 200 in the second from bottom line and so on. A dark rectangle indicates that the integer represented by this rectangle is a prime. Note: the column with column-index 10, 20, ... is to the left of the vertical line |. For example, the top line contains the nine prime numbers 9901, 9907, 9923, 9929, 9931, 9941, 9949, 9967, and 9973.

Table of Prime Numbers

2	3	5	7	11	13	17	19	23	29
31	37	41	43	47	53	59	61	67	71
73	79	83	89	97	101	103	107	109	113
127	131	137	139	149	151	157	163	167	173
179	181	191	193	197	199	211	223	227	229
233	239	241	251	257	263	269	271	277	281
283	293	307	311	313	317	331	337	347	349
353	359	367	373	379	383	389	397	401	409
419	421	431	433	439	443	449	457	461	463
467	479	487	491	499	503	509	521	523	541
547	557	563	569	571	577	587	593	599	601
607	613	617	619	631	641	643	647	653	659
661	673	677	683	691	701	709	719	727	733
739	743	751	757	761	769	773	787	797	809
811	821	823	827	829	839	853	857	859	863
877	881	883	887	907	911	919	929	937	941
947	953	967	971	977	983	991	997	1009	1013
1019	1021	1031	1033	1039	1049	1051	1061	1063	1069
1087	1091	1093	1097	1103	1109	1117	1123	1129	1151
1153	1163	1171	1181	1187	1193	1201	1213	1217	1223
1229	1231	1237	1249	1259	1277	1279	1283	1289	1291
1297	1301	1303	1307	1319	1321	1327	1361	1367	1373
1381	1399	1409	1423	1427	1429	1433	1439	1447	1451
1453	1459	1471	1481	1483	1487	1489	1493	1499	1511
1523	1531	1543	1549	1553	1559	1567	1571	1579	1583
1597	1601	1607	1609	1613	1619	1621	1627	1637	1657
1663	1667	1669	1693	1697	1699	1709	1721	1723	1733
1741	1747	1753	1759	1777	1783	1787	1789	1801	1811
1823	1831	1847	1861	1867	1871	1873	1877	1879	1889
1901	1907	1913	1931	1933	1949	1951	1973	1979	1987
1993	1997	1999	2003	2011	2017	2027	2029	2039	2053
2063	2069	2081	2083	2087	2089	2099	2111	2113	2129
2131	2137	2141	2143	2153	2161	2179	2203	2207	2213
2221	2237	2239	2243	2251	2267	2269	2273	2281	2287
2293	2297	2309	2311	2333	2339	2341	2347	2351	2357
2371	2377	2381	2383	2389	2393	2399	2411	2417	2423
2437	2441	2447	2459	2467	2473	2477	2503	2521	2531
2539	2543	2549	2551	2557	2579	2591	2593	2609	2617
2621	2633	2647	2657	2659	2663	2671	2677	2683	2687
2689	2693	2699	2707	2711	2713	2719	2729	2731	2741
2749	2753	2767	2777	2789	2791	2797	2801	2803	2819
2833	2837	2843	2851	2857	2861	2879	2887	2897	2903
2909	2917	2927	2939	2953	2957	2963	2969	2971	2999
3001	3011	3019	3023	3037	3041	3049	3061	3067	3079
3083	3089	3109	3119	3121	3137	3163	3167	3169	3181
3187	3191	3203	3209	3217	3221	3229	3251	3253	3257
3259	3271	3299	3301	3307	3313	3319	3323	3329	3331
3343	3347	3359	3361	3371	3373	3389	3391	3407	3413
3433	3449	3457	3461	3463	3467	3469	3491	3499	3511
3517	3527	3529	3533	3539	3541	3547	3557	3559	3571
3581	3583	3593	3607	3613	3617	3623	3631	3637	3643
3659	3671	3673	3677	3691	3697	3701	3709	3719	3727
3733	3739	3761	3767	3769	3779	3793	3797	3803	3821
3823	3833	3847	3851	3853	3863	3877	3881	3889	3907
3911	3917	3919	3923	3929	3931	3943	3947	3967	3989
4001	4003	4007	4013	4019	4021	4027	4049	4051	4057

T a b l e I.2. Prime Numbers below 4058

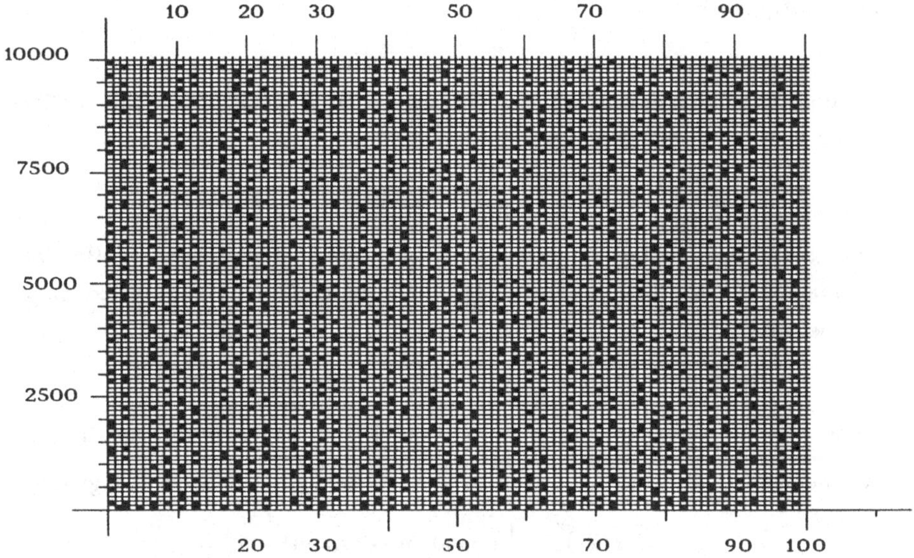

F i g u r e I.10

Characteristic Function of the Primes below 10.000

1.7. CHARACTERS, L-FUNCTIONS, PRIMES IN ARITHMETIC PROGRESSIONS

For a treatment of primes in arithmetic progressions (primes p in residue-classes $p \equiv a \mod q$, where $\gcd(a,q) = 1$) it is necessary to have functions that single out the elements of one residue-class. Such functions, which are, in addition, multiplicative and periodic, are the DIRICHLET characters, defined on $\mathbb{N}$ or $\mathbb{Z}$. Characters will be introduced in a more general setting: we assume that $\mathcal{G}$ is a topological group, which, in addition, is also abelian.

A **character** χ on $\mathcal{G}$ is a *continuous* homomorphism from $\mathcal{G}$ into the circle group

$$\mathbb{T} = \{ z \in \mathbb{C} \; ; \; |z| = 1 \} \tag{7.1}$$

by multiplication,

$$\chi : (\mathcal{G}, \cdot) \to (T, \cdot).$$

By pointwise multiplication the characters on $\mathcal{G}$ again form an abelian group, i. e. the character group

(7.2) $$\mathcal{G}^{\wedge} = \{\chi : \mathcal{G} \to T,\ \chi \text{ continuous homomorphism}\}.$$

This group $\mathcal{G}^{\wedge}$ can be converted to a *topological group* in the following manner: a basis of neighbourhoods of the unit element e of $\mathcal{G}$ consists of the sets

$$U(\varepsilon,K) = \{ \chi \in \mathcal{G} ,\ |\chi(x) - 1| < \varepsilon \text{ for all } x \text{ in } K \},$$

where ε is any positive real number, and $K \subset \mathcal{G}$ any *compact* set in $\mathcal{G}$.

As mentioned already in section 3, a character χ defined on the group

$$(\mathbb{Z}/m\mathbb{Z})^{\times} = \{a \bmod m,\ \gcd(a,m) = 1\}$$

of residue-classes a mod m, prime to m, with discrete topology, in other words a group-homomorphism

$$\chi : (\mathbb{Z}/m\mathbb{Z})^{\times} \to (\mathbb{C}, \cdot),\ |\chi(n)| = 1,$$

induces a completely multiplicative, m-periodic function

(7.3) $$\chi : \mathbb{N} \to \{ z \in \mathbb{C},\ |z| = 1 \text{ or } z = 0 \},$$

defined by $\chi(n) = \chi(n \bmod m)$, if $\gcd(n,m) = 1$, and $\chi(n) = 0$ otherwise. The unit element of the character group induces the so-called *principal character* χ_0 mod m, with values $\chi_0(n) = 1$, if $\gcd(n,m) = 1$, $\chi_0(n) = 0$ otherwise. The character group of $(\mathbb{Z}/m\mathbb{Z})^{\times}$ has $\varphi(m)$ elements.

These functions (7.3), called DIRICHLET characters, allow the construction of **DIRICHLET L-functions**

(7.4) $$\begin{cases} L(s,\chi) = \sum_{n=1}^{\infty} \chi(n)\cdot n^{-s} = \prod_p \left(1 + \chi(p)\cdot p^{-s} + \chi(p^2)\cdot p^{-2s} + \dots\right) \\ \qquad = \prod_p \left(1 - \chi(p)\cdot p^{-s}\right)^{-1}. \end{cases}$$

The series and products in (7.4) are absolutely convergent in Re s > 1. Moreover the series $\sum_{n=1}^{\infty} \chi(n)\cdot n^{-s}$ is conditionally convergent in Re s > 0 if χ is **not** the principal character. If χ **is** the principal character χ_0, then

(7.5) $$L(s, \chi_0) = \prod_{p \nmid m} \left(1 - p^{-s}\right)^{-1} = \prod_{p | m} \left(1 - p^{-s}\right) \cdot \zeta(s).$$

DIRICHLET characters χ satisfy (like characters in *locally compact* topological abelian groups in general, where summation is replaced by integration with respect to the HAAR measure on $\mathcal{G}$) the *orthogonality relations*:

If a *runs through a full set of representatives* mod m (*for example,* a = 1, 2, ..., m), *then*

(7.6) $$\sum_{a \bmod m} \chi(a) = \begin{cases} \varphi(m), & \textit{if } \chi \textit{ is the principal character,} \\ 0 & \textit{otherwise.} \end{cases}$$

If χ *runs through all the* $\varphi(m)$ *DIRICHLET characters* mod m, *then*

(7.7) $$\sum_{\chi} \chi(a) = \begin{cases} \varphi(m), & \textit{if } a \equiv 1 \bmod m, \\ 0 & \textit{otherwise.} \end{cases}$$

Corollary *(Orthogonality Relations for DIRICHLET Characters).*

(7.8) $$\sum_{a \bmod m} \chi_1(a) \cdot \overline{\chi_2(a)} = \begin{cases} \varphi(m), & \textit{if } \chi_1 = \chi_2, \\ 0 & \textit{otherwise,} \end{cases}$$

and

(7.9) $$\sum_{\chi} \chi(a_1) \cdot \overline{\chi(a_2)} = \begin{cases} \varphi(m), & \textit{if } a_1 \equiv a_2 \bmod m \textit{ and } \gcd(a_1 a_2, m) = 1, \\ 0 & \textit{otherwise.} \end{cases}$$

These relations allow specific residue-classes mod m to be singled out:

If f *is an arithmetical function and* gcd(a,m) = 1, *and if* n_ι, $\iota = 1, \ldots, J$, *are positive integers, then*

(7.10) $$\sum_{\iota \le J,\ n_\iota \equiv a \bmod m} f(n_\iota) = \frac{1}{\varphi(m)} \cdot \sum_{\chi} \overline{\chi(a)} \sum_{\iota \le J} f(n_\iota) \cdot \chi(n_\iota).$$

Since

$$- L'(s,\chi) / L(s,\chi) = \sum_{n=1}^{\infty} \chi(n) \cdot \Lambda(n) \cdot n^{-s} ,$$

one finds results on **primes in arithmetic progressions** in the same way as is possible for ordinary primes (for example, using the method of "complex integration"). DIRICHLET L-functions have properties similar to those of the RIEMANN zeta-function, and so, using the functions

$$(7.11)\qquad \begin{cases} \pi(x;a,q) = \sum_{p\le x,\ p\equiv a \bmod q} 1, \\ \vartheta(x;a,q) = \sum_{p\le x,\ p\equiv a \bmod q} \log p, \\ \psi(x;a,q) = \sum_{n\le x,\ n\equiv a \bmod q} \Lambda(n)\ , \end{cases}$$

one obtains the following theorem.

Theorem 7.1. *If* $\gcd(a,q) = 1$, *then, with some positive constant* γ, *depending on* a *and* q, *the following asymptotic formulae hold :*

$$(7.12)\qquad \begin{cases} \psi(x;a,q) = \frac{1}{\varphi(q)}\, x + \mathcal{O}\left(x\cdot\exp(- \gamma\ (\log\ x)^{\frac{1}{2}})\right), \\ \vartheta(x;a,q) = \frac{1}{\varphi(q)}\, x + \mathcal{O}\left(x\cdot\exp(- \gamma\ (\log\ x)^{\frac{1}{2}})\right), \\ \pi(x;a,q) = \frac{1}{\varphi(q)}\, \mathrm{li}\ x + \mathcal{O}\left(x\cdot\exp(- \gamma\ (\log\ x)^{\frac{1}{2}})\right), \end{cases}$$

with the integral-logarithm $\mathrm{li}\ x = \mathrm{li}\ e + \int_e^x (\log\ u)^{-1}\, du.$

It is sometimes important to have <u>uniform</u> estimates on $\pi(x;a,q)$ in $1 \le a \le q$, with q restricted to some range, depending on x. An important result of this kind is provided by the following theorem.

Theorem 7.2 (Prime Number Theorem of PAGE-SIEGEL-WALFISZ**).** *If* $1 \le a \le q$, *if* $\gcd(a,q) = 1$, *and if* $1 \le q \le (\ \log\ x\ )^A$ *with some fixed constant* A, *then, as* x *tends to infinity, the asymptotic formula*

$$(7.13)\qquad \pi(x;a,q\ ) = \frac{1}{\varphi(q)}\ \mathrm{li}\ x + \mathcal{O}_A\left(x\cdot\exp\left\{ - \gamma\, (\log\ x)^{\frac{1}{2}} \right\} \right)$$

holds uniformly in a *and* q. *As indicated, the constant implicit in the* $\mathcal{O}$*-symbol may depend on* A.

For a proof see, for example, PRACHAR [1957] or ESTERMANN [1952]. For some applications the range of admissible values for q in (7.13) is not sufficient, for example when consideration of larger values of q is unavoidable; this occurs in problems from the additive theory of numbers. The sieve method (V. BRUN, A. SELBERG) or the *"large sieve"* easily gives the upper estimate

$$(7.14)\qquad \pi(x;a,q) < \gamma \cdot x \Big/ \left(\varphi(q)\cdot\log(x/q) \right)$$

with some constant γ (which, in fact, may be taken to be 2, as long as $q < \delta \cdot x$ for some positive constant δ). See, for example, MONTGOMERY [1971], HALBERSTAM & RICHERT [1974], or SCHWARZ [1974].

Another deep and extremely useful result is the prime number theorem of E. BOMBIERI and A. I. VINOGRADOV, which says that on average (7.13) is true [with a better remainder term] for a much larger range of values of q.

Theorem 7.3. (**BOMBIERI–VINOGRADOV's Prime Number Theorem**). *For any positive constant* A *there is a [positive] constant* B, *for which the estimate*

$$\sum_{q \le \sqrt{x} \cdot \log^{-B} x} \ \max_{y \le x} \ \max_{a \bmod q,\ \gcd(a,q)=1} \left| \pi(y;q,a) - \frac{\operatorname{li} y}{\varphi(q)} \right| = \mathcal{O}_A\left(x \cdot \log^{-A} x\right)$$

is true.

For a proof, see, for example, DAVENPORT [1967] or HUXLEY [1972].

I.8. EXERCISES

Remark. Many similar exercises may be found, for example in T. APOSTOL [1976].

1) Deduce higher EULER summation formulae such as

$$\sum_{a<n\le x} g(n) = \int_a^x g(u)du + \int_a^x B_1(u) \cdot g''(u)du - g(x) \cdot B_0(x) - \tfrac{1}{2} g(a) + B_1(x) \cdot g'(x),$$

where a is an integer, and $B_1(x) = \int_a^x B_0(u)du = \frac{1}{2}\left(x-[x]\right) \cdot \left(x-[x]-1\right)$.

2) Prove (by partial summation) for every real $s > 1$ and for $x \ge 1$

$$\left| \sum_{n \le x} n^{-s} - (s-1)^{-1} \cdot (1-x^{s-1}) - \tfrac{1}{2} + s \int_1^\infty u^{-(s+1)} \cdot B_0(u)\, du \right| \le x^{-s}.$$

Prove, for $s > 0$ and $x \ge 1$,

$$\left| \sum_{n \le x} n^s - (x^{s+1} - 1) \cdot (s+1)^{-1} \right| \le x^s.$$

3) Exhibit infinitely many functions, linearly independent in the vector-space $\mathbb{C}^{\mathbb{N}}$. Prove orthogonality relations for the functions g_d, $d|r$, defined in I.2.

4) Denote by $\mu_r(n)$ the characteristic function of the set of r-free integers (r = 2, 3, ...), so that

$\mu_r(n) = 0$, if there is some prime for which p^r divides n, and
$\mu_r(n) = 1$ otherwise.

Prove: The function μ_r is multiplicative and $\mu_r(n) = \sum_{d^r|n} \mu(d)$.

5) Denote by $\rho(n)$ the number of solutions d mod n of the congruence $f(d) \equiv 0 \bmod n$, where $f(X)$ is a non-constant polynomial with integer coefficients. Prove that ρ is a multiplicative function.

6) Prove $1 * \mu^2 = 2^\omega$, and $\Lambda * \mu = -\mu \log$, $\Lambda * 1 = \log$, where Λ is VON MANGOLDT's function (see I.6).

7) Show that $\dfrac{n}{\varphi(n)} = \sum_{d|n} \dfrac{\mu^2(d)}{\varphi(d)}$.

8) The LIOUVILLE function $\lambda: n \mapsto (-1)^{\Omega(n)}$ has the convolution inverse $\lambda^{-1(*)} = \mu\cdot\lambda$, more generally a completely multiplicative function h has convolution inverse $h^{-1(*)} = \mu\cdot h$. The function $\sigma_1: n \mapsto \sum_{d|n} d$ has the inverse $\mu * (\mu\cdot\mathrm{id})$.

9) If f_1 is completely multiplicative, and $f_2 > 0$ is multiplicative and integer-valued, then $f_1 \circ f_2$ is multiplicative.

10) If f is a multiplicative solution of the functional equation $f^2 = 2^\omega * f$, then f is integer-valued.

11) Prove or disprove the following:

(a) If f is r-periodic, $f(1) = 1$, then $f^{-1(*)}$ is r-periodic.
(b) If r is r-even, $f(1) = 1$, then $f^{-1(*)}$ is r-even.
(c) Every r-periodic function is s-even for some positive integer s.
(d) If f is strongly multiplicative, then $f^{-1(*)}$ is 2-multiplicative.
(e) If f is 2-multiplicative, then $f^{-1(*)}$ is strongly multiplicative.

12) The RAMANUJAN sum $n \mapsto c_r(n)$ is multiplicative if and only if $\mu(r) = 1$ [so that r = 1 or r is a product of an even number of different primes].

13) (HÖLDER 1936). Put $n' = r/\gcd(n,r)$. Then $c_r(n) = \mu(n')\cdot\big(\varphi(r)/\varphi(n')\big)$.

14) (RADEMACHER 1925). Denote by f(n,r) the number of solutions of the linear congruence

$$x_1 + x_2 + \dots + x_s \equiv n \bmod r$$

in vectors $(\,x_\rho \bmod r\,)_{1 \le \rho \le s}$ with the additional condition $\gcd(\,x_\rho, r\,) = 1$. Prove:

(a) $f(n, r_1 r_2) = f(n, r_1) \cdot f(n, r_2)$, if $\gcd(r_1, r_2) = 1$.

(b) If $p \nmid n$ then

$$f(n, p^k) = p^{ks-k-s} \cdot \{ (p-1)^s + (-1)^{s-1} \},$$

and for $p|n$

$$f(n,p^k) = p^{ks-k-s} \cdot (p-1) \cdot \{ (p-1)^{s-1} + (-1)^s \}.$$

15) The vector-space $\mathbb{C}^{\mathbb{N}}$ with multiplication

$$f \perp g:\ n \mapsto \sum_{d|n,\gcd(d,n/d)=1} f(d) \cdot g(n/d)$$

("unitary convolution") becomes a commutative algera with unit element ε.

16) Dualize (4.7), which means: prove for $x \ge 2$ and any complex numbers w_n the inequality

$$\sum_{p \le x} p^{-1} \Big| \frac{p}{x} \sum_{n \le x, p|n} w_n - x^{-1} \cdot \sum_{n \le x} w_n \Big|^2 \le (4C_1+2) \cdot x^{-1} \sum_{n \le x} |\, w_n \,|^2.$$

17) [DABOUSSI]. Prove

$$\sum_{p^k \le x} p^{-k} \cdot \Big| p^k/x \cdot \sum_{n \le x, p^k \| n} w_n - x^{-1} \sum_{n \le x, p \nmid n} w_n \Big|^2 \le 2(C_1+1) \cdot x^{-1} \sum_{n \le x} |w_n|^2.$$

18) Assume that $\sum_{n=1}^{\infty} n^{-s} \cdot f(n)$ is absolutely convergent at the point $s = \sigma_1 + it_1$. Prove that this DIRICHLET series is absolutely convergent for every $s = \sigma + it$ if $\sigma \ge \sigma_1$. This result is not true, if the assumption of absolute convergence is weakened to convergence. In this case, prove convergence of $\sum_{n=1}^{\infty} n^{-s} \cdot f(n)$ in the region $\mathrm{Re}\, s > \sigma_1$.

19) Prove Lemma 5.1. Hint:

$$\Big| \sum_{n=1}^{\infty} f(n) - \prod_{p \le x} (\, 1 + f(p) + f(p^2) + \dots \,) \Big| \le \sum_{n > x} |f(n)|.$$

20) For integers k, in Re s > 1, prove

$$\sum_{n=1}^{\infty} c_n(k) \cdot n^{-s} = \tau_{s-1}(k) \,/\, (k^{s-1} \cdot \zeta(s)).$$

21) (SIERPINSKI, 1952). Let $p_1 < p_2 < p_3 < \dots$ be the ordered sequence of all primes. Prove:

a) $\sum_{n=1}^{\infty} p_n \cdot 10^{-2^n}$ has a limit, say c.

b) The formula $p_n = \left[10^{2^n} \cdot c\right] - 10^{2^{n-1}} \left[10^{2^{n-1}} c\right]$
holds for n = 1, 2, ...

22) Define the polynomial p(x) by

$$p(x) = \prod_{\substack{1 \le s \le n \\ (s,n)=1}} (x - e^{2\pi i \cdot (s/n)}).$$

Prove $p(x) = \prod_{d|n} (x^{n/d} - 1)^{\mu(d)}$.

23) Give the proof of EULER's summation formula (Theorem 1.2) in detail.

24) Define D(f) by D(f): $n \mapsto f(n) \cdot \log n$. Then the map D is a derivation (so that $D: \mathbb{C}^{\mathbb{N}} \to \mathbb{C}^{\mathbb{N}}$ is linear, $D\varepsilon = 0$, and

$$D(f * g) = f * D(g) + D(f) * g).$$

25) g is completely additive if and only if the map $f \mapsto f \cdot g$ is a derivation. Note that many properties of derivations are dealt with in T. APOSTOL [1976], §2.18.

26) Prove: For every positive integer k,

$$\sum_{d|n} d^k = \zeta(k+1) \cdot n^k \cdot \sum_{r \ge 1} c_r(n) \cdot r^{-(k+1)},$$

and this series is absolutely convergent.

Paul Erdös

J. Kubilius

S. Ramanujan
(1887–1920)

P. Turán
(1910–1976)

A. Wintner
(1903–1958)

Turán's photo, given to the first–named author by Prof. Dr. K. Jacobs, was already used in an article in *"The Development of Mathematics from 1900 to 1950"*, Birkhäuser Verlag (forthcoming 1994), edited by J. P. Pier. Birkhäuser Verlag has kindly given permission to use this photograph again.

J. P. L. DIRICHLET
(1805–1859)

A. F. MÖBIUS
(1790–1868)

G. H. HARDY
(1877–1947)

J. E. LITTLEWOOD
(1885–1977)

H. DAVENPORT
(1907–1969)

Chapter II

Mean-Value Theorems and Multiplicative Functions, I

Abstract. *This chapter mainly deals with estimates of sums over multiplicative functions and with asymptotic formulae for these sums. Rather simple, elementary methods lead to the mean-value theorems of* WINTNER *and* AXER*, in which multiplicativity does not play any rôle. Next, inequalities for sums over prime-powers are shown to be sufficient to obtain upper bounds for sums over non-negative multiplicative functions; lower bounds for such sums may be obtained under stronger assumptions. The* HARDY-LITTLEWOOD-KARAMATA *Tauberian Theorem is employed to prove a useful theorem by E.* WIRSING *with some applications. Finally, following* DABOUSSI*'s proofs, an elementary proof of the prime number theorem is given, and* SAFFARI*'s result on direct decompositions of the set of positive integers is proved.*

II.1. MOTIVATION

Given an arithmetical function $f\colon \mathbb{N} \to \mathbb{C}$, the **mean-value** $M(f)$ of the function f is defined to be the limit

$$(1.1) \qquad M(f) = \lim_{x\to\infty} x^{-1} \cdot \sum_{n\le x} f(n)$$

if this limit exists. In case $f = 1_{\mathcal{A}}$ is the characteristic function of some set $\mathcal{A}$ of integers ($f(n) = 1$ if $n \in \mathcal{A}$, $f(n) = 0$ otherwise), the mean-value (1.1) is also the **density** $\delta(\mathcal{A})$ of the set $\mathcal{A}$:

$$(1.2) \qquad \delta(\mathcal{A}) = \lim_{x\to\infty} x^{-1}\cdot\sum_{n\le x} 1_{\mathcal{A}}(n).$$

If f is a *real-valued* function, the *upper* [resp. *lower*] *mean-value* $M^{-}(f)$ [resp. $M_{-}(f)$] of f, defined as

$$M^{-}(f) = \limsup_{x\to\infty} x^{-1}\cdot \sum_{n\le x} f(n), \qquad M_{-}(f) = \liminf_{x\to\infty} x^{-1}\cdot\sum_{n\le x} f(n),$$

always exists, and so the upper density $\delta^{-}(\mathcal{A})$ and lower density $\delta_{-}(\mathcal{A})$ of a subset $\mathcal{A}$ of $\mathbb{N}$ always exist. The density $\delta(\mathcal{A})$ exists if and only if the upper and lower density of $\mathcal{A}$ are equal.

More generally, often the asymptotic behaviour of the mean-value-function $x \mapsto M(f,x)$ is required, where

$$(1.3) \qquad M(f,x) = \sum_{n\le x} f(n),$$

for example, if one is interested in results beyond the pure existence of $M(f) = \lim_{x\to\infty} x^{-1}\cdot M(f,x)$, or if the mean-value (1.1) does not exist (this is the case, for example, for $f = \tau$, the divisor function).

The **existence** of the limit (1.1) is often a disguised form of some other arithmetical statement, and thus it is of considerable importance to obtain results on the existence (and the value) of the limit (1.1).

For **example**:

- The assertion $M(\mu) = 0$ is equivalent to the prime number theorem in the form $\psi(x) = x \cdot (1 + o(1))$, this being nothing more than $M(\Lambda) = 1$.

- The result $M(\mu^2) = 6\cdot\pi^{-2}$ is a result on the density of the set of squarefree integers.
- The result $M(a) = \prod_{2\le k\le\infty} \zeta(k)$, where $a(n)$ denotes the number of non-isomorphic abelian groups of order n, provides some information on algebraic objects (much more precise information is available, also for other types of algebraic objects; see, for example, J. KNOPFMACHER [1975]).
- Knowledge of $\sum_{n\le x} \tau(n)$ provides information on the number of lattice-points (these are points in $\mathbb{R}^2$ with integer coordinates) in the planar region between the hyperbola $\xi\cdot\eta \le x$ and the axis. Denoting by r(n) the number of representations of the integer n as a sum of two squares, then the behaviour of $\sum_{n\le x} r(n)$ contains results on the number of lattice points in the disc $B(0,x^{\frac{1}{2}})$ [see I.2, Theorem 2.7, and II.4].
- If $\rho(n)$ denotes the number of solutions of the congruence $P(u) \equiv 0$ mod n, where P(X) is a monic, irreducible polynomial with integer coefficients, then the existence of $M(\rho)$ is a non-trivial result concerning polynomial congruences.
- FOURIER coefficients $\hat{f}(\alpha) = M(f\cdot e_{-\alpha})$ of arithmetical functions and RAMANUJAN[-FOURIER] coefficients $a_r(f) = \{\varphi(r)\}^{-1}\cdot M(f\cdot c_r)$ are defined via the notion of mean-value.
- In the theory of sieve methods and their applications, estimates for sums such as $\sum_{n\le x,\gcd(n,k)=1} \mu^2(n)\cdot\{\varphi(n)\}^{-1}$ are useful.
- The question of the existence of a **limit distribution**

(1.4) $$\lim_{N\to\infty} N^{-1}\cdot \#\Big\{ n \le N;\ g(n) \le x \Big\}$$

for a *real-valued* arithmetical function g is, according to the *continuity theorem for characteristic functions*, connected with the existence of the mean-value

(1.5) $$M\Big(n \mapsto \exp\{ 2\pi i\cdot t\cdot g(n)\}\Big).$$

- In the theory of **uniform distribution** modulo one, the foundations having been layed by HERMANN WEYL in 1916, the uniform distribution of a real-valued sequence $\{x_n\}_{n=1,2,\dots}$ depends on the existence of the mean- values $M_k = M(n \mapsto \exp\{2\pi i\cdot k\cdot x_n\})$ for every k in

$\mathbb{Z}$. The condition $M_k = 0$ for every $k \neq 0$ is necessary and sufficient for the uniform distribution of $\{x_n\}_{n=1,2,\dots}$.

Figure II.1 shows the characteristic function of the **squarefree numbers** in the range 1, ..., 10.000. The small squares indicate the integers 1, ..., 100 (line at the bottom), 101,...,200 (next line), ..., 9901,...,10.000 (top line).

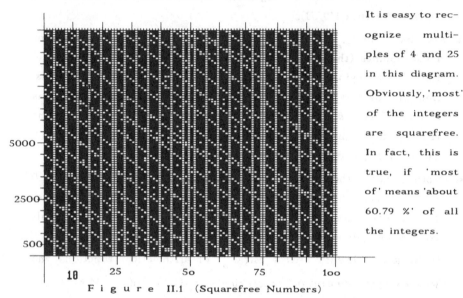

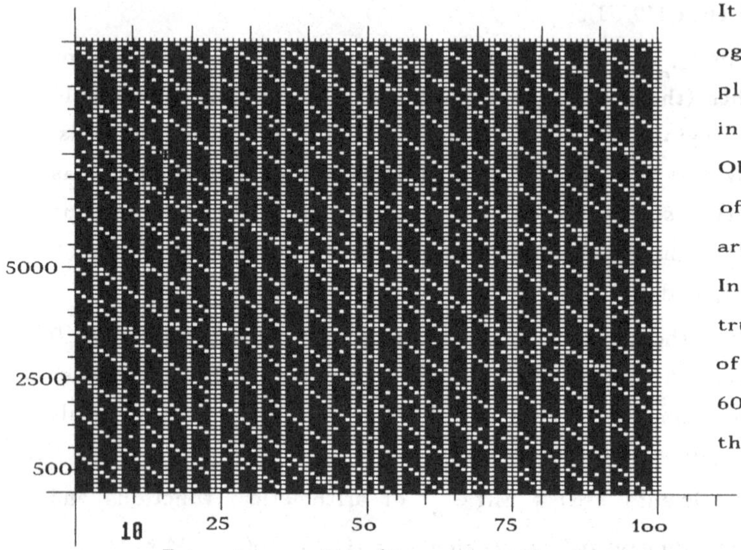

It is easy to recognize multiples of 4 and 25 in this diagram. Obviously, 'most' of the integers are squarefree. In fact, this is true, if 'most of' means 'about 60.79 %' of all the integers.

F i g u r e II.1 (Squarefree Numbers)

Numerous investigations dealt with the determination of mean-values for special arithmetical functions. The aim of this chapter, however, is, to provide general theorems which secure the existence of mean-values for large classes of arithmetical functions, or, at least, to provide estimates for the sum $\sum_{n\le x} f(n)$. Of course, it is easier to obtain results for classes of arithmetical functions which have some kind of arithmetical structure, and thus the functions most frequently dealt with are multiplicative.

Multiplicative functions, according to I. (2.6'), are determined by their values on the prime-powers. Higher prime-powers are rather rare, their number up to x being

(1.6) $$\sum_{p^k\le x,\ k\ge 2} 1 \le \sum_{p\le\sqrt{x}} 1 + \sum_{3\le k\le \log x/\log 2} x^{\frac{1}{3}} = \mathcal{O}\left(x^{\frac{1}{2}} \right).$$

Therefore, it is reasonable to assume that the behaviour of $\sum_{n\le N} f(n)$

is intimately connected with the behaviour of $\sum_{p \le x} f(p)$. An important example here is E. WIRSING's theorem, which is introduced in II, § 4. Further examples are provided by DELANGE's theorem and the theorems of P. D. T. A. ELLIOTT and H. DABOUSSI (see Chapter VI).

The HALÁSZ Theorem (see section 5 of this chapter) deals with complex-valued multiplicative functions that are restricted in size (the condition $|f| \le 1$ is assumed), but there is no assumption on $\sum_{p \le x} f(p)$. Multiplicativity is treated as a Tauberian condition. E. WIRSING published his similar theorems with other restrictions on the values of f just one year before G. HALÁSZ. In particular, his results (and, of course, the HALÁSZ Theorem, too) contain a proof of the famous ERDÖS-WINTNER conjecture: *Any multiplicative function* f, *assuming only the values* 1, 0 , *and* -1, *possesses a mean-value.* Choosing $f = \mu$, the MÖBIUS function, the truth of this conjecture includes the prime number theorem.

II.2. ELEMENTARY MEAN-VALUE THEOREMS (WINTNER, AXER)

Chapter I, section 2, discussed, for a given completely multiplicative function h, the linear transformation

$$T_h : \mathbb{C}^{\mathbb{N}} \to \mathbb{C}^{\mathbb{N}}, \ T_h : f \mapsto f * h,$$

of the vector-space $\mathbb{C}^{\mathbb{N}} = \left\{ f\colon \mathbb{N} \to \mathbb{C} \right\}$ of complex-valued arithmetical functions onto itself. The map T_h is bijective, and its inverse is

$$T_h^{-1}\colon f \mapsto (\mu \cdot h) * f.$$

Theorem 2.1. *Assume that* h *is a fixed* <u>*completely*</u> *multiplicative arithmetical function with mean-value* $M(h) = H$, *and that the arithmetical function* $f\colon \mathbb{N} \to \mathbb{C}$ *has the property*

$$(2.1) \qquad \sum_{n=1}^{\infty} n^{-1} \cdot |(T_h^{-1}(f))(n)| = \sum_{n=1}^{\infty} n^{-1} \cdot \Big| \sum_{d|n} \mu(d) \cdot h(d) \cdot f(n/d) \Big| < \infty.$$

Then the mean-value M(f) *exists and is equal to*

(2.2) $$M(f) = H \cdot \sum_{n=1}^{\infty} n^{-1} \cdot (T_h^{\ -1}(f))(n) .$$

Choosing h = 1 (a constant function), we obtain the following result.

Corollary 2.2 (A. WINTNER). *If*

(2.3) $$\sum_{n=1}^{\infty} n^{-1} \cdot \Big| \sum_{d|n} \mu(d)\cdot f(n/d) \Big| < \infty ,$$

then the mean-value

(2.4) $$M(f) = \sum_{n=1}^{\infty} n^{-1} \cdot \sum_{d|n} \mu(d)\cdot f(n/d)$$

exists.

Examples.

(1) Consider EULER's function $\varphi = \mu * id_{\mathbb{N}}$. Then

$$n^{-1}\cdot\varphi(n) = \sum_{d|n} d^{-1}\cdot\mu(d), \text{ and so } (id)^{-1}\cdot\varphi = 1 * (id^{-1}\cdot\mu).$$

Obviously the series $\sum_{n=1}^{\infty} \left|n^{-1}\cdot\mu(n)\right|\cdot n^{-1}$ is convergent; hence Corollary 2.2 gives

$$M((id)^{-1}\cdot\varphi\ ) = \sum_{n=1}^{\infty} n^{-2}\cdot \mu(n) = (\zeta(2))^{-1} = 6 \cdot \pi^{-2}.$$

(2) If $\sigma(n) = \sum_{d|n} d$ denotes the *sum of the divisors* of n, then $\sigma = 1 * id$, and $(id)^{-1}\cdot\sigma = (id)^{-1} * 1$. Arguing as before, Corollary 2.2 leads to

$$M(\ (id)^{-1}\cdot\sigma\ ) = \sum_{n=1}^{\infty} n^{-2} = \frac{1}{6} \cdot \pi^2.$$

(3) If h is completely multiplicative , then $h(n) = h(d)\cdot h(n/d)$ for any d dividing n. Assume that $M(h) = H$ exists and that $|h(n)| \le K$ is bounded. Then, after a short calculation, Theorem 2.1, applied with $f = h\cdot\varphi/id$, gives the following mean-value result:

$$M(h\cdot\varphi/id) = H \cdot \sum_{n=1}^{\infty} n^{-1}\cdot \sum_{d|n} \mu(d)\cdot h(d)\cdot h(n/d)\cdot\varphi(n/d)\Big/(n/d)$$

$$= H \cdot \sum_{n=1}^{\infty} n^{-2}\cdot h(n)\cdot\mu(n) = H \cdot \prod_p \left(1 - p^{-2}\cdot h(p) \right).$$

(4) If h is the characteristic function of the set of integers coprime with some fixed integer m, then $H = M(h) = m^{-1}\cdot\varphi(m)$, and so

$$\sum_{n\le x,\ \gcd(n,m)=1} n^{-1}\cdot\varphi(n) = \frac{\varphi(m)}{m} \cdot \prod_{p\nmid m} \left(1 - p^{-2} \right) .$$

Proof of Theorem 2.1. Abbreviate $T_h^{-1}(f)$ by f'. Then $f = h * f'$, and

$$\sum_{n\le x} f(n) = \sum_{n\le x} \sum_{d|n} f'(d)\cdot h\left(\frac{n}{d}\right)$$
$$= \sum_{d\le x} f'(d) \cdot \sum_{n\le x,\ n\equiv 0 \bmod d} h\left(\frac{n}{d}\right)$$
$$= \sum_{d\le x} f'(d) \cdot \left(H\cdot\frac{x}{d} + \Theta\left(\frac{x}{d}\right)\right),$$

where $\Theta\left(\frac{x}{d}\right) = o\left(\frac{x}{d}\right)$, as $\frac{x}{d} \to \infty$; therefore $\left|\Theta\left(\frac{x}{d}\right)\right| \le \varepsilon\cdot\frac{x}{d}$, if $\frac{x}{d} \ge K(\varepsilon)$, and $\left|\Theta\left(\frac{x}{d}\right)\right| \le L(\varepsilon)$ is bounded, if $\frac{x}{d} \le K(\varepsilon)$. Thus

$$\sum_{n\le x} f(n) = \sum_{1\le d<\infty} f'(d)\cdot H\cdot\frac{x}{d} + E_1(x) + E_2(x),$$

with two error terms; the first one is

$$E_1(x) = - H\cdot x \cdot \sum_{d>x} d^{-1}\cdot f'(d) = o(x), \text{ as } x \to \infty,$$

due to the absolute convergence of $\sum_d d^{-1}\cdot f'(d)$. The second error term is

$$E_2(x) = \sum_{d\le x} f'(d) \cdot\Theta\left(\frac{x}{d}\right),$$

and so

$$\left| E_2(x)\right| \le \sum_{d\le x/K(\varepsilon)} | f'(d)| \cdot\varepsilon\cdot \left(\frac{x}{d}\right) + \sum_{x/K(\varepsilon)<d\le x}|f'(d)|\cdot L(\varepsilon).$$

The first sum is $\le \varepsilon\cdot x\cdot\sum d^{-1}\cdot|f'(d)|$, and is thus sufficiently small [using (2.1)]. The second sum is

$$\le L(\varepsilon) \cdot \sum_{x/K(\varepsilon)<d\le x}|f'(d)|\cdot \frac{x}{d} \le L(\varepsilon)\cdot x \cdot \sum_{x/K(\varepsilon)<d} d^{-1}\cdot|f'(d)|,$$

and so is $< \varepsilon\cdot x$, if x is sufficiently large. □

If the function f in Corollary 2.2 is *multiplicative*, then a nicer-looking result is available.[1)]

Corollary 2.3. Let $f\colon \mathbb{N} \to \mathbb{C}$ *be a multiplicative function satisfying*

$$\sum_p p^{-1}\cdot\left| f(p) - 1 \right| < \infty, \tag{2.5}$$

---

1) This theorem will be sharpened later (DELANGE, ELLIOTT, WIRSING Theorems, see II.4 and VII).

and

(2.6) $$\sum_p \sum_{k\ge 2} p^{-k} \cdot \left| f(p^k) \right| < \infty.$$

Then the mean-value

(2.7) $$M(f) = \prod_p \left(1 + p^{-1}\cdot(f(p)-1) + p^{-2}\cdot(f(p^2)-f(p)) + \dots \right)$$

exists.

Remark 1. (a) If, in addition, f is *strongly multiplicative* (recall: this means that $f(p^k) = f(p)$ for any p and any $k = 1,2 \dots$), then

(2.7') $$M(f) = \prod_p \left(1 + p^{-1}\cdot(f(p)-1) \right).$$

(b) If, in addition, f is *completely multiplicative*, then

(2.7") $$M(f) = \prod_p \left(1 + p^{-1}\cdot(f(p)-1)\cdot(1 - p^{-1}\cdot f(p))^{-1} \right).$$

Note that, in this case, $| p^{-1}\cdot f(p)| < 1$ for any prime p by condition (2.6).

Proof of Corollary 2.3. The function $f' = \mu * f$ is multiplicative, and so

$$\sum_{n\le x} n^{-1}\cdot |f'(n)| \le \prod_{p\le x} \left(1 + p^{-1}\cdot|f'(p)| + p^{-2}\cdot|f'(p^2)| + \dots \right).$$

Using the values $f'(p) = f(p) - 1$, and $f'(p^k) = f(p^k) - f(p^{k-1})$, and utilizing the inequality $1 + \beta \le \exp(\beta)$ [if $\beta \ge 0$], we obtain

$$\sum_{n\le x} n^{-1}\cdot|f'(n)| \le \prod_{p\le x} \left(1 + p^{-1}\cdot|f(p)-1| + \sum_{k\ge 2} p^{-k}\cdot\left| f(p^k) - f(p^{k-1})\right| \right)$$

$$\le \exp\left\{ \sum_{p\le x} \left(p^{-1}\cdot|f(p)-1| + \sum_{k\ge 2} p^{-k}\cdot\left| f(p^k) - f(p^{k-1})\right|\right)\right\}.$$

The convergence of the series $\sum_p p^{-2}\cdot|f(p)|$ follows from (2.5) and the convergence of $\sum_p p^{-2}$. Equations (2.5) and (2.6) imply the boundedness of the series $\left(\sum_{p\le x}(\dots) \right)$ in the argument of the exponential function; thus the assumptions of Corollary 2.2 are verified, and Corollary 2.3 is proven. □

The method used for the proof of Theorem 2.1 also gives the following result.

Theorem 2.4. *Let* g, *and* h *be arithmetical functions. Suppose that the mean-value* M(g) *exists and that the series* $\sum_{n=1}^{\infty} n^{-1}\cdot|h(n)|$ *is con-*

vergent. Then the function $f = g * h$ *has the mean-value*

(2.8) $$M(f) = \sum_{n=1}^{\infty} n^{-1} \cdot h(n) \cdot M(g).$$

The **proof** is left as an exercise (Exercise 2).

It is possible to weaken the assumptions in Corollary 2.2: absolute convergence of the series (2.3) is replaced by mere convergence; but an additional condition (2.10), that is weaker than absolute convergence is unavoidable. A sufficient condition is displayed in A. AXER's Theorem, as follows.

Theorem 2.5. *Assume that* f *is an arithmetical function, and put* $f' = \mu * f$. *Suppose that*

(2.9) *the series* $\sum_{n=1}^{\infty} n^{-1} \cdot f'(n)$ *is convergent,*

and that

(2.10) $$\sum_{n \le N} |f'(n)| = \mathcal{O}(N) \text{ for } N \to \infty.$$

Then the mean-value $M(f)$ *exists and equals*

(2.4) $$M(f) = \sum_{n=1}^{\infty} n^{-1} \cdot f'(n).$$

Examples. (5) The convolution formulae $\mu^2 = ((\mu \circ \surd) \cdot sq) * 1$, $id/\varphi = 1 * (\mu^2/\varphi)$, where sq is the characteristic function of the set of squares of integers, lead to $M(\mu^2) = \sum_{n=1}^{\infty} n^{-2} \cdot \mu(n) = 6 \cdot \pi^{-2}$, and to $x^{-1} \cdot \sum_{n \le x} (n/\varphi(n)) \to \prod (1 + \{(p-1)p\}^{-1})$.

(6) The function $\frac{1}{4} r(n)$, where $r(n)$ is the number of representations of n as a sum of two squares, is multiplicative and representable as $\sum_{d|n} \chi(d)$, where χ is the non-principal character modulo 4; therefore $\frac{1}{4} r = 1 * \chi$, with convergent sum $\sum d^{-1} \cdot \chi(d)$. So AXER's result gives the mean-value result

$$M(\tfrac{1}{4} r) = 1 - \frac{1}{3} + \frac{1}{5} - + \ldots = \tfrac{1}{4} \pi.$$

Remark 2. WINTNER's Theorem (Corollary 2.2) follows from AXER's Theorem: The absolute convergence of $\sum_{n=1}^{\infty} n^{-1} \cdot |f'(n)|$ implies the following by partial summation (see I.1):

$$\sum_{n \le x} n^{-1} \cdot |f'(n)| \cdot n = x \cdot \sum_{n \le x} n^{-1} \cdot |f'(n)| - \int_1^x \sum_{n \le u} n^{-1} \cdot |f'(n)| \, du \le C \cdot x.$$

In fact, the same idea and sensitive handling of the sum and integral yields the stronger result

$$\sum_{n\le x} |f'(n)| = o(x).$$

Remark 3. Condition (2.9) *alone* is not sufficient for the existence of M(f). See, for example, A. WINTNER [1943].

Remark 4. *If* M(f) *exists and if* $\sum_n n^{-1} \cdot f'(n)$ *is convergent, then*

$$M(f) = \sum_n n^{-1} \cdot f'(n).$$

Proof. The existence of M(f) implies $M(f,x) = \sum_{n\le x} f(n) = M(f)\cdot x + o(x)$. Partial summation gives

$$\mathcal{D}(f,s) = \sum_n f(n)\cdot n^{-s} \sim M(f) \cdot (s-1)^{-1}, \text{ as } s \to 1+.$$

Therefore,

$$M(f) = \lim_{s\to 1+} \mathcal{D}(f,s)\cdot\zeta^{-1}(s) = \lim_{s\to 1+} \sum (\mu*f)(n)\cdot n^{-s}.$$

But the convergence of $\sum (\mu*f)(n)\cdot n^{-1}$ implies, by the *continuity theorem for* DIRICHLET *series* (or, what amounts to the same, by partial summation), that $\lim_{s\to 1+} \sum (\mu*f)(n)\cdot n^{-s} = \sum (\mu*f)(n)\cdot n^{-1}$, and Remark 4 is proved. □

Proof of AXER's Theorem. Abbreviating $\beta - [\beta]$ by $\{\beta\}$, a routine calculation (change of the order of summation) gives

$$\sum_{n\le N} f(n) = \sum_{n\le N} \sum_{d|n} f'(d) = \sum_{d\le N} f'(d) \cdot \left[\frac{N}{d}\right]$$

$$= N \cdot \sum_{d\le N} d^{-1}\cdot f'(d) - \sum_{d\le N} f'(d) \cdot \left\{\frac{N}{d}\right\}$$

$$= N \cdot \sum_{d=1}^{\infty} d^{-1}\cdot f'(d) + o(N) + R(N),$$

with the remainder term

$$R(N) = -\sum_{d\le N} f'(d)\cdot\left\{\frac{N}{d}\right\}. \tag{2.11}$$

It must be shown that R(N) is not larger than $o(N)$. The summatory function

$$M(f', x) = \sum_{n\le x} f'(n)$$

has the following properties:

(i) $|M(f',x)| \le \sum_{n\le x} |f'(n)| \le C\cdot x$ with some constant C by (2.10),

(ii) $|M(f',x)| = o(x)$, as $x \to \infty$.

Statement (ii) is obtained from (2.9) and partial summation:

$$M(f',x) = \sum_{n\le x} (n^{-1}\cdot f'(n)) \cdot n = \sum_{n\le x} n^{-1}\cdot f'(n) \cdot x - \int_1^x \sum_{n\le u} n^{-1}\cdot f'(n)\, du,$$

and, inserting $\sum_{n\le x} n^{-1}\cdot f'(n) = \gamma + o(1)$, (ii) is easily proved [in order to show $\int_1^x o(1)\,du = o\left(\int_1^x 1\cdot du\right)$, split the integral into $\int_1^{\sqrt{x}} + \int_{\sqrt{x}}^x$].

Next we consider the sum

$$R(M,N) = -\sum_{M<d\le N} f'(d) \cdot \left\{\frac{N}{d}\right\}.$$

ABEL summation [I. 1, Theorem 1.3; Theorem 1.1, which is easier, is not applicable, the function $x \mapsto \{x\}$ being not differentiable] gives

$$-R(M,N) = (M(f',N) - M(f',M))\cdot\left\{\frac{N}{N}\right\}$$

$$- \sum_{M<n\le N-1} (M(f',n) - M(f',M))\cdot\left(\left\{\frac{N}{n+1}\right\} - \left\{\frac{N}{n}\right\}\right).$$

Therefore,

$$|R(M,N)| \le 2\cdot \max_{M<n\le N-1} |M(f',n)| \cdot \sum_{M<n\le N-1} \left|\left\{\frac{N}{n+1}\right\} - \left\{\frac{N}{n}\right\}\right|.$$

Now divide the integers n in $M < n \le N-1$ into two disjoint classes $\mathcal{P}$ and $\mathcal{T}$: n is in $\mathcal{P}$ if $\left[\frac{N}{n+1}\right] = \left[\frac{N}{n}\right]$, otherwise it is in $\mathcal{T}$. If n is in $\mathcal{P}$, then

$$\left|\left\{\frac{N}{n+1}\right\} - \left\{\frac{N}{n}\right\}\right| = \left|\frac{N}{n+1} - \frac{N}{n}\right| \le \frac{N}{n\cdot(n+1)},$$

and so

$$\left|\sum_{M<n\le N-1,\ n\in\mathcal{P}} (M(f',n) - M(f',M))\cdot\left(\left\{\frac{N}{n+1}\right\} - \left\{\frac{N}{n}\right\}\right)\right|$$

$$\le \max_{M<n\le N} |M(f',n)| \cdot \sum_{n>M} \frac{N}{n\cdot(n+1)} \le \frac{N}{M}\cdot \max_{M<n\le N} |M(f',n)|.$$

For integers $n \in \mathcal{T}$ the inequality

$$1 \le \left[\frac{N}{n+1}\right] < \left[\frac{N}{n}\right] \le \frac{N}{M}$$

implies that there are at most $\frac{N}{M}$ elements in $\mathcal{T}$; thus

$$\sum_{n\in\mathcal{T}} \left| \left\{\frac{N}{n+1}\right\} - \left\{\frac{N}{n}\right\} \right| \le 2 \cdot \frac{N}{M} .$$

Collating our estimates, we obtain

$$|R(N)| \le \left|\sum_{d\le M} f'(d) \cdot \left\{ \frac{N}{d}\right\} \right| + \left| \sum_{M<d\le N} f'(d) \cdot \left\{ \frac{N}{d}\right\} \right|$$

$$\le C\cdot M + 6 \cdot \frac{N}{M} \cdot \max_{M<n\le N} |M(f', n)| + o(M).$$

Choose $\varepsilon > 0$, then fix $M = \varepsilon\cdot N$, and finally choose N so large (this is possible by (ii)) that $\max_{M<n\le N} |M(f', n)| < \varepsilon^2\cdot N$. Then

$$| R(N) | \le (C + 6) \cdot \varepsilon \cdot N. \qquad \Box$$

II.3. ESTIMATES FOR SUMS OVER MULTIPLICATIVE FUNCTIONS (RANKIN'S TRICK)

This section deals with sums over non-negative multiplicative functions; the results given connect [estimates for] the size of values of f at prime-powers in mean with the estimates of the sum $\sum_{n\le N} f(n)$ from above (this is rather easy) and from below (this is more difficult). The ideas involved are not too difficult: to obtain an upper estimate for $\sum_{n\le N} f(n)$, where $f \ge 0$, an additional weight factor $g(n) \ge 1$ is introduced (for example, $g(n) = \log n$ if $n \ge 3$, or $g(n) = (N/n)^{\beta}$ if $n \le N$), which makes the treatment of the sum easier (for example, log n can be split additively, or other factors g(n) make it possible to remove some troublesome conditions of summation without increasing the sum under consideration too much); the factor g(n) has to be chosen in such a way that the new sum $\sum_{n\le N} f(n)\cdot g(n)$ can be dealt with in a simpler way. Surprisingly that this simple method ("RANKIN's trick") is often very effective.

Theorem 3.1. *Suppose that for a non-negative, multiplicative arithmetical function* $f: \mathbb{N} \to [0,\infty[$, *for every* $y \ge 1$, *the upper estimate*

(3.1) $$\sum_{p^k \le y} f(p^k) \cdot \log p^k \le c_1 \cdot y \cdot (\log y)^\alpha$$

is true with some $\alpha \ge 0$ *and some positive constant* c_1. *Then there are constants* c_2, c_3, *which depend only on* c_1 *so that for all* $x \ge 2$

(3.2) $$\sum_{n \le x} f(n) \le c_2 \cdot x\,(\log x)^{\alpha-1} \exp\Big\{\sum_{p \le x} p^{-1} \cdot f(p) + \sum_{p \le x}\sum_{k \ge 2} p^{-k} f(p^k)\Big\},$$

(3.3) $$\sum_{n \le x} f(n) \le c_3 \cdot x \cdot \log^\alpha x \cdot \exp\Big\{\sum_{p \le x} p^{-1} \cdot (f(p)-1) + \sum_{p \le x}\sum_{k \ge 2} p^{-k} \cdot f(p^k)\Big\}.$$

Remark. If, for every prime-power p^k, $0 \le f(p^k) \le \gamma_1 \cdot \gamma_2^{\,k}$, where $\gamma_2 < 2$, then (3.1) holds with $\alpha = 0$.

Proof. The function $n \mapsto \log n$ is additive, and so $\log n = \sum_{p^k \| n} \log p^k$. Inserting this into $\sum_{n \le x} f(n) \log n$, inverting the order of summation, and using the multiplicativity of f, we obtain

$$\sum_{n \le x} f(n) \cdot \log n = \sum_{p^k \le x} \log p^k \cdot \sum_{m \le x/p^k,\ p \nmid m} f(m) \cdot f(p^k).$$

Inverting the order of summation again, and neglecting the condition $p \nmid m$, this leads to the estimate

$$\sum_{n \le x} f(n) \cdot \log n \le \sum_{m \le x} f(m) \cdot \sum_{p^k \le x/m} f(p^k) \cdot \log p^k \le c_1 \cdot x \cdot (\log x)^\alpha \cdot \sum_{m \le x} m^{-1} \cdot f(m).$$

Assumption (3.1) was used in the last line. The multiplicativity of f and the inequality $1 + y \le \exp(y)$ imply

$$\sum_{m \le x} m^{-1} \cdot f(m) \le \prod_{p \le x} \big(1 + p^{-1} \cdot f(p) + p^{-2} \cdot f(p^2) + \dots\big)$$

$$\le \exp\Big(\sum_{p \le x} p^{-1} \cdot f(p) + \sum_{p \le x}\sum_{k \ge 2} p^{-k} \cdot f(p^k)\Big) = E(x),$$

for the moment. Then, for $x \ge 2$, $\sum_{n \le x} f(n) \cdot \log(n) \le c_1 \cdot x \cdot \log^\alpha x \cdot E(x)$, and

$$\sum_{2 \le n \le x} f(n) \le \sum_{n \le \sqrt{x}} f(n) \cdot (\log n / \log 2) + \sum_{\sqrt{x} < n \le x} f(n) \cdot \log n / (\tfrac{1}{2} \log x)$$

$$\le c_1 \cdot x \cdot \log^{\alpha-1} x \cdot E(x) \cdot \big\{x^{-\frac{1}{2}} \cdot \log x / \log 2 + 2\big\} \le 4\, c_1 \cdot x \cdot \log^{\alpha-1} x \cdot E(x).$$

In our argument we used the fact that $y \mapsto E(y)$ is monotonically increasing, and that the maximum of $x^{-\frac{1}{2}} \cdot \log x / \log 2$ in $x \ge 2$ is attained at $x = e^2$ and equals 1.06 … .

The estimate $\sum_{p \le x} p^{-1} \le \log\log(x) + c_4$ (see I.7) implies the second assertion. □

Another form of this theorem is given in G. HALL & G. TENENBAUM [1988], Theorem 0.1.

Theorem 3.2. *Let* f *be a non-negative multiplicative function satisfying*

(3.4) $$\sum_{p\le x} f(p)\cdot\log p \le c_1\cdot x \quad (x \ge 1),$$

and

(3.5) $$\sum_p \sum_{k\ge 2} p^{-k}\cdot f(p^k)\cdot \log(p^k) \le c_2.$$

Then, for any $x \ge 1$,

$$\sum_{n\le x} f(n) \le (c_1 + c_2 + 1)\cdot x\cdot \log^{-1}x \cdot \sum_{n\le x} n^{-1}\cdot f(n).$$

Remark. WIRSING's condition $f(p^k) \le \gamma_1\cdot\gamma_2{}^k$ for $k \ge 2$, where $0 \le \gamma_2 < 2$, is stronger than condition (3.5) for the higher prime-powers.

Proof. As before, we begin with

$$\sum_{n\le x} f(n)\cdot\log(n) = \sum_{p^k\le x} \log (p^k) \cdot \sum_{m\le x/p^k,\ p\nmid m} f(m)\cdot f(p^k)$$

$$\le \sum_{m\le x} f(m) \sum_{p\le x/m} f(p) \log p + \sum_{p^k\le x, k\ge 2} f(p^k) \log p^k \cdot \sum_{m\le x/p^k} f(m).$$

The obvious estimate

$$\Sigma_{m\le x/p^k} f(m) \le x \cdot p^{-k} \cdot \Sigma_{m\le x/p^k} m^{-1} \cdot f(m)$$

implies, together with the assumptions of Theorem 3.2,

$$\sum_{n\le x} f(n)\cdot\log(n) \le (c_1 + c_2) \cdot x \cdot \sum_{m\le x} m^{-1} \cdot f(m),$$

and so, noting that $(\log n + x/n) \ge \log x$ in $1 \le n \le x$,

$$\sum_{n\le x} f(n) \le \frac{1}{\log x} \cdot \sum_{n\le x} f(n) \log n + \frac{x}{\log x} \cdot \sum_{n\le x} \frac{f(n)}{n}$$

$$\le (c_1 + c_2 + 1) \cdot \frac{x}{\log x} \cdot \sum_{n\le x} \frac{f(n)}{n}. \qquad \square$$

RANKIN's trick is applied to the proof of the following theorem.

Theorem 3.3. *Assume that* f *is a non-negative multiplicative function, and denote the maximal prime divisor of* n *by* $P(n) = p_{max}(n)$. *Suppose that* $z > y$, *and that for some* $\Delta > 0$ *the series*

(3.6) $$\sum_{p} \sum_{k\ge 2} p^{-k} \cdot f(p^k) \cdot p^{k\Delta}$$

is convergent. If

(3.7) $$\log\Big(\log z/\log y \Big) \Big/ \log y \to 0$$

as $y \to \infty$, *then*

(3.8) $$\sum_{n>z, P(n)\le y} n^{-1} \cdot f(n) \ll \exp\Big(\frac{\log z}{\log y} \cdot \Big[\sum_{p\le y} p^{-1} \cdot f(p) - \log\Big(\frac{\log z}{\log y}\Big)\Big]\Big).$$

Remark. *The series*

$$\sum_{n, P(n)\le y} n^{-1} \cdot f(n) \quad \textit{is equal to} \quad \prod_{p\le y} \Big(1 + p^{-1} \cdot f(p) + \ldots \Big),$$

and so, in some cases, Theorem 3.3 also implies results on

$$\sum_{n\le z, P(n)\le y} n^{-1} \cdot f(n).$$

Proof of Theorem 3.3. With some parameter δ, $0 \le \delta \le \Delta$, we use a weight-function, printed in bold-face, to obtain

$$\begin{aligned}
\sum_{n>z, P(n)\le y} n^{-1} \cdot f(n) &\le \sum_{n>z, P(n)\le y} \mathbf{(n/z)^{\delta}} \cdot n^{-1} \cdot f(n) \\
&\le z^{-\delta} \cdot \sum_{n, P(n)\le y} n^{-1} \cdot f(n) \cdot n^{\delta} \\
&\le z^{-\delta} \cdot \prod_{p\le y} \Big(1 + p^{-1} \cdot f(p) \cdot p^{\delta} + p^{-2} \cdot f(p^2) \cdot p^{2\delta} + \ldots \Big) \\
&\ll \exp\Big\{ -\delta \cdot \log(z) + \sum_{p\le y} p^{-1} \cdot f(p) \cdot p^{\delta} \Big\}.
\end{aligned}$$

A good choice for δ is $\delta = (\log y)^{-1} \cdot \log\Big(\frac{\log z}{\log y}\Big)$. Owing to condition (3.7) this expression is $< \Delta$ if y is sufficiently large. Using this choice of δ, the assertion of Theorem 3.3 is proved. □

Lower bounds for non-negative multiplicative functions are accessible only with greater effort. First we deduce an auxiliary result on "rather small" non-negative multiplicative functions, where, for simplicity, the summation is extended only over squarefree integers. We need some elementary results on prime numbers with explicit constants. The inequality

(3.9) $$\prod_{p\le n} p < 4^n$$

(valid for $n = 1, 2, \ldots$) is easily proved by induction (see Exercise 17). DENIS HANSON, Canad. Math. Bull. 15, 33-37 (1972), gives a stronger result). Equation (3.9) implies

$$\vartheta(n) < 2\cdot\log 2 \cdot n.$$

Partial summation leads to

$$\sum_{p\le x} p^{-1}\cdot\log p < 2\cdot\log 2 \cdot \log x + 2\cdot\log 2 \cdot (1-\log 2)$$

$$\le 1.4 \cdot \log x + 0.45 \quad < 1.55 \cdot \log x,$$

if $x \ge e^3$. In passing we mention that, according to ROSSER-SCHOENFELD [1962], $\sum_{p\le x} p^{-1}\cdot\log p < \log x$ for $x > 1$ (and therefore $x \ge e^3$ could be replaced by $x \ge e$).

Figure II.2 shows the rather smooth function $(\log x)^{-1}\cdot \sum_{p\le x} p^{-1}\cdot\log p$ in the range $1 \le x \le 600$.

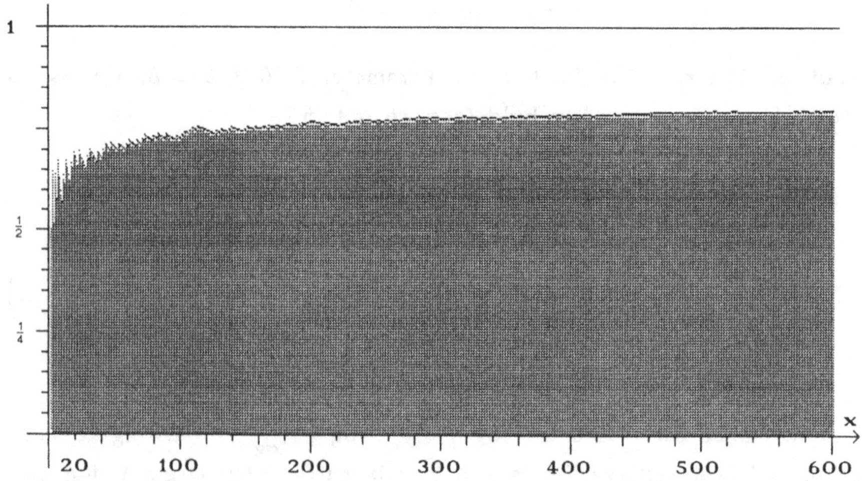

F i g u r e II.2.

Lemma 3.4. *Let* h *be a multiplicative, non-negative arithmetical function satisfying*

$$0 \le h(p) \le 0.1 \textit{ for every prime } p.$$

Then, for $x \ge e^3$,

$$\sum_{n\le x} \mu^2(n) \cdot n^{-1}\cdot h(n) \ge 0.8 \cdot \prod_{p\le x} \left(1 + p^{-1}\cdot h(p) \right). \tag{3.10}$$

Proof. Denote by P(n) the maximal prime divisor of n, as before. Consider the difference between the sum under consideration and the expected approximation $\prod_{p\le x}\left(1+p^{-1}\cdot h(p)\right)$,

$$\prod_{p\le x}\left(1+p^{-1}\cdot h(p)\right)-\sum_{n\le x}\mu^2(n)\cdot n^{-1}\cdot h(n)=\sum_{n>x,P(n)\le x}\mu^2(n)\cdot n^{-1}\cdot h(n).$$

Then, applying RANKIN's idea again, with the weight-function **log n /log x**, this difference is (writing n = p·d, neglecting the condition $d \ge x/p$ and using $h(p) \le 0.1$)

$$\le \sum_{n>x,P(n)\le x}\mu^2(n)\cdot n^{-1}\cdot h(n)\cdot\frac{\log n}{\log x}$$

$$=\sum_{n>x,P(n)\le x}\mu^2(n)\cdot n^{-1}\cdot h(n)\cdot\sum_{p|n}\log p\,/\log x$$

$$\le(\log x)^{-1}\cdot\sum_{p\le x}p^{-1}\cdot h(p)\cdot\log p\cdot\sum_{P(d)\le x,\ d\ge x/p}\mu^2(d)\cdot d^{-1}\cdot h(d)$$

$$\le(\log x)^{-1}\cdot\sum_{p\le x}p^{-1}\cdot h(p)\cdot\log p\cdot\prod_{p\le x}\left(1+p^{-1}\cdot h(p)\right)$$

$$\le 0.1\cdot(\log x)^{-1}\cdot(1.55\cdot\log x)\cdot\prod_{p\le x}\left(1+p^{-1}\cdot h(p)\right)$$

if $x \ge e^3$, and the assertion will follow, after replacing 1.55 by 2. □

Theorem 3.5 [BARBAN]. *Let g be a non-negative multiplicative arithmetical function bounded at the primes,*

$$0\le g(p)\le C_1.$$

Then there is some positive constant C_2, depending only on C_1, such that the inequality

$$\sum_{n\le N}\mu^2(n)\cdot n^{-1}\cdot g(n)\ge C_2\cdot\exp\left(\sum_{p\le N}p^{-1}\cdot g(p)\right)\tag{3.11}$$

holds as soon as N is sufficiently large.

Proof. In order to apply Lemma 3.4, choose an integer m so large that $m^{-1}\cdot g(p)$ is ≤ 0.1, for example $m = [10\cdot C_1] + 1$, and put $z = N^{1/m}$. Define *completely* multiplicative functions g^* and H_o by

$$\begin{cases} g^*(p^k)=\{g(p)\}^k,\ k=1,2,\ldots,\\ H_o(n)=m^{-\Omega(n)}\cdot n^{-1}\cdot g^*(n).\end{cases}\tag{3.12}$$

If H is any non-negative, completely multiplicative function, then

$$\Big\{\sum_{n\le z} H(n)\Big\}^m = \sum_{n_1\le z,\dots,n_m\le z} \cdots \sum H(n_1\cdot\ldots\cdot n_m)$$
$$\le \sum_{r\le z^m} H(r)\cdot\tau_m(r)$$

with the divisor function $\tau_m(r)$ counting the number of representations of r as a product of m factors. The values of τ_m at primes p are $\tau_m(p) = m$, and thus $\tau_m(n) = m^{\Omega(n)}$ if n is squarefree. Using the representation $\tau_m = 1 * \ldots * 1 = 1 * \tau_{m-1}$, the relation

$$\tau_m(p^k) = \binom{k+m-1}{m-1},\ k = 1, 2, \ldots,\ m = 1, 2, \ldots, \tag{3.13}$$

is easily proved by induction (Exercise 16).

Write r in the form $r = r'\cdot d$, where d is squarefree, r' is 2-full (this means that $p|r'$ implies $p^2|r'$), and $\gcd(r',d) = 1$. Then, neglecting the condition $\gcd(r',d) = 1$, we obtain

$$\Big\{\sum_{n\le z} H(n)\Big\}^m \le \sum_{r'\le z^m,\, r'\text{ 2-full}} H(r')\cdot\tau_m(r')\cdot\sum_{d\le z^m}\mu^2(d)\cdot H(d)\cdot\tau_m(d)$$
$$\le \prod_{p\le z^m}\Big\{1 + H(p^2)\cdot\tau_m(p^2) + H(p^3)\cdot\tau_m(p^3) + \ldots\Big\}$$
$$\times \sum_{d\le z^m}\mu^2(d)\cdot H(d)\cdot\tau_m(d).$$

With the choice $H = H_0$ given above (see (3.12)), and paying attention to $g^*(p) \le C_1$, the product

$$P_0 = \prod_p \Big\{1 + H_0(p^2)\cdot\tau_m(p^2) + H_0(p^3)\cdot\tau_m(p^3) + \ldots\Big\}$$

is convergent (for this, an estimate such as $\tau_m(p^k) \ll_{m,\varepsilon} p^{k\varepsilon}$ is useful). The sum

$$G(N,m) = \Big\{\sum_{n\le N^{1/m}}\mu^2(n)\cdot n^{-1}\cdot g^*(n)\cdot m^{-\Omega(n)}\Big\}^m = \Big\{\sum_{n\le N^{1/m}}\mu^2(n)\cdot H_0(n)\Big\}^m$$

satisfies

$$G(N,m) \le \Big\{\sum_{n\le N^{1/m}} H_0(n)\Big\}^m \le P_0\cdot\sum_{d\le N}\mu^2(d)\cdot d^{-1}\cdot g(d)$$

on the one hand; on the other hand we obtain, from Lemma 3.4,

$$G(N,m) \ge (0.8)^m\cdot\prod_{p\le N^{1/m}}\Big(1 + \frac{g(p)}{p\cdot m}\Big)^m.$$

These two estimates imply the relation

$$\sum_{d\le N} \mu^2(d)\cdot d^{-1}\cdot g(d) \ge P_o^{-1}\cdot(0.8)^m\cdot\prod_{p\le N^{1/m}}\Big(1+\frac{g(p)}{p\cdot m}\Big)^m,$$

for every $N \ge e^{3m}$. For $0 \le x < 1$ the inequality

$$(1+x) = (1-x^2)\cdot(1-x)^{-1} \ge (1-x^2)\cdot\exp(x)$$

is valid. Thus we obtain (with $x = g(p)/mp$)

$$P_1 := \prod_p\Big(1-\big((m\cdot p)^{-1}\cdot g(p)\big)^2\Big) \ge \prod_p\Big(1-0.01\cdot p^{-2}\Big) > 0.9$$

(by an easy numerical estimate), and, finally,

$$\sum_{d\le N}\mu^2(d)\cdot d^{-1}\cdot g(d) \ge P_o^{-1}\cdot P_1^{m}\cdot(0.8)^m\cdot\exp\Big(-C_1\cdot\sum_{N^{1/m}<p\le N}p^{-1}\Big)$$

$$\times\exp\Big(\sum_{p\le N} m\cdot(m\cdot p)^{-1}\cdot g(p)\Big).$$

The well-known asymptotic formula for $\sum_{p\le x} p^{-1}$ (I.6, (6.7)) yields

$$\sum_{N^{1/m}<p\le N} p^{-1} = \log(m) + o(1),\ N\to\infty,$$

and thus Theorem 3.5 is proved. □

BARBAN's Theorem dealt with functions of order n^{-1} "in mean". For multiplicative functions of order 1 "in mean", removing the restriction of summation over squarefree integers, one gets the following theorem.

Theorem 3.6. *Let f be a non-negative multiplicative arithmetical function, satisfying*

(3.14) $$f(p) \ge \gamma_1 > 0 \text{ for all primes } p \ge P_o,$$

and

$$f(p^k) \le \gamma_2 \text{ for every prime } p \text{ and every } k = 1,2,\ldots .$$

Then, with some positive constant γ depending only on γ_1, γ_2 and P_o, the inequality

(3.15) $$\sum_{n\le x} f(n) \ge \gamma\cdot x\cdot\exp\Big(\sum_{p\le x}p^{-1}\cdot(f(p)-1)\Big)$$

holds for every $x > x_o$.

Proof. As in the proof of Theorem 3.1 we begin with

$$L_f(x) := \sum_{n\le x} f(n)\cdot \log n = \sum_{m\le x} f(m)\cdot \sum_{p^k\le x/m,\ p\nmid m} f(p^k)\cdot\log(p^k)$$

$$\ge \sum_{m\le x} f(m)\cdot \sum_{p\le x/m,\ p\nmid m} f(p)\cdot\log p.$$

Assumption (3.14) on the values f(p), and a TCHEBYCHEFF-estimate for $\vartheta(x)$, yield

$$L_f(x) \ge \sum_{m\le\sqrt{x},\ m \text{ squarefree}} f(m)\cdot\left\{\tfrac{1}{2}\gamma_1\cdot\frac{x}{m} + \mathcal{O}(1) - \gamma_2\cdot\omega(m)\cdot\log x\right\};$$

the constant hidden in the $\mathcal{O}$-notation depends only on P_0, γ_1 and γ_2. Since $\omega(m)\cdot\log x \le \tau(m)\cdot\log x = \mathcal{O}(m^{\frac{1}{4}}\cdot\log x) = o(x/m)$ for $m \le x^{\frac{1}{2}}$, we obtain

$$L_f(x) \ge \gamma_3\cdot \sum_{m\le\sqrt{x}} f(m)\cdot\frac{x}{m}\cdot\mu^2(m),$$

and BARBAN's Theorem 3.5 gives

$$\sum_{m\le\sqrt{x}} \mu^2(m)\cdot m^{-1}\cdot f(m) \ge \gamma_3'\cdot\exp\{\sum_{p\le\sqrt{x}} p^{-1}\cdot f(p)\}.$$

Since $\sum_{p\le x} p^{-1} = \log\log x + \gamma_4 + o(1)$, and

$$\sum_{\sqrt{x}<p\le x} p^{-1} f(p) \le \gamma_2\cdot\log 2 + o(1)$$

{for these results, the prime number theorem is needed} we obtain finally (for $x \ge x_0$)

$$\sum_{n\le x} f(n) \ge (\log x)^{-1}\cdot L_f(x) \ge \gamma_5\cdot\frac{x}{\log x}\cdot\exp\left\{\sum_{p\le x} p^{-1}\cdot f(p)\right\}$$

$$\ge \gamma\cdot x\cdot\exp\left\{\sum_{p\le x} p^{-1}\cdot(f(p)-1)\right\}.$$

This is the result we looked for. □

II.4. WIRSING'S MEAN-VALUE THEOREM FOR SUMS OVER NON-NEGATIVE MULTIPLICATIVE FUNCTIONS

A rather general, and very useful, result concerning the behaviour of sums $\sum_{n\le x} f(n)$ over non-negative multiplicative arithmetical functions is a E. WIRSING's Mean-Value Theorem [1961]. The main assumption is one specifying the asymptotic behaviour of the sum $\sum_{p\le x} f(p)$. The HARDY-LITTLEWOOD-KARAMATA Tauberian Theorem allows deduction of a result on $\sum_{n\le x} n^{-1}\cdot f(n)$ (in fact, in his paper [1961] WIRSING used elementary arguments, but in [1969] he sketched the method used here in [1969]). Elementary arguments establish a connection between the sum $\sum_{n\le x} f(n)\cdot \log n$ and $\sum_{n\le x} n^{-1}\cdot f(n)$.

Theorem 4.1 [E. WIRSING]. *Let* $f\colon \mathbb{N} \to [0,\infty[$ *be a non-negative multiplicative function. Assume that with some constant* $\tau > 0$ *the asymptotic relation*

$$\sum_{p\le x} f(p)\cdot\log p = (\ \tau + o(1)\ )\cdot x,\quad x\to\infty \tag{4.1}$$

holds, and that for every prime p *and* $k = 2, 3, \ldots,$ *the values of* f *at prime-powers are 'small',*

$$f(p^k) \le \gamma_1\, \gamma_2^{\,k},\ \textit{where } 0 < \gamma_2 < 2. \tag{4.2}$$

Then, as $x \to \infty$, *the asymptotic formula*

$$\sum_{n\le x} f(n) = (1 + o(1))\cdot \frac{x}{\log x}\cdot\frac{e^{-\mathscr{C}\tau}}{\Gamma(\tau)}\cdot \prod_{p\le x}\left(1 + \frac{f(p)}{p} + \frac{f(p^2)}{p^2} + \ldots\right) \tag{4.3}$$

holds. $\mathscr{C}$ *denotes* EULER*'s constant,* $\Gamma(.)$ *the gamma-function.*

Remark 1. (4.2) may be replaced by weaker assumptions, e.g.

$$\sum_p \sum_{k\ge 2,\ p^k\ge x} p^{-k}\cdot f(p^k) = o(\ (\log x)^{-1}\ ),$$

and

$$f(p) = \mathcal{O}(p^{1-\delta}\ )\ \text{for some}\ \delta > 0.$$

Using the Relationship Theorem from Chapter III (Theorem 2.1), these assumptions can be weakened further.

Remark 2. (i) Starting with (4.1), partial summation gives

$$\sum_{p \le x} p^{-1} \cdot f(p) \cdot \log p = (\tau + o(1)) \cdot \log x, \tag{4.4}$$

and, furthermore, the convergence of the series

$$\sum_{p} (p \cdot \log p)^{-1} \cdot f(p) = \int_2^{\infty} (\tau + o(1)) \cdot \frac{2 + \log u}{u \cdot \log^3 u} \, du . \tag{4.5}$$

(ii) (4.1) gives $f(p) \cdot \log p = \mathcal{O}(p)$, and this estimate, together with (4.5), implies the convergence of

$$\sum_p p^{-2} \cdot f^2(p) \ll \sum_p (p \cdot \log p)^{-1} \cdot f(p).$$

Lemma 4.2. *Suppose that the assumptions of* WIRSING*'s Theorem* 4.1 *are valid. Abbreviate*

$$\sum_{n \le x} f(n) \quad \text{by} \quad \mathbf{M}(f,x), \qquad \text{and} \qquad \sum_{n \le x} n^{-1} \cdot f(n) \quad \text{by} \quad \mathbf{m}(f,x). \tag{4.6}$$

Then, as $x \to \infty$,

$$\mathbf{M}(f,x) = (\tau + o(1)) \cdot \frac{x}{\log x} \cdot \mathbf{m}(f,x). \tag{4.7}$$

Proof. Put $\ell(f,x) = \sum_{n \le x} f(n) \cdot \log n$. Then, making use of the multiplicativity of f (as in the last section, II.3, but with a little bit more care):

$$\ell(f,x) = \sum_{n \le x} f(n) \cdot \sum_{p^k \| n} \log p^k = \sum_{m \cdot p^k \le x, \ p \nmid m} f(m) \cdot f(p^k) \cdot \log p^k$$

$$= \sum_{m \cdot p^k \le x} f(m) \cdot f(p^k) \cdot \log p^k - \sum_{m \cdot p^k \le x, p | m} f(m) \cdot f(p^k) \cdot \log p^k$$

$$= \sum_{m \le x} f(m) \cdot \Big\{ \sum_{p^k \le x/m} f(p^k) \cdot \log p^k - \sum_{\substack{p^{k+r} \le x/m \\ p \nmid m, r \ge 1, k \ge 1}} f(p^k) \cdot f(p^r) \cdot \log p^k \Big\}.$$

Using assumption (4.1), the first sum $\sum_{p^k \le x/m} f(p^k) \cdot \log(p^k)$ equals

$$\tau \cdot \frac{x}{m} + o\Big(\frac{x}{m} \Big) + \sum_{p^k \le x/m, k \ge 2} f(p^k) \cdot \log(p^k).$$

Thus the problem is reduced to an estimation of remainder terms. Having proved the formulae

$$S_1 := \sum_{p^k \le x, k \ge 2} f(p^k) \cdot \log(p^k) = o(x), \tag{4.8'}$$

(4.8")
$$S_2 := \sum_{\substack{p^{k+r}\le x\\ r\ge 1, k\ge 1}} f(p^k)\cdot f(p^r)\cdot \log(p^k) = o(x),$$

we obtain

(4.9)
$$\begin{cases} \ell(f,x) = \sum_{m\le x} \left\{ \tau\cdot \frac{x}{m}\cdot f(m) + o\left(\frac{x}{m}\cdot f(m)\right)\right\} \\ \quad = \tau\cdot x\cdot \mathbf{m}(f,x) + \sum_{m\le x} o\left(\frac{x}{m}\cdot f(m)\right) \\ \quad = \tau\cdot x\cdot \mathbf{m}(f,x) + o(x\cdot \mathbf{m}(f,x)). \end{cases}$$

Partial summation (see I. Theorem 1.1) then gives the assertion [the integral $\int_2^x \tau\cdot \mathbf{m}(f,u)\cdot \log^{-2}u\cdot du$ is easily shown to be of lower order, using the trivial estimate $\mathbf{m}(f,u) \le \mathbf{m}(f,x)$].

In order to substantiate (4.9) one has to show that, for some [fixed, large] constant K, the estimate

$$x\cdot \sum_{x/K<m\le x} m^{-1}\cdot f(m) = o(x\cdot \mathbf{m}(f,x))$$

holds. But this sum is [the bold-face factor in the next formula is ≥ 1, as long as $x/K \le m$]

$$\le K\cdot \sum_{x/K<m\le x} f(m) \le K\cdot \sum_{m\le x} f(m)\cdot \mathbf{\log(m)/\log(x/K)}$$
$$= O_K(\ell(f,x)/\log x) = O_K(x\cdot \mathbf{m}(f,x)/\log x).$$

For S_1, the sum up to $x\cdot(\log x)^{-2}$ is easily estimated by $O(x/\log x)$, using the weaker condition of Remark 1. The remaining sum, where $k\ge 2$ and $x\cdot(\log x)^{-2} < p^k \le x$ is less than

$$\sum_{x\cdot(\log x)^{-2}<p^k\le x, k\ge 2} \sum f(p^k)\cdot \log x\cdot \frac{x}{p^k} = o\left(x\cdot \log x\cdot \left(\log\left(\frac{x}{\log^2 x}\right)\right)^{-1}\right) = o(x).$$

The sum

$$S_2 := \sum_{\substack{p^{k+r}\le x\\ r\ge 1, k\ge 1}} f(p^k)\cdot f(p^r)\cdot \log(p^k)$$

is more difficult. Split this sum into $S' + 2S'' + S'''$, with the following summation conditions:

$$S': k = r = 1,\ p^2 \le x,$$

$$S'': r = 1,\ k \ge 2,\ p^{1+k} \le x,$$
$$S''': k \ge 2,\ r \ge 2,\ p^{r+k} \le x.$$

Since $f(p) \cdot \log p = o(p)$,

$$S' = \sum_{p\le\sqrt{x}} o(p) \cdot f(p) = \mathcal{O}(\ x \cdot (\log\ x)^{-1}\ ).$$

For S", fix an integer K so large that $(1+K)^{-1} + (\log \gamma_2/\log 2) < 1$; this is possible by the condition $\gamma_2 < 2$. Then split S" into $S''_1 + S''_2$, where $2 \le k \le K$ in S''_1, and $k > K$ in S''_2, and use $f(p^k) \le \gamma_1\gamma_2{}^k$. Then

$$S''_2 \le \sum_{p\le x^{1/(1+K)}} f(p)\log p \cdot \gamma_1 \cdot \sum_{k\le \log x/\log 2} k \cdot \gamma_2{}^k$$

$$\ll \sum_{p\le x^{1/(1+K)}} f(p) \cdot \log^2(x) \cdot x^{\log\gamma_2/\log 2} \ll x^{1-\delta}$$

with some $\delta > 0$. More easily, we obtain

$$S''_1 \ll \sum_{p\le x^{1/3}} f(p)\ \log\ p \cdot \sum_{2\le k\le K} k \cdot \gamma_2{}^k \ll x^{1-\delta}.$$

The treatment of S''' with the result $S''' \ll x^{1-\delta}$ is simpler than that of S" and is left as an exercise (see Exercise 12). □

By Lemma 4.2 it is sufficient to prove an asymptotic formula for $m(f,x) = \sum_{n\le x} n^{-1} \cdot f(n)$. The following will be proved, applying the HARDY-LITTLEWOOD-KARAMATA Tauberian Theorem.

Theorem 4.3 (E. WIRSING). *Denote by* $g: \mathbb{N} \to [0,\infty[$ *a non-negative multiplicative function satisfying*

(4.10) $$\sum_{p\le x} g(p) \cdot \log(p) = (\ \tau + o(1)\ ) \cdot \log\ x, \quad \textit{where } \tau > 0,$$

(4.11) $$\sum_{p} g^2(p) < \infty,$$

and, for all primes p *and* $k = 2, 3, \dots,$

(4.12) $$g(p^k) \le \gamma_1 \cdot (\gamma_2/p)^k, \quad \textit{where } \gamma_2 < 2.$$

Then

(4.13) $$\sum_{n\le x} g(n) = (1 + o(1)) \cdot \frac{e^{-\mathcal{C}\tau}}{\Gamma(\tau+1)} \cdot \prod_{p\le x} (\ 1 + g(p) + g(p^2) + \dots\ )$$
$$= (\ 1 + o(1)\ ) \cdot \frac{e^{-\mathcal{C}\tau}}{\Gamma(\tau+1)} \cdot P \cdot \exp\Big(\sum_{p\le x} g(p)\Big),$$

with a convergent product

$$P = \prod_p \exp\bigl(-g(p)\bigr) \cdot \bigl(1 + g(p) + g(p^2) + \dots\bigr). \tag{4.14}$$

Remark 3. Condition (4.12) may be replaced by the weaker assumption

$$\sum_p \sum_{k\ge 2} g(p^k) < \infty. \tag{4.12'}$$

Proof of Theorem 4.3. Consider the generating DIRICHLET series

$$\mathcal{D}(g,\sigma) = \Sigma_{n=1}^{\infty} g(n) \cdot n^{-\sigma}.$$

The following conditions hold:

(i) The product P is convergent.

(ii) $\mathcal{D}(g,\sigma)$ is convergent in $\sigma > 0$.

(iii) $\mathcal{D}(g,\sigma) \sim P \cdot \sigma^{-\tau} \cdot \exp\ (\ H(\sigma)\ ),\ \sigma \to 0+$,

where

$$H(\sigma) = \sum g(p)\cdot p^{-\sigma} - \tau \cdot \log(\sigma^{-1}).$$

(iv) The function $x \mapsto L(x) := \exp(\,H(1/x))$ is slowly oscillating.

(v) $\sum_p g(p) \cdot p^{-\sigma} = \sum_{p\le\exp(1/\sigma)} g(p) - \mathscr{C}\cdot\tau + o(1)$, as $\sigma \to 0+$.

(vi) For any real r, $0 < r < 1$,

$$\sum_{y^r<p\le y} g(p) = \tau\cdot\log(\tfrac{1}{r}) + o(1),\ y \to \infty.$$

Taking these assertions for granted, (iii) and (iv), with $g \ge 0$, permit application of the HARDY-LITTLEWOOD-KARAMATA Tauberian Theorem (see Appendix A.4) to the DIRICHLET series $\mathcal{D}(g,\sigma)$; thus we obtain [using (v) in the 2nd line]

$$\sum_{n\le x} g(n) \sim \frac{P}{\Gamma(\tau+1)} \cdot (\log x)^{\tau} \cdot \exp\Bigl\{ \sum_p g(p)\cdot p^{-1/\log(x)} - \tau\cdot\operatorname{loglog} x\Bigr\}$$

$$\sim \frac{P}{\Gamma(\tau+1)} \cdot e^{-\mathscr{C}\tau} \cdot \exp\Bigl\{ \sum_{p\le x} g(p) \Bigr\}. \qquad \square$$

Assertions (i) to (vi) remain to be proven.

(i) First abbreviate $\exp(-g(p))\cdot(1 + g(p) + g(p^2) + \dots)$ to $w(p)$, and choose p_0 so large that $g(p) \le \frac{1}{2}$ for $p \ge p_0$ (this is possible; $g(p)$ tends to zero). For $0 \le \varepsilon \le 1$ [for example, by TAYLOR's formula] the inequality

$$1 - \varepsilon \le \exp(-\varepsilon) \le 1 - \varepsilon + \tfrac{1}{2}\varepsilon^2$$

is valid; therefore, with some $\Theta = \Theta_p$, $0 \le \Theta_p \le 1$,

$$\left|w(p) - 1\right| = \left|\left(1 - g(p) + \tfrac{1}{2}\Theta\cdot g^2(p)\right)\cdot\left(1 + g(p) + g(p^2) + \dots\right) - 1\right|$$
$$\le \left|-g^2(p) + (1-g(p))\cdot(g(p^2)+\dots) + \tfrac{1}{2}\Theta\cdot g^2(p)\cdot(1+g(p)+g(p^2)+\dots)\right|.$$

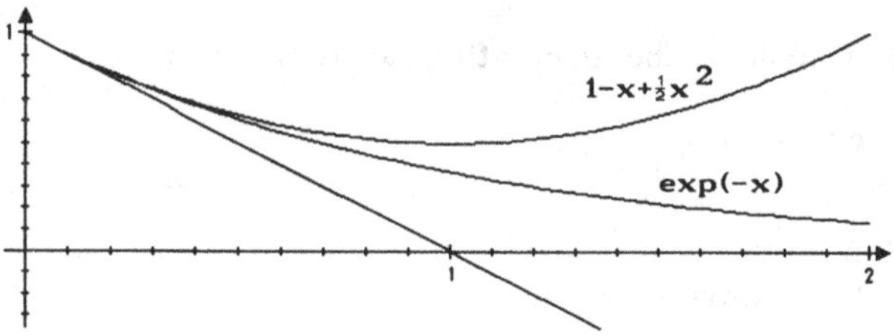

Figure II.3.

Figure II.3 shows the functions $1-x$, $\exp(-x)$, and $1-x+\frac{1}{2}x^2$ for $0 \le x \le 2$. The diagram was produced using the computer algebra system RIEMANN II.

The convergence of the series $\sum_p g^2(p)$ and $\sum_p \sum_{k\ge2} g(p^k)$ and the inequality $|g(p)| \le \frac{1}{2}$ [for large p] imply the absolute convergence of the product

$$P = \prod_p \left(1 + (w(p)-1)\right).$$

(ii) Because of the inequality (here multiplicativity is exploited !)

$$\sum_{n\le x} g(n)\cdot n^{-\sigma} \le \prod_{p\le x}\left(1 + g(p)\cdot p^{-\sigma} + g(p^2)\cdot p^{-2\sigma} + \dots\right)$$

the convergence of the series $\sum g(p)\cdot p^{-\sigma}$ (for $\sigma > 0$, deduced by partial summation from (4.7)!) and of $\sum_p\sum_{k\ge2} g(p^k)$ imply the convergence of $\mathcal{D}(g,\sigma)$ in $\sigma > 0$. So (ii) is true.

(vi) is proved by partial summation:

$$\sum_{y^r<p\le y} g(p) = \sum_{y^r<p\le y} g(p)\cdot\log(p)\cdot\{\log p\}^{-1}$$
$$= (\tau+o(1))\cdot\left(\log(y) - \log(y^r)\right)\cdot\{\log y\}^{-1}$$
$$+ \tau\cdot\int_{y^r}^{y}\{\log u - \log y^r + o(\log u)\}\cdot\{u\log^2 u\}^{-1}du,$$

and, performing an easy calculation, this equals

$$= -\tau\cdot\log r + o(1).$$

Next we come to (iii), (iv) and (v). As long as $\sigma > 0$,

$$\log \mathcal{D}(g,\sigma) = \sum_p g(p)\cdot p^{-\sigma} + \sum_p h_p(\sigma) = \Sigma_1(\sigma) + \Sigma_2(\sigma),$$

where

$$h_p(\sigma) = \Big\{\log\Big(1 + g(p)\cdot p^{-\sigma} + g(p^2)\cdot p^{-2\sigma} + \dots\Big) - g(p)\cdot p^{-\sigma}\Big\}.$$

TAYLOR's formula gives $|\log(1+x+y) - x| \le y + \frac{1}{2}(x+y)^2$ if $x \ge 0$, $y \ge 0$.

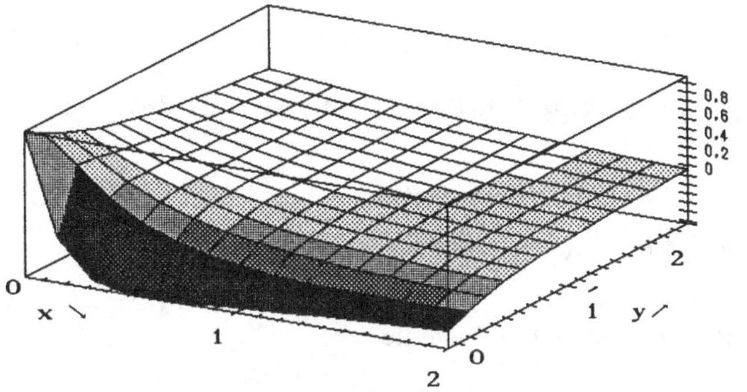

F i g u r e II.4.

Figure II.4 gives the function

$$\frac{\log(1+x+y) - x}{y + \frac{1}{2}(x+y)^2}$$

in the region $x,y > 0.01$. It was produced using the computer algebra system RIEMANN II (Begemann & Niemeyer, Detmold).

Therefore,

$$|\, h_p(\sigma) \,| \le \sum_{k\ge2} g(p^k) + \tfrac{1}{2}\Big(\sum_{k\ge1} g(p^k) \Big)^2.$$

By the WEIERSTRASS criterion, $\Sigma_2(\sigma)$ is uniformly convergent in $\sigma \ge 0$, and so

$$\Sigma_2(\sigma) = \Sigma_2(0) + o(1), \quad \sigma \to 0+.$$

$\Sigma_1(\sigma)$ is to be treated by partial summation. Split this sum into

$$\Sigma_1(\sigma) = \sum_{p\le\exp(1/\sigma)} g(p)\cdot p^{-\sigma} + \sum_{p>\exp(1/\sigma)} g(p)\cdot p^{-\sigma} = \Sigma_{11}(\sigma) + \Sigma_{12}(\sigma).$$

The second sum is easier to handle:

$$\Sigma_{12}(\sigma) = \int_{e^{1/\sigma}}^{\infty} \Big\{ \sum_{e^{1/\sigma}<p\le u} g(p)\cdot\log(p) \Big\} \cdot \frac{1+\sigma\cdot\log u}{u^{1+\sigma}\cdot\log^2 u}\, du.$$

By assumption (4.7) the expression in braces $\{\dots\}$ is

$$\tau \cdot \Big(\log u - \log\exp(\sigma^{-1})\Big) + o(\log u),$$

and thus, using the substitution $w = \sigma\cdot\log u$, a straightforward calculation results in

$$\Sigma_{12}(\sigma) = (\, \tau + o(1)\,) \cdot \int_1^\infty e^{-w}\cdot w^{-1}\, dw.$$

For $\Sigma_{11}(\sigma)$, we apply partial summation to the difference:

$$\Sigma_{11}(\sigma) - \Sigma_{p\le\exp(1/\sigma)} g(p) = \Sigma_{p\le\exp(1/\sigma)} g(p)\cdot\log p \cdot \left(\frac{p^{-\sigma}-1}{\log p}\right)$$

$$= \sum_{p\le e^{1/\sigma}} g(p)\cdot\log p\cdot(e^{-1}-1)\cdot\sigma - \int_2^{e^{1/\sigma}} \sum_{p\le u} g(p)\cdot\log p\cdot\frac{d}{du}\left(\frac{u^{-\sigma}-1}{\log u}\right) du.$$

Using the asymptotic formula (4.7) again for $\Sigma_{p\le x}\, g(p)\cdot\log p$, and substituting $w = \sigma\cdot\log(u)$, we obtain

$$\Sigma_{11}(\sigma) - \Sigma_{p\le\exp(1/\sigma)} g(p) = o(1) - \tau \int_0^1 w^{-1}\cdot(1-e^{-w})\, dw.$$

Summarizing, using the integral representation

$$\mathscr{C} = \int_0^1 w^{-1}\cdot(1-e^{-w})\, dw - \int_1^\infty e^{-w}\cdot w^{-1}\, dw,$$

for EULER's constant, we obtain the formula

$$\Sigma_{p\le\exp(1/\sigma)} g(p)\cdot p^{-\sigma} = \Sigma_{p\le\exp(1/\sigma)} g(p) - \tau\cdot\mathscr{C} + o(1),$$

and (v) is proved.

The function $x \mapsto L(x) = \exp\left(H(x^{-1})\right)$ is slowly oscillating: without loss of generality, $c \le 1$; then, using the definition of $H(.)$, (v) and (vi),

$$\log L(c\cdot x) - \log L(x) = H\left((cx)^{-1}\right) - H\left(x^{-1}\right)$$

$$= \sum_{\exp(x)<p\le\exp(cx)} g(p) - \tau\cdot\log x + o(1)$$

$$= \tau\cdot\log c - \tau\cdot\log c + o(1) = o(1).$$

So, finally, as $\sigma \to 0+$, collating our results,

$$\mathscr{D}(g,\sigma) = \exp\left(\Sigma_1(\sigma) + \Sigma_2(\sigma)\right) \sim P\cdot\exp\left(\Sigma_1(\sigma) - \tau\cdot\log(\sigma^{-1})\right)\cdot\sigma^{-\tau}$$

$$= P\cdot\sigma^{-\tau}\cdot\exp\left(H(\sigma)\right),$$

where $H(\sigma) = \Sigma_p\, g(p)\cdot p^{-\sigma} - \tau\cdot\log(\sigma^{-1})$ is slowly oscillating. The proof of Theorem 4.3 and 4.1 is now complete. □

We note that E. WIRSING also proved results on complex-valued functions and quote his result without proof.

Theorem 4.4. *Assume that* f *satisfies the conditions of Theorem* 4.1 *with* $\tau \neq 0$; *let* $f^*: \mathbb{N} \to \mathbb{C}$ *be a second multiplicative function,* $|f^*| \leq f$, *satisfying*

$$(4.15) \qquad \sum_{p \leq x} f^*(p) \cdot \log p \sim \tau^* \cdot x, \quad x \to \infty.$$

and

$$(4.16) \qquad \prod_{p \leq x} \left(1 + p^{-1} \cdot f^*(p) + p^{-2} \cdot f^*(p^2) + \dots \right)$$
$$= o\left(\prod_{p \leq x} \left(1 + p^{-1} \cdot f(p) + p^{-2} \cdot f(p^2) + \dots \right) \right).$$

Then

$$(4.17) \qquad \sum_{n \leq x} f^*(n) = o\left(\sum_{n \leq x} f(n) \right).$$

Remark 4. Condition (4.16), concerning the product $\prod_{p \leq x} (\dots)$, follows from the simpler condition

$$(4.16^*) \qquad \sum_{p} p^{-1} \cdot \left(f(p) - \mathrm{Re}\{f^*(p)\} \right) = \infty.$$

There are numerous possible **Applications of WIRSING's Theorem**, some of which are presented below.

(1) Denote by b the characteristic function of the set $\mathcal{B}$ of positive integers, representable as a sum of two squares:

$$\mathcal{B} = \{ n \in \mathbb{N};\ \exists\ x,y \in \mathbb{Z}:\ n = x^2 + y^2 \},$$

$$b(n) = 1, \text{ if } n \in \mathcal{B},\ b(n) = 0 \text{ otherwise}.$$

Then b is multiplicative by virtue of the identity

$$(x^2 + y^2) \cdot (\xi^2 + \eta^2) = (x \cdot \xi + y \cdot \eta)^2 + (x \cdot \eta - \xi \cdot y)^2.$$

Obviously, $b(2^k) = 1$, $b(p^{2k}) = 1$ for any prime p and any $k \in \mathbb{N}$. From elementary number theory it is known that $b(p) = 0$, and $b(p^{2k+1}) = 0$ if $p \equiv 3 \bmod 4$, and $b(p^k) = 1$ for any $k \in \mathbb{N}$ if $p \equiv 1 \bmod 4$. Thus,

$$\sum_{p \leq x} b(p) \cdot \log p = \mathcal{O}(1) + \sum_{p \leq x,\ p \equiv 1 \bmod 4} \log p = \tfrac{1}{2} x + o(x),$$

by the prime number theorem (see I.7 (7.12)). Now WIRSING's Theorem 4.1 is applicable and yields

$$\sum_{n\le x} b(n) = (1+o(1))\cdot e^{-\frac{1}{2}\mathscr{C}}\cdot\frac{1}{\Gamma(\frac{1}{2})}\cdot\frac{x}{\log x}\cdot\prod_{\substack{p\le x\\ p\equiv 1(4)}}\left(1+p^{-1}+p^{-2}+\dots\right)\cdot 2\prod_{\substack{p\le x\\ p\equiv 3(4)}}\left(1+p^{-2}+p^{-4}+\dots\right).$$

Note that $\Gamma(\frac{1}{2}) = \sqrt{\pi}$. The product over the primes $p \equiv 3 \bmod 4$ is convergent, and thus

$$\sum_{n\le x} b(n) \sim \gamma\cdot\frac{x}{\log x}\cdot\prod_{p\le x, p\equiv 1 \bmod 4}(1+p^{-1})$$

$$\sim \gamma^{*}\cdot\frac{x}{\log x}\cdot\exp\Big(\sum_{p\le x, p\equiv 1 \bmod 4} p^{-1}\Big) \sim \gamma^{**}\cdot x\cdot(\log x)^{-\frac{1}{2}}.$$

Figure II.5 shows the function $x^{-1}\cdot\sum_{n\le x} b(n)$ (upper curve) and the approximation, given by the right-hand side of the formula at the top of this page, in the range $1 \le x \le 900$. Only every third value is plotted.

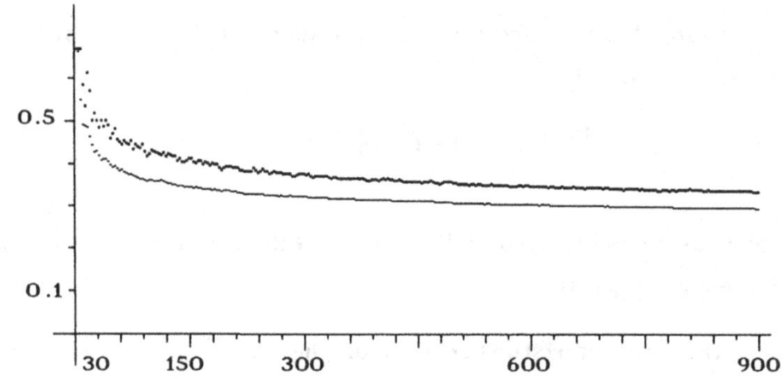

F i g u r e II.5.

(2) Given a monic, irreducible polynomial $P(X) \in \mathbb{Z}[X]$ of degree ≥ 1, denote by $\rho(n)$ the number of solutions of the congruence

$$P(m) \equiv 0 \bmod n,\ 1 \le m \le n.$$

Then $\rho \ge 0$, ρ is multiplicative, $0 \le \rho(p) \le \deg(P)$, $\rho(p^k)$ is bounded by a result of T. NAGEL(L) ([1919], [1923], see [1956], p.90) and the prime ideal theorem implies

$$\sum_{p\le x} \rho(p)\cdot\log p \sim x$$

(see ERDÖS, [1952]). Thus, again, WIRSING's theorem is applicable, giving

$$\sum_{n\le x} \rho(n) \sim e^{-\mathscr{C}}\cdot\frac{x}{\log x}\cdot\prod_{p\le x}\left(1+p^{-1}\cdot\rho(p)+p^{-2}\cdot\rho(p^2)+\dots\right).$$

(3) The *divisor-function* τ is multiplicative, $\tau(p^k) = k+1$, and for any real $r \ge 0$

$$\sum_{p\le x} \tau^r(p)\cdot \log p \sim 2^r \cdot x.$$

Thus, WIRSING's theorem gives a much sharper result than the upper estimate deduced in Chapter I (Theorem 2.7 (c)):

$$\sum_{n\le x} \tau^r(n) \sim e^{-2^r\cdot\mathcal{C}}\cdot(\Gamma(2^r))^{-1}\cdot \frac{x}{\log x}\cdot\prod_{p\le x}\left(1+p^{-1}\cdot 2^r + p^{-2}\cdot 3^r+\ldots\right)$$

$$\sim \gamma' \cdot \frac{x}{\log x}\cdot \exp\left(\sum_{p\le x} 2^r\cdot p^{-1}\right) \sim \gamma''\cdot x\cdot(\log x)^{2^r-1}.$$

(4) Given some fixed set $\mathcal{T}$ of primes with the property

$$\sum_{p\le x, p\in\mathcal{T}} \log p \sim \tau\cdot x,\ \tau > 0,$$

we denote by $\mathcal{M}_0$ the set of positive integers solely composed from primes in $\mathcal{T}$,

$$n \in \mathcal{M}_0 \Leftrightarrow (\ p|n \Rightarrow p\in\mathcal{T}\ ),$$

and the characteristic function of this set by $f_{\mathcal{T}}: n \mapsto 1$, if $n\in\mathcal{M}_0$, 0 otherwise. Then there is an asymptotic formula for $\sum_{n\le x} f_{\mathcal{T}}(n)$, and, for any fixed integer m, the values $\omega(n)$, where $n\in\mathcal{M}_0$, are relatively uniformly distributed modulo m; this means that the number $M_{\omega,r,m,\mathcal{T}}(x)$ of integers $n\le x$, n in $\mathcal{T}$, with the property $\omega(n)\equiv r \bmod m$, is asymptotically equal to $m^{-1}\cdot \#\{\ n\le x,\ n\in\mathcal{M}_0\ \}$.

Denote by ε an m-th root of unity (abbreviated **m-r.u**). Put $f^*(n) = f_{\mathcal{T}}(n)\cdot \varepsilon^{\omega(n)}$; then the conditions of Theorem 4.3 are fulfilled, $\tau^* = \varepsilon\cdot\tau$, and

$$\sum_p p^{-1}\cdot\left(\ f_{\mathcal{T}}(p) - \mathrm{Re}(\ f^*(p)\ )\ \right) = \infty, \text{ if } \varepsilon \ne 1.$$

WIRSING's Theorem 4.1 gives an asymptotic formula for

$$M_{\mathcal{T}}(x) = \sum_{n\le x} f_{\mathcal{T}}(n),$$

and Theorem 4.4 (which is not proved in this Chapter) gives

$$\sum_{n\le x} f_{\mathcal{T}}(n)\cdot \varepsilon^{\omega(n)} = o\left(\ M_f(x)\ \right), \text{ if } \varepsilon \ne 1.$$

But

$$M_{\omega,r,m,\mathcal{T}}(x) = \sum_{n\le x} f_{\mathcal{T}}(n)\cdot m^{-1}\cdot \sum_{\varepsilon\ \text{m-r.u}} \varepsilon^{\omega(n)-r}$$

$$= m^{-1} \cdot M_f(x) + o\left(M_f(x) \right),$$

and so the result is proved. □

II.5. THE THEOREM OF G. HALÁSZ ON MEAN-VALUES OF COMPLEX-VALUED MULTIPLICATIVE FUNCTIONS

E. WIRSING's Theorem 4.1 is not sufficiently strong to give a solution for the old conjecture of ERDÖS and WINTNER: *any multiplicative function taking only the values* +1, −1 *and* 0 *possesses a mean-value.* But WIRSING believed that an assumption such as

$$\sum_{p\le x} p^{-1}\cdot f(p)\cdot \log p \sim \tau \cdot \log x$$

instead of (4.1) ought to be sufficient for deducing [some] results on the mean-value of a multiplicative, real-valued function f. One of his results, 1967, is quoted below. One year later G. HALÁSZ dealt with complex- valued multiplicative functions of modulus less than or equal to one. However, for complex-valued functions difficulties may occur. The example

$$f(n) = \exp(\, i\cdot t\cdot \log n \,),$$

with

$$\sum_{n\le x} f(n) = (\, 1+it \,)^{-1} \cdot x \cdot x^{it} + \mathcal{O}(\, \log x \,)$$

shows that for $t \neq 0$ the mean-value of a complex-valued function of modulus 1 need not exist (see Exercise IX, 1).

E. WIRSING [1967] proved the following results.

Theorem 5.1. *Let* f *be a non-negative multiplicative arithmetical function, satisfying*

(5.1) $$|f(p)| \le G \textit{ for all primes } p,$$

(5.2) $$\sum_{p\le x} p^{-1} \cdot f(p) \cdot \log p \sim \tau\cdot \log x,$$

with some constants $G > 0$, $\tau > 0$, *and*

$$\sum_p \sum_{k \ge 2} p^{-k} \cdot |f(p^k)| < \infty; \tag{5.3}$$

if $0 < \tau \le 1$, *then, in addition, the condition*

$$\sum_p \sum_{k \ge 2,\ p^k \le x} |f(p^k)| = \mathcal{O}\left(x/\log x \right). \tag{5.4}$$

is assumed to hold. Then

$$\sum_{n \le x} f(n) = (1 + o(1)) \cdot \frac{x}{\log x} \cdot \frac{e^{-\mathscr{C}\tau}}{\Gamma(\tau)} \cdot \prod_{p \le x} \left(1 + \frac{f(p)}{p} + \frac{f(p^2)}{p^2} + \ldots \right). \tag{5.5}$$

Theorem 5.2. *Assume that* $f\colon \mathbb{N} \to \mathbb{C}$ *is multiplicative and satisfies assumptions* (5.1) *to* (5.4) *of Theorem* 5.1, *and, moreover, that*

$$\sum_p p^{-1} \cdot \left(|f(p)| - \operatorname{Re}\,(f(p)) \right) < \infty. \tag{5.6}$$

Then assertion (5.5) *of Theorem* 5.1 *is true.*

Some preparations are helpful for the formulation of the next result. Let

$$\mathcal{g} = \{ z = \rho \cdot e^{i\varphi} \in \mathbb{C};\ 0 \le \varphi < 2\pi,\ 0 \le \rho \le r(\varphi) \}$$

be some region in the complex plane $\mathbb{C}$ containing 0. Define its "average radius" by

$$\tilde{r}(\mathcal{g}) = (2\pi)^{-1} \cdot \int_0^{2\pi} r(\varphi)\, d\varphi.$$

Theorem 5.3. *Assume that* $f\colon \mathbb{N} \to [0,\infty[$ *is multiplicative and satisfies the assumptions of Theorem* 5.1. *Let* $f^*\colon \mathbb{N} \to \mathbb{C}$ *be multiplicative,* $|f^*| \le f$. *Suppose there is a convex region* $\mathcal{g} \subset \mathbb{C}$ *with average radius* $\tilde{r}(\mathcal{g}) < 1$, *containing* 0, *such that* $\tau \cdot \mathcal{g}$ *contains all values* $f^*(p)$. *Then*

$$\sum_{n \le x} f^*(n) = \frac{x}{\log x} \cdot \frac{e^{-\mathscr{C}\tau}}{\Gamma(\tau)} \cdot \prod_{p \le x} \left(1 + \frac{f^*(p)}{p} + \frac{f^*(p^2)}{p^2} + \ldots \right) + o\Big(\sum_{n \le x} f(n) \Big).$$

We do not prove these theorems here, but refer to WIRSING's paper [1967]; a proof by A. HILDEBRAND for a special version of WIRSING's Theorem for real-valued functions is given in Chapter IX, and, in the same chapter, we give a result due to G. HALÁSZ[1)] [1968] for complex-valued multiplicative functions of modulus $|f| \le 1$; the proof given there will be "elementary" and follows H. DABOUSSI and K.-H. INDLEKOFER [1990].

Theorem 5.4 (G. HALÁSZ). *Let* f *be a multiplicative arithmetical function of modulus* $|f| \le 1$. Then *there exist a real constant* α, *a complex constant* C *and a slowly oscillating, continuous function* $L: [1,\infty[\to \mathbb{C}$, $|L| = 1$, *for which the asymptotic relation*

$$\sum_{n\le x} f(n) = C \cdot L(\log x) \cdot x^{1+i\alpha} + o(x)$$

is true.

The function L and the constants α, C may be given explicitly.[2)] The proof of parts of Theorem 5.4 is postponed until Chapter IX.

II.6. THE THEOREM OF DABOUSSI AND DELANGE ON THE FOURIER-COEFFICIENTS OF MULTIPLICATIVE FUNCTIONS

In 1974 H. DABOUSSI and H. DELANGE announced the result that, for *irrational* values of α, FOURIER-coefficients $f^\wedge(\alpha)$ of multiplicative arithmetical functions f of absolute value $|f| \le 1$ are zero. DABOUSSI and DELANGE [1982] proved the following stronger result.

Theorem 6.1. *Let* f *be a multiplicative arithmetical function for which the semi-norm*

$$\|f\|_2^2 := \limsup_{x\to\infty} x^{-1} \cdot \sum_{n\le x} |f(n)|^2 \tag{6.1}$$

---

1) In the authors' opinion, GABOR HALÁSZ's method, a skilful variant of the method of *complex integration*, seems to be definitely simpler than WIRSING's method of dealing with convolution integrals.

2) See also the paper by K.–H. INDLEKOFER [1981a].

is finite. Then, for every irrational α, *the mean-value* (FOURIER-*coefficient)*

(6.2) $$\hat{f}(\alpha) = M(f\cdot e_\alpha) = \lim_{x\to\infty} x^{-1}\cdot\sum_{n\le x} f(n)\cdot e^{2\pi i\alpha n}$$

is zero.

We do not give DABOUSSI and DELANGE's proof, but sketch a proof of a result which is a little weaker - the relationship theorem of Chapter III allows the deduction of Theorem 6.1. The result is as follows.

Theorem 6.2. *Denote by* $\mathcal{F}_A$ *the set of multiplicative functions with the properties*

(6.3') $$|f(p)| \le A \text{ for all primes } p,$$

(6.3") $$\sum_{n\le N} |f(n)|^2 \le A^2\cdot N \text{ for all integers } N \ge 1.$$

Abbreviate by $S_f(\alpha)$ *the exponential-sum*

$$S_f(\alpha) = \sum_{n\le N} f(n)\cdot e^{2\pi i\alpha n}.$$

Then $S_f(\alpha) = o(N)$, *as* $N\to\infty$, *if* $f\in\mathcal{F}_A$ *and* α *is irrational.*

Remark. Based on the "Large Sieve", H. L. MONTGOMERY and R. C. VAUGHAN [1977] prove a stronger result:

If f *is in* $\mathcal{F}_A$, *then, for* $q\le N$ *and* $\gcd(a,q) = 1$,

(6.4) $$S_f(a/q) \ll_A N\cdot(\log 2N)^{-1} + N\cdot(\varphi(q))^{-1} + (qN)^{\frac12}\,(\log(2N/q))^{1+\frac12}$$

uniformly for all functions f *in* $\mathcal{F}_A$; *the implied* $\mathcal{O}$*-constant depends only on* A.

They deduce the

Corollary 6.3. *Let* $\left|\alpha - \frac{a}{q}\right| \le q^{-2}$, *where* $\gcd(a,q) = 1$. *If* $2 \le R \le q \le N/R$, *then*

$$S_f(\alpha) \ll_A N\cdot(\log 2N)^{-1} + N\cdot(\log R)^{1+\frac12}\cdot R^{-\frac12}.$$

Proof of Theorem 6.2. The "large sieve inequality" [see Appendix] gives

$$\sum_{q\le Q}\sum_{a,\ (a,q)=1}\Big|\sum_{M+1\le n\le M+N} a_n\cdot e\Big(n\cdot\frac{a}{q}\Big)\Big|^2$$

$$\ll (N + Q^2) \cdot \sum_{M+1\le n\le M+N} |a_n|^2,$$

for complex sequences a_n. This implies

$$\sum_{p\le Q} p \cdot \Big| p^{-1}\cdot\sum_{n\le N} f(n)\, e(\alpha n) - \sum_{m\le N, m\equiv 0\ (p)} f(m)\cdot e(\alpha m)\Big|^2$$

$$= \sum_{p\le Q} p \cdot\Big| p^{-1}\cdot\sum_{n\le N} f(n)e(\alpha n) - \sum_{m\le N} p^{-1}\cdot\sum_{1\le b\le p} e(b/p)\cdot f(m)\cdot e(\alpha m)\Big|^2$$

$$= \sum_{p\le Q} p\cdot p^{-2}\Big|\sum_{1\le b\le p} e(b/p)\cdot \sum_{m\le N} f(m)\cdot e(\alpha m)\Big|^2.$$

Using the CAUCHY-SCHWARZ inequality for $| \dots |^2$, this is

$$\le \sum_{p\le Q} p^{-1} \sum_{1\le b\le p} 1 \cdot \sum_{1\le b\le p} \Big|\sum_{m\le N} f(m)\cdot e(\alpha m)\Big|^2,$$

and the sieve inequality gives the estimate

$$\ll (N + Q^2) \cdot \sum_{m\le N} |f(m)|^2 \ll_A N^2 \quad \text{if } Q \le N^{\frac12}.$$

Summarizing, we obtain

$$\sum_{p\le Q} p \cdot \Big| p^{-1}\cdot N^{-1}\cdot S_f(\alpha) - N^{-1}\cdot \sum_{m\le N, m\equiv 0\ (p)} f(m)\cdot e(\alpha m)\Big|^2 \ll 1$$

if $Q \le N^{\frac12}$. Next, replacing $\sum_{m\le N, m\equiv 0 (p)} f(m)$ by $\sum_{m\le N/p} f(p)\cdot f(m)$, with an error of $\mathcal{O}(p^{-2}\cdot N)$, we obtain, as long as $Q \le \sqrt{N}$,

$$\sum_{p\le Q} p \cdot \Big| p^{-1}\cdot N^{-1}\cdot S_f(\alpha) - N^{-1}\cdot \sum_{m\le N/p} f(p)\cdot f(m)\cdot e(\alpha m p)\Big|^2 \ll 1.$$

The CAUCHY-SCHWARZ inequality now implies

$$\Big\{ \sum_{p\le Q} \Big| p^{-1}\cdot N^{-1}\cdot S_f(\alpha) - N^{-1}\cdot \sum_{m\le N/p} f(p)\cdot f(m)\cdot e(\alpha m p)\Big| \Big\}^2 \ll P(Q),$$

where $P(Q) := \sum_{p\le Q} p^{-1}$. Abbreviating $|N^{-1}\cdot S_f(\alpha)|$ by $D(\alpha)$, and writing

$$D(\alpha) \le \Big\{ \Big|\sum_{p\le Q} \dots \Big| \Big\} + \Big|\sum_{p\le Q} N^{-1}\cdot \sum_{m\le N/p} \dots \Big|,$$

we obtain

$$P(Q)\cdot D(\alpha) \ll (P(Q))^{\frac12} + N^{-1} \cdot \Big|\sum_{p\le Q} \sum_{m\le N/p} f(p)\cdot f(m)\cdot e(\alpha m p)\Big|.$$

The square of the last double-sum over p, m is dealt with by the CAUCHY-SCHWARZ inequality:

$$\left\{ \left| \sum_{p\le Q} \sum_{m\le N/p} f(p)\cdot f(m)\cdot e\,(\alpha mp) \right| \right\}^2$$
$$= \left\{ \sum_{m\le N} \sum_{p\le Q, p\le N/m} f(p)\cdot f(m)\cdot e\,(\alpha mp)\right\}^2$$
$$\le A^2\cdot N \cdot \sum_{m\le N} \left| \sum_{p\le Q, p\le N/m} f(p)\cdot e(\alpha mp) \right|^2 = N \sum_m \sum_p \ldots \overline{\left\{\sum_{p'} \ldots\right\}}$$
$$= N \cdot \sum_{p\le Q} \sum_{p'\le Q} f(p)\cdot \overline{f(p')}\cdot \sum_{m\le \min(N/p, N/p')} e(\ \alpha m\cdot(p-p')\ ).$$

The contribution of the diagonal elements $(p = p')$ to this sum is

$$\le N \cdot \sum_{p\le Q} |f(p)|^2 \cdot N/p \ll N^2 \cdot P(Q).$$

The non-diagonal elements $p \ne p'$ contribute

$$\ll N \cdot Q^2 \cdot m,$$

where, obtained by summing the geometric series,

$$m = \min \left\{N,\ \max_{1\le h\le Q} \left\{ \ |\sin \pi\alpha\cdot h|\ \right\}^{-1}\right.$$.

Summarizing, for $Q \le N^{\frac{1}{2}}$, with some constant γ, depending on A only,

$$D(\alpha) \le \gamma \cdot \{ P(Q) \}^{-\frac{1}{2}} \cdot \{ 1 + (m\cdot Q^2)^{\frac{1}{2}} \}.$$

Given a large K, choose Q with the property $P(Q) \ge K^2\cdot\gamma^2$. Then $m\cdot Q^2 \le 1$ if N is large; the irrationality of α is exploited here. Our estimates then give

$$D(\alpha) < K^{-1}\cdot 2.$$ □

II.7. APPLICATION OF THE DABOUSSI - DELANGE THEOREM TO A PROBLEM OF UNIFORM DISTRIBUTION

A sequence $\{x_n\}$ of real numbers is termed *uniformly distributed modulo 1*, abbreviated u.d.mod 1, if for any pair α, β of real numbers, $0 \le \alpha < \beta \le 1$,

$$\lim_{N\to\infty} N^{-1} \cdot \# \left\{ n \le N;\ \alpha \le \{x_n\} \le \beta \right\} = \beta - \alpha,$$

where $\{x\} = x - [x]$ denotes the fractional part of x. According to H. WEYL's criterion, a sequence $\{x_n\}$ of real numbers is u.d.mod 1, if and only if

$$\lim_{N \to \infty} N^{-1} \cdot \sum_{n \le N} \exp (2\pi ik \cdot x_n) = 0$$

for every integer $k \neq 0$.

Theorem 7.1. *If* g *is a real-valued additive function, and if* α *is an irrational real number, then the sequence*

$$x_n = g(n) - \alpha \cdot n$$

is uniformly distributed modulo 1.

Proof. Given an integer $k \neq 0$, the multiplicative (!) function

$$f: n \mapsto \exp\left\{2\pi ik \cdot g(n)\right\}$$

fulfils the assumptions of Theorem 6.1 (or 6.2). Therefore, the mean-value $M(f \cdot e_{-k\alpha}) = 0$. Written in full,

$$\lim_{x \to \infty} x^{-1} \cdot \sum_{n \le x} \exp (2\pi ik \cdot g(n) - 2\pi ik\alpha \cdot n) = 0,$$

and so, according to WEYL, the sequence $g(n) - \alpha n$ is u.d. mod 1. □

II.8. THE THEOREM OF SAFFARI AND DABOUSSI, I

Let the semi-group $\mathbb{N}$ of positive integers be represented as a direct product of two subsets $\mathcal{A}$ and $\mathcal{B}$,

$$\mathbb{N} = \mathcal{A} \times \mathcal{B},$$

so that every n is uniquely representable as a product

$$n = a \cdot b, \ a \in \mathcal{A}, \ b \in \mathcal{B}.$$

In terms of the characteristic functions α of $\mathcal{A}$, β of $\mathcal{B}$ ($\alpha(n) = 1$ if $n \in \mathcal{A}$, $\alpha(n) = 0$ otherwise), this property is equivalent to

(8.1) $$1 = \alpha * \beta.$$

The question as to whether a "direct factor" $\mathcal{A}$ of $\mathbb{N}$ possesses a density was answered in the affirmative by B. SAFFARI [1976a,b] and P. ERDÖS, B. SAFFARI and R. C. VAUGHAN [1979]. SAFFARI proved the following theorem.

Theorem 8.1. *Assume that* $\mathbb{N} = \mathcal{A} \times \mathcal{B}$ *is a direct product, and*

(8.2) $$\sum_{b\in\mathcal{B}} b^{-1} < \infty.$$

Then $\mathcal{A}$ *has a density*

$$\delta(\mathcal{A}) = \Big(\sum_{b\in\mathcal{B}} b^{-1} \Big)^{-1}.$$

Of course, the density of $\mathcal{B}$ is zero. The case where $\sum_{b\in\mathcal{B}} b^{-1} = \infty$ was treated in the [1979] paper quoted above. In this case the density $\delta(\mathcal{A})$ also exists and is zero. We give DABOUSSI's proof [1979] of Theorem 8.1.

Given $y \ge 2$, define completely multiplicative functions ℓ_y and s_y, the characteristic functions of integers composed of $\underline{\ell}$arge, resp. $\underline{s}$mall, primes, by

(8.3) $$\ell_y(p) = \begin{cases} 1, & \text{if } p > y \\ 0, & \text{if } p \le y \end{cases}, \qquad s_y(p) = \begin{cases} 1, & \text{if } p \le y \\ 0, & \text{if } p > y \end{cases}.$$

Then $\ell_y * s_y = 1$, $\sum_n n^{-1} \cdot s_y(n) = \prod_{p\le y} \big(1 + p^{-1} + p^{-2} + \dots \big) < \infty$, and the mean-value

$$M(\ell_y) = M(1 * s_y^{-1(*)}) = M(1 * (\mu s_y)) = \big(\sum n^{-1} \cdot s_y(n) \big)^{-1}$$

exists. This is proved using WINTNER's Corollary 2.3, for example.

Define $\alpha_y = (s_y \cdot \alpha) * \ell_y : n \mapsto \sum_{d|n} s_y(d) \cdot \alpha(d) \cdot \ell_y(n/d)$. Then the mean-value

(8.4) $$M(\alpha_y) = \big(\sum_{b\in\mathcal{B}} b^{-1} \cdot s_y(b) \big)^{-1}$$

exists for every $y \ge 2$. Since $\sum n^{-1} \cdot (\alpha s_y)(n) < \infty$, this follows from Theorem 2.4 for example, and, by multiplicativity, the result

$$\big(\sum n^{-1} \cdot s_y(n) \big)^{-1} \cdot \sum n^{-1} \cdot (\alpha \cdot s_y)(n)$$

is easily transformed into (8.4). Next we show that

(8.5) $$M(\alpha,x) := \sum_{n\le x} \alpha(n) \le M(\alpha_y,x), \text{ for } x \ge 1.$$

Taking this for granted, the theorem is proved as follows: $\alpha*\beta = 1$ implies $\sum_{b\in\mathcal{B},\ b\le x} M(\alpha, x/b) = \sum_{n\le x} 1 = [x]$, and so

$$M(\alpha,x) + \sum_{b\in\mathcal{B},\,1<b\le x} M(\alpha, x/b) = [x].$$

Therefore, making use of (8.5),

(8.6) $$[x] - \sum_{b\in\mathcal{B},\,1<b\le x} M(\alpha_y, x/b) \le M(\alpha,x) \le M(\alpha_y,x).$$

In the case where $\sum_{b\in\mathcal{B}} b^{-1} = \infty$, by (8.4), we obtain

$$\limsup_{x\to\infty} x^{-1}\cdot M(\alpha,x) \le \limsup_{x\to\infty} x^{-1}\cdot M(\alpha_y,x) = \Big(\sum_{b\in\mathcal{B}} b^{-1}\cdot s_y(b)\Big)^{-1},$$

and, letting $y\to\infty$, $\delta(\mathcal{A}) = 0$.

If $\sum_{b\in\mathcal{B}} b^{-1} < \infty$, then (8.6) gives

$$1 - \Big(\sum_{b\in\mathcal{B},b>1} b^{-1}\Big)\cdot \lim_{x\to\infty} x^{-1}\cdot M(\alpha_y,x) \le \liminf_{x\to\infty} x^{-1}\cdot M(\alpha,x)$$

$$\le \limsup_{x\to\infty} x^{-1}\cdot M(\alpha,x) \le \lim_{x\to\infty} x^{-1}\cdot M(\alpha_y,x).$$

Using $\lim_{x\to\infty} x^{-1}\cdot M(\alpha_y,x) = \big(\sum_{b\in\mathcal{B}} b^{-1}\cdot s_y(b)\big)^{-1}$, and then letting $y\to\infty$, we obtain

$$\Big(\sum_{b\in\mathcal{B}} b^{-1}\Big)^{-1} \le \liminf_{x\to\infty} x^{-1}\cdot M(\alpha,x) \le \limsup_{x\to\infty} x^{-1}\cdot M(\alpha,x) \le \Big(\sum_{b\in\mathcal{B}} b^{-1}\Big)^{-1}.$$

It remains to prove $\sum_{n\le x} \alpha(n) \le \sum_{n\le x} \alpha_y(n)$ (this is (8.5)). Since, by complete multiplicativity, the relation $s_y\cdot(\alpha*\beta) = s_y\cdot\alpha * s_y\cdot\beta$ holds, we obtain $s_y\cdot\alpha * s_y\cdot\beta * \ell_y = 1$. Therefore

$$\sum_{n\le x} \alpha(n) = \sum_{n\le x} \alpha(n)\cdot(s_y\cdot\alpha*s_y\cdot\beta*\ell_y)(n)$$

$$= \sum_{abc\le x,a\in\mathcal{A},b\in\mathcal{B}} s_y(a)\cdot s_y(b)\cdot\ell_y(c)\cdot\alpha(abc)$$

$$= \sum_{a\le x,a\in\mathcal{A}} s_y(a) \sum_{c\le x/a} \ell_y(c)\cdot \sum_{b\le x/ac,b\in\mathcal{B}} s_y(b)\cdot\alpha(abc)$$

$$\le \sum_{a\le x,a\in\mathcal{A}} s_y(a) \sum_{c\le x/a} \ell_y(c) = \sum_{n\le x}(s_y\cdot\alpha*\ell_y)(n) = M(\alpha_y,x).$$

Here

$$\sum_{b \le x/ac, b \in \mathcal{B}} s_y(b) \cdot \alpha(abc) \le \sum_{b \in \mathcal{B}} \alpha(abc) \le 1,$$

was used. This is true, since, given a number of the form ac, there is at most **one** $b \in \mathcal{B}$ for which $b \cdot ac = A \in \mathcal{A}$ [otherwise, if $b' \cdot ac = A' \in \mathcal{A}$, $b' \neq b$, then $A \cdot b' = A' \cdot b$, which is impossible, since $\mathbb{N}$ is a *direct* product of $\mathcal{A}$ and $\mathcal{B}$]. □

II.9. DABOUSSI'S ELEMENTARY PROOF OF THE PRIME NUMBER THEOREM

The ideas used in II.8, worked out in greater detail, make it possible to give an elementary proof of the prime number theorem (DABOUSSI [1984]).

Theorem 9.1. *Put* $M(x) = \sum_{n \le x} \mu(n)$. *Then*

$$\lim_{x \to \infty} x^{-1} \cdot M(x) = 0.$$

Proof. With some [large] parameter y we use the completely multiplicative functions ℓ_y, s_y defined in II.8, where $\ell_y(p) = 1$ for "large" primes $p > y$ and $s_y(p) = 1$ for "small" primes $p \le y$. For brevity, write $M_y(x) = \sum_{n \le x} \mu(n) \cdot s_y(n)$. Then $\mu = \mu\ell_y * \mu s_y$, and so

$$M(x) = \sum_{n \le x} \mu(n) \cdot \ell_y(n) \, M_y(x/n). \tag{9.1}$$

Denote by $d_1 = 1 < d_2 < \dots < d_q$ the finite sequence of squarefree integers composed only of prime-factors $p \le y$; in $x/d_{j+1} < n \le x/d_j$ the function $M_y\left(\frac{x}{n}\right)$ is constant. Therefore, with $d_{q+1} = \infty$, $x/d_{q+1} = 0$, we obtain, using (9.1),

$$M(x) = \sum_{1 \le j \le q} M_y(d_j) \cdot \sum_{x/d_{j+1} < n \le x/d_j} \ell_y(n) \cdot \mu(n). \tag{9.2}$$

The mean-value $M(\ell_y) = \prod_{p \le y} \left(1 - p^{-1}\right)$ exists. Dividing (9.2) by x, we obtain

$$(9.3)\quad \begin{cases} \limsup_{x\to\infty} x^{-1}\cdot|M(x)| \le \prod_{p\le y}(1-p^{-1})\cdot \sum_{1\le j\le q} |M_y(d_j)|\cdot\{1/d_j - 1/d_{j+1}\} \\ \qquad = \prod_{p\le y}(1-p^{-1})\cdot \int_1^\infty t^{-2}\cdot|M_y(t)|\,dt, \end{cases}$$

again using the remark preceding (9.2). Put

$$C = \lim_{y\to\infty} (\log y)^{-1}\cdot \prod_{p\le y}(1-p^{-1})^{-1},$$

and $\alpha = \limsup_{x\to\infty} x^{-1}\cdot|M(x)|$. Then $0 \le \alpha \le 1$. We are going to prove the following.

(i) *There exists a constant* $\delta \ge 1$ *such that for any* β *in* $\alpha < \beta < 2$ *the inequality*

$$(9.4\text{i})\qquad \int_1^y t^{-2}\cdot|M_y(t)|\,dt \le \beta\cdot\delta^{-1}\cdot\log y + O(1)$$

is true. If $\alpha > 0$, *then* $\delta > 1$.

$$(9.4\text{ii})\qquad \int_y^\infty t^{-2}\cdot|M_y(t)|\,dt \le \beta\cdot(C-1)\cdot\log y + o(\log y).$$ [1]

Having proved the inequalities (9.4i), (9.4ii), and making use of inequality (9.3), we obtain

$$\alpha \le \beta\cdot(1-C^{-1}(1-\delta^{-1})),$$

and so, for $\beta \to \alpha+$, $\alpha = 0$. Therefore, the prime number theorem is proved and so it remains to prove (9.4i) and (9.4ii).

Proof of (9.4i). It is known, that for any a, $1 \le a < b$, the integral

$$\left|\int_a^b t^{-2}\cdot M(t)\,dt\right| \le 6$$

is bounded. [The value 6 is of no importance; any fixed bound M is sufficient. See Exercise 18.] If $M(x)$ does not change sign in $[1, y]$, then

$$\int_1^y t^{-2}\cdot|M_y(t)|\,dt = \int_1^y t^{-2}\cdot|M(t)|\,dt \le 6,$$

and we are ready. In the other case, we prove (9.4i) using

$$(9.5)\qquad \delta = \min\left(2,\ 1+\alpha^2/(24)\right).$$

[1] In fact, it is easy to improve the remainder term in (9.4ii) to $O(1)$.

Given β, $\alpha < \beta < 2$, then $|M(x)| \le \beta x$, if $x \ge x_\beta$. It suffices to show

(9.6) $$\int_a^b t^{-2} \cdot |M(t)|\, dt \le (\beta/\delta) \cdot \log(b/a)$$

for two consecutive zeros a, b of M(x) in $x_\beta \le a < b \le y$. According to the size of b/a we distinguish three cases:

1st case: $\log(b/a) \ge 6 \cdot (\delta/\beta)$. Then, trivially,

$$\int_a^b t^{-2} \cdot |M(t)|\, dt \le 6 \le (\beta/\delta) \cdot \log(b/a).$$

2nd case: $\log(b/a) < 6\ (\delta/\beta)$ *and* $(b/a) \le (1-\frac{1}{2}\beta)^{-1}$. Since $M(a) = 0$, for every t in [a,b]

(9.7) $$|M(t)| = |M(t) - M(a)| \le |t-a| \le \tfrac{1}{2}\beta \cdot t.$$

Again, this estimate implies

$$\int_a^b t^{-2} \cdot |M(t)|\, dt \le \tfrac{1}{2}\beta \int_a^b t^{-1}\, dt \le \tfrac{1}{2}\beta \cdot \log(b/a) \le (\beta/\delta) \cdot \log(b/a),$$

which is (9.6) in the second case.

3rd case: $\log(b/a) < 6\ \delta/\beta$ *and* $b/a > (1-\frac{1}{2}\beta)^{-1}$.
We apply inequality (9.7) in the interval $a \le t \le a \cdot (1-\frac{1}{2}\beta)^{-1}$, and we apply the estimate $|M(t)| \le \beta\cdot t$ in $[x_\beta \le]\ a\cdot(1-\frac{1}{2}\beta)^{-1} \le t \le b$. This leads to

$$\int_a^b t^{-2} \cdot |M(t)|\, dt \le \tfrac{1}{2}\beta \log(1-\tfrac{1}{2}\beta)^{-1} + \beta \log\big((b/a)\cdot(1-\tfrac{1}{2}\beta)\big)$$

$$= \beta \log(b/a) + \tfrac{1}{2}\beta \log(1-\tfrac{1}{2}\beta).$$

Using the definition of δ and

$$\tfrac{1}{2}\beta \log(1-\tfrac{1}{2}\beta) \le -\tfrac{1}{4}\beta^2 \le -\tfrac{1}{4}\alpha^2,$$

we see that

$$\tfrac{1}{2}\beta \log(1-\tfrac{1}{2}\beta) \le -6\ (\delta-1) \le -(1-\delta^{-1})\ .\ \beta \cdot \log(b/a),$$

and so we obtain (9.6) in the third case also.

Proof of (9.4 ii). The following *auxiliary functions* are needed:

(9.8) $$F(x) = \int_0^x u^{-1} \cdot (1-e^{-u})\, du, \text{ where } x > 0,$$

$$(9.9)\qquad k(s) = \int_0^\infty e^{-sx} \cdot e^{F(x)}\, dx, \text{ where } s > 0.$$

It is obvious that the function $s \mapsto k(s)$ is positive, decreasing and continuously differentiable. Furthermore,

$$(9.10)\qquad s \cdot k(s) - \int_s^{s+1} k(u)\, du = 1 \text{ for any } s > 0.$$

[The left-hand side of (9.10) equals

$$- \int_0^\infty e^{F(x)-sx} \cdot (F'(x) - s)\, dx = - \int_0^\infty \frac{d}{dx}\left(e^{F(x)-sx} \right) dx = 1. \,]$$

For fixed $y \geq 2$ we consider the function h, defined in $x > 1$:

$$(9.11)\qquad h(x) = (\log y)^{-1} \cdot k\left(\frac{\log x}{\log y} \right).$$

Then, for any $x > 1$, (9.10) leads to the equation

$$(9.12)\qquad \log x \cdot h(x) = 1 + \int_x^{yx} \upsilon^{-1} h(\upsilon)\, d\upsilon.$$

Partial integration gives

$$(9.13)\qquad \int_1^y x^{-1} \cdot \left(\int_x^{yx} \upsilon^{-1} h(\upsilon)\, d\upsilon \right) dx = \int_1^2 (2-u) \cdot k(u)\, du \cdot \log y.$$

Lemma 9.2. *Denote by* C *the limit given in the displayed formula immediately following* (9.3). *Then*

$$\int_1^2 (2-u) \cdot k(u)\, du = C - 1.$$

Proof. Starting with the convolution relations

$$\log = \Lambda * 1, \text{ and } s_y \cdot \log = (s_y \Lambda) * s_y ,$$

and using the abbreviation $S_y(x) = \sum_{n \leq x} s_y(n)$, we obtain

$$\sum_{n \leq x} s_y(n) \cdot \log n = \sum_{d \leq x} s_y(d)\, \Lambda(d) \cdot S_y\left(\frac{x}{d}\right)$$

This relation implies

$$S_y(x) \cdot \log x = \sum_{n \leq x} s_y(n) \log n + \sum_{n \leq x} s_y(n) \log(x/n)$$

$$= \sum_{p \leq y, p^k \leq x} \log p \cdot S_y\left(\frac{x}{p^k}\right) + \sum_{n \leq x} s_y(n) \log \frac{x}{n}.$$

Therefore,

$$(9.14)\quad \int_y^\infty S_y(x)\cdot\frac{\log x}{x^2}\cdot h(x)\,dx = \int_y^\infty \sum_{p\le y}\log p\cdot S_y\Big(\frac{x}{p}\Big)\,\frac{h(x)}{x^2}\,dx + R_1 + R_2,$$

with the remainder terms

$$(9.14.1)\qquad R_1 = \int_y^\infty \sum_{\substack{p\le y\\ k\ge 2,\, p^k\le x}} \log p\cdot S_y\Big(\frac{x}{p^k}\Big)\,\frac{h(x)}{x^2}\,dx,$$

$$(9.14.2)\qquad R_2 = \int_y^\infty \sum_{n\le x} s_y(n)\cdot \log\Big(\frac{x}{n}\Big)\,\frac{h(x)}{x^2}\,dx.$$

The error terms R_1 and R_2 are bounded: the estimate

$$\int_1^\infty u^{-2}\cdot S_y(u)\,du \le \sum_{n\le x} n^{-1}\, s_y(n) = \prod_{p\le y}\left(1-p^{-1}\right)^{-1} = \mathcal{O}(\log y)$$

implies

$$R_1 \le h(y)\cdot \sum_{p,\,k\ge 2} p^{-k}\log p\cdot \int_1^\infty u^{-2}\cdot S_y(u)\,du = \mathcal{O}(1),$$

and

$$R_2 = \mathcal{O}\Big(h(y)\cdot \textstyle\sum_{n=1}^\infty n^{-1}\, s_y(n)\Big) = \mathcal{O}(1).$$

Starting with the elementary relation $\sum_{n\le y} p^{-1}\log p = \log y + \mathcal{O}(1)$ [see I. (6.6)], partial summation yields the formulae

$$(9.15.1)\qquad \sum_{p\le y} p^{-1}\log p\cdot h(pt) = \int_t^{yt} \upsilon^{-1}\, h(\upsilon)\,d\upsilon + \mathcal{O}\big(h(y)\big),\ \text{ for } t\ge y,$$

$$(9.15.2)\qquad \sum_{y/t<p\le y} p^{-1}\log p\cdot h(pt) = \int_y^{yt} \upsilon^{-1}\, h(\upsilon)\,d\upsilon + \mathcal{O}\big(h(y)\big),\ \text{ for } t>1.$$

The integral on the right-hand-side of (9.14) equals $\sum_{p\le y}\log p\cdot\int_y^\infty \dots\, dx$, and using the substitution $x\mapsto p\cdot t$, we obtain

$$\sum_{p\le y}\log p\cdot \int_{y/p}^\infty S_y(t)\cdot t^{-2}\, h(pt)\cdot p^{-1}\,dt$$

for this integral. Interchanging summation and integration, we arrive at

$$\int_y^\infty S_y(x)\cdot\frac{\log x}{x^2}\cdot h(x)\,dx = \int_1^y t^{-2}\cdot S_y(t)\cdot \sum_{y/t<p\le y} p^{-1}\cdot\log p\cdot h(pt)\,dt$$

$$+\int_y^\infty t^{-2}\cdot S_y(t)\cdot \sum_{p\le y} p^{-1}\cdot\log p\cdot h(pt)\,dt + \mathcal{O}(1).$$

Using (9.15.1) and (9.15.2), the right-hand-side changes into

$$\int_1^y t^{-2}\cdot S_y(t)\cdot\Big(\int_y^{yt} u^{-1}\, h(u)\, du\Big)\, dt + \int_y^\infty t^{-2}\cdot S_y(t)\cdot\Big(\int_t^{yt} u^{-1}\, h(u)\, du\Big)\, dt + \mathcal{O}(1).$$

Using (9.12), we finally obtain

$$(9.16)\quad \int_y^\infty S_y(x)\cdot\frac{\log x}{x^2}\, dx = \int_1^y x^{-2}\, S_y(x)\cdot\Big(\int_y^{yx} u^{-1}\, h(u)\, du\Big)\, dx + \mathcal{O}(1).$$

The left-hand-side of (9.16) becomes

$$\int_1^\infty x^{-2}\, S_y(x)\, dx - \int_1^y x^{-2}\, S_y(x)\, dx = \sum_{n=1}^\infty n^{-1}\, s_y(n) - \sum_{n\le y} n^{-1}\, s_y(n) + \mathcal{O}(1)$$

$$(9.17.1)\qquad = \prod_{p\le y}\big(1-p^{-1}\big)^{-1} - \sum_{n\le y} n^{-1} + \mathcal{O}(1)$$

$$= \big(C + o(1) - 1\big)\cdot \log y + \mathcal{O}(1).$$

Using $S_y(x) = x + \mathcal{O}(1)$ for all $x \le y$, and (9.13), the right-hand-side of (9.16) becomes

$$\int_1^y x^{-1}\Big(\int_y^{yx} u^{-1}\, h(u)\, du\Big)\, dx + \int_1^y \mathcal{O}\Big(x^{-2}\int_y^{yx} u^{-1}\, h(u)\, du\Big)\, dx + \mathcal{O}(1)$$

$$(9.17.2)\qquad = \int_1^2 (2-u)\, k(u)\, du\cdot \log y + \mathcal{O}(1).$$

Dividing (9.17.1) and (9.17.2) by log y, then as $y\to\infty$, Lemma 9.2 is proved. □

The proof of (9.4ii) can now be concluded, using very similar ideas. Starting with the convolution relations $\Lambda * \mu = -\mu\cdot\log$, and $(s_y\,\Lambda) * (s_y\,\mu) = -\, s_y\cdot\mu\cdot\log$, we obtain

$$|M_y(x)|\cdot\log x \le \sum_{n\le x} s_y(n)\,\Lambda(n)\cdot|M_y(x/n)| + \sum_{n\le x} s_y(n)\log(x/n),$$

and so

$$\int_y^\infty |M_y(x)|\cdot\log x\cdot x^{-2}\, h(x)\, dx = \int_y^\infty \sum_{p\le x}\log p\cdot|M_y(x/p)|\cdot x^{-2}\, h(x)\, dx + \mathcal{O}(1),$$

and

$$\int_y^\infty |M_y(x)|\cdot x^{-2}\, dx = \int_1^y x^{-2}\, |M_y(x)|\cdot\Big(\int_y^{yx} u^{-1}\, h(u)\, du\Big)\, dx + \mathcal{O}(1).$$

Using $M_y(x) = M(x)$ for $x \le y$, and estimating $|M(x)|$ by βx, if $x \ge x_\beta$, we finally obtain

$$\int_y^\infty |M_y(x)| \cdot x^{-2}\,dx = \beta \int_1^y x^{-1} \left(\int_y^{yx} \upsilon^{-1}\, h(\upsilon)\, d\upsilon \right) dx + \mathcal{O}(1)$$

$$= \beta \int_1^2 (2-u)\, k(u)\, du \cdot \log y + \mathcal{O}(1) = \beta\,(C-1) \log y + \mathcal{O}(1),$$

and so (9.4ii) is proved. □

II.10. MOHAN NAIR'S ELEMENTARY METHOD IN PRIME NUMBER THEORY

The prime number theorem implies the asymptotic formula

(10.1) $$\psi_1(x) \sim \tfrac{1}{2}\, x^2, \text{ as } x \to \infty,$$

where $\psi_1(x) = \sum_{n\le x} \psi(n)$. On the other hand, by elementary arguments, (10.1) implies the prime number theorem $\psi(x) \sim x$. M. NAIR [1982] gave a simple method for obtaining good lower estimates for $\psi_1(x)$.

Theorem 10.1. *If x is sufficiently large, then*

(10.2) $$\psi_1(x) \ge 0.47459\ \ldots\ \cdot\ x^2.$$

By a more elaborate calculation, NAIR was able to improve the constant 0.47459 to 0.49517, which is rather near the best-possible constant $\frac{1}{2}$. We do not deal with this improvement here.

Lemma 10.2 (CAUCHY). *The determinant D_n of the n-rowed matrix $\left(\alpha_{i,j}\right)_{i,j=1,\ldots,n}$ with elements*

$$\alpha_{i,j} = (a_i + b_j)^{-1},$$

where a_i, b_j are given real, positive numbers, has the value

$$D_n = \prod_{1\le j<i\le n} (a_i - a_j)\,(b_i - b_j) \left(\prod_{i,j=1}^{n} (a_i + b_j) \right)^{-1}.$$

The **Proof** is given in POLYA-SZEGÖ, "*Aufgaben und Lehrsätze aus der Analysis*", II, 7. Abschnitt, Aufgabe 3 (p.299). It is by inductive arguments (see Exercise 19). □

Lemma 10.2 is used with $a_i = m + i$, $b_j = j$ for some positive integer m. Then

$$D_n = \prod_{1\le j<i\le n} (i - j)^2 \Big(\prod_{i,j=1}^{n} (m + i + j) \Big)^{-1}.$$

A good approximation for D_n may be obtained with the aid of STIRLING's formula

$$(10.3)\quad \begin{cases} \log D_n = -\frac{1}{2}(2n+m)^2 \cdot \log(2n+m) + (m+n)^2 \cdot \log(m+n) \\ \qquad + n^2 \cdot \log n - \frac{1}{2} m^2 \cdot \log m + \mathcal{O}\big((n+m)\log(n+m)\big) \end{cases}$$

(see exercise 20). Denote by d_n the least common multiple of 1, 2, ..., n:

$$d_n = \operatorname{lcm}[1, 2, \dots, n].$$

The connection with prime number theory is due to the relation

$$\psi(n) = \log d_n.$$

Multiplying D_n by the product $\prod_{1\le i\le n} d_{n+m+i}$, we obtain a determinant with integer entries and a positive value. Thus

$$\log D_n + \sum_{1\le i\le n} \psi(m+n+i) \ge 0,$$

which may be written as

$$\psi_1(m+2n) - \psi_1(m+n) \ge -\log D_n.$$

In the special case where $m = [s\cdot n]$ with some constant $s > 0$, one obtains for large n the relation

$$\psi_1((2+s)\cdot n) - \psi_1((1+s)\cdot n) \ge \eta(s) \cdot n^2 + \mathcal{O}(n \log n),$$

with the function

$$\eta(s) = \tfrac{1}{2}(2+s)^2 \log(2+s) - (1+s)^2 \log(1+s) + \tfrac{1}{2} s^2 \cdot \log s.$$

Therefore,

$$\psi_1(x) - \psi_1\big((1+s)x \cdot (2+s)^{-1}\big) \ge \eta(s)\cdot(2+s)^{-2} \cdot x^2 + \mathcal{O}(x \log x),$$

and splitting the interval $[1,x]$ into $\mathcal{O}(\log x)$ intervals of the shape $[(1+s)(2+s)^{-1}\cdot\rho,\ \rho]$, we obtain by summation

$$\psi_1(x) \geq \eta(s) \cdot (2s+3)^{-1} \cdot x^2 + \mathcal{O}(\,x \log^2 x\,).$$

The choice $s \to 0+$ leads to the lower bound (note that $\frac{1}{3} \cdot 2 \log 2 \approx 0.462$)

$$\psi_1(x) \geq 0.46 \cdot x^2, \text{ if } x \text{ is large.}$$

Some numerical calculations (see the graphical illustration in Figure II.8) show that the maximum of the function $\eta(s) \cdot (2s+3)^{-1}$ occurs near $s = 0.22$, with the "good" value $\eta(s) \cdot (2s+3)^{-1} = 0.47459 \ldots$. □

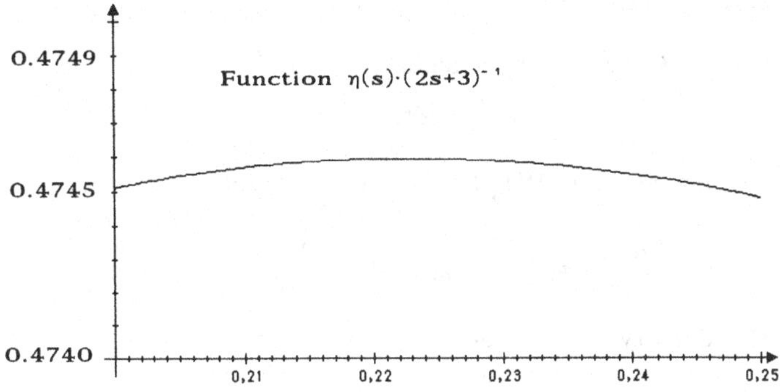

Figure II.8 was produced using the program system RIEMANN II.

Figure II.8

II.11. EXERCISES

1) Let h be a completely multiplicative arithmetical function with mean-value $M(h)$, which is bounded at the primes, $|h(p)| \leq \gamma$ for every prime p. If f is a multiplicative function with the property that the two series

$$\sum_p p^{-1} \cdot |\, f(p) - h(p)\,| < \infty, \text{ and } \sum_p \sum_{k\geq 2} p^{-k} \cdot |\, f(p^k)\,| < \infty$$

are convergent, then the mean-value $M(f)$ exists and is equal to

$$M(f) = M(h) \cdot \prod_p \Big(1 + \sum_{k\geq 1} p^{-k} \cdot (f(p^k) - h(p)\, f(p^{k-1})) \Big).$$

Hint: use Theorem 2.1.

2) Prove Theorem 2.4 in detail.

Hint: $x^{-1} \cdot \sum_{n\leq x} f(n) = \sum_{n=1}^{\infty} n^{-1} \cdot h(n) \cdot M(g) - \sum_{n>x} n^{-1} \cdot h(n) \cdot M(g)$
$+ \sum_{n\leq x} n^{-1} \cdot h(n) \cdot o(x/n).$

3) Let h be a bounded, completely multiplicative function with mean-value M(h). Then

$$M\big(\mathrm{id}^{-1} \cdot h \cdot \tau\big) = M(h) \sum_{n=1}^{\infty} n^{-2} \cdot h(n) = M(h) \prod_p \big(1 - p^{-2} \cdot h(p)\big)^{-1}.$$

4) Define the sets $\mathcal{F}$ and $\mathcal{C}$ by

$$\mathcal{F} = \{ n \in \mathbb{N},\ n \text{ squarefree} \},$$
$$\mathcal{C} = \{ (m,n) \in \mathbb{N}^2,\ \gcd(m,n) = 1 \}.$$

Prove that the mean-value $M(1_{\mathcal{F}})$ exists and equals

$$\lim_{x \to \infty} x^{-2} \sum\sum_{m,n\leq x} 1_{\mathcal{C}}(m,n).$$

5) Prove that the density of the set

$$\mathcal{A} = \{ n \in \mathbb{N}\text{: the number of primes with } p^2 | n \text{ is even} \}$$

exists and is equal to $\frac{1}{2} \prod_p (1 - 2 \cdot p^{-2}) + \frac{1}{2}$.

Hint: the function $2 \cdot 1_{\mathcal{A}} - 1$ is multiplicative. Use Corollary 2.3.

6) Prove that for any non-negative integer ℓ the density of the set

$$D_\ell = \{ n \in \mathbb{N},\ \omega(n) \leq \ell \}$$

is zero.

7) Prove that the frequency with which the digit $m \in \{ 1, 2, ..., 9 \}$ occurs as first number in the sequence 2^n (written in decimal scale), is $\log_{10} (1 + m^{-1})$.

8) For every direct product $\mathbb{N} = \mathcal{A} \times \mathcal{B}$, both the series

$$\sum_{a\in\mathcal{A}} a^{-1},\ \sum_{b\in\mathcal{B}} b^{-1}$$

cannot be convergent Is it possible that both series are divergent?

9) Let a, $K_o \in \mathbb{N}$. Does there exist a direct product $\mathbb{N} = \mathcal{A} \times \mathcal{B}$, such that

(a) $\mathcal{A} = \{ a^k, k = 0, 1, 2, \dots \}$?

(b) $\mathcal{A} = \{ a^k, 0 \le k < K_o \}$?

10) Prove that there exists a constant $c > 0$, such that for every real number $\alpha > 0$ and every integer $N > 1$ there exists a multiplicative function f, where $|f| \le 1$, satisfying

$$| \sum_{n \le N} f(n) \cdot e(n\alpha) | \ge c \cdot N \cdot (\log N)^{-1}.$$

11) a) Define

$$g(n) = \tfrac{1}{4} \# \{ (x,y) \in \mathbb{Z}^2, x^2 + y^2 = n \}, \ g' = \mu * g.$$

It is known that g' is the non-principal character modulo 4. Use this to prove $M(g) = \frac{1}{4} \pi$.

(b) Let $f = \mu^2 \cdot g$. Prove that f is multiplicative, and has mean-value

$$M(f) = \frac{3}{2\pi} \cdot \prod_{p \equiv 1 \ (4)} \left(1 - 2 (p(p+1))^{-1} \right).$$

12) Prove the estimate for S''' given in section (II.4).

13) Assume that f is a multiplicative arithmetical function, satisfying $f(p) \ge c_1 \, p^{3/2+\varepsilon}$ for every prime p, with some positive constant c_1. Prove that

$$\sum_{n \ge x} \mu^2(n) \cdot (f(n))^{-1} << x^{-1}.$$

Hint: apply RANKIN's trick.

14) Let α be real and $N \ge 2$ a positive integer. Define a completely multiplicative function f by: $f(p) = c$ if $p \le \frac{1}{2} N$, $f(p) = e(-\alpha p)$ if $\frac{1}{2} N < p \le N$, $f(p) = 1$ if $p > N$. Prove that there exists a complex constant c , $|c| = 1$, such that f is in $\mathcal{F}_1$ and has the property $|S_f(\alpha)| >> N \cdot \log^{-1} N$ [see Theorem 6.2].

15) Prove:

a) $\varphi^{-1} \cdot \mathrm{id} = (\varphi^{-1} \cdot \mu^2) * 1$.

b) $\sum_{n \le x} (\varphi(n)^{-1} \cdot n)^2 = \mathcal{O}(x)$.

c) $\sum_{n \le x} n \cdot \varphi^{-2}(n) = \mathcal{O}(\log x)$.

16) Prove the formula $\tau_m(p^k) = \binom{k+m-1}{m-1}$, k, m = 1, 2, ...

17) Prove by mathematical induction the inequality $\prod_{p \le N} p < 4^N$.

Hint: if $N = 2 \cdot k + 1$ is odd, use $\prod_{p \le N} p \le \binom{2k+1}{k} \cdot \prod_{p \le k+1} p$.

18) In $1 \le a < b$, prove $\left| \int_a^b t^{-2} \cdot \sum_{n \le t} \mu(n)\, dt \right| \le 6$.

Hint: use I. Cor. 2.5.

19) For positive real numbers a_i, b_j, where $1 \le i, j \le n$, calculate the determinant

$$D_n = \det\left(\frac{1}{a_i + b_j}\right) = \prod_{1 \le j < i \le n} (a_i - a_j)(b_i - b_j) \left(\prod_{i,j=1}^{n} (a_i + b_j) \right)^{-1}.$$

Hint: subtract the n-th row from another one, and extract suitable factors. Then subtract the n-th column from another one. Proceed by induction on n.

20) Prove formula (10.3), using STIRLING's formula

$$\log(n!) = n \cdot \log n - n + O(n).$$

Chapter III

Related Arithmetical Functions

Abstract. *The simple fact that multiplicative functions are determined by their values at the primes leads to the idea that multiplicative functions which do not differ "too much" at the primes behave similarly. The aim of this chapter is to render these vague idea more precise and to provide a universally applicable result in order to reduce proofs to the simplest possible assumptions. The notion of "relatedness" is a measure for "not differing too much at the primes". Our result states that two related functions* f *and* g, *which are not too large, are connected by a convolution formula* $g = f * h$, *where the function* h *is small in the sense that the series* $\sum |h(n)| \cdot n^{-1}$ *is convergent.*
This chapter is close to the paper HEPPNER & SCHWARZ *[1983].*

III.1. INTRODUCTION, MOTIVATION

Multiplicative functions are determined by their values at the prime-powers p^k for the relation

$$f\Big(\prod_{p^k \| n} p^k \Big) = \prod_{p^k \| n} f(p^k).$$

Higher prime-powers p^k, where $k \geq 2$, are rare: the number of these up to x is

$$\sum_{p^k \leq x,\ k \geq 2} 1 = \mathcal{O}(x^{\frac{1}{2}}),$$

and so one is inclined to conjecture that multiplicative arithmetical functions behaving similarly at primes have similar properties[1)]. This chapter aims to give an exact meaning to these vague formulations. The theorem we are going to prove will be important for simplifying proofs by reduction of these to special cases which are easier to handle (for example, multiplicative functions may be replaced by com-

---

1) For example, given two multiplicative arithmetical functions f and g, which behave similarly at the primes, one might ask for conditions that ensure one or more of the following assertions:

- If f has a mean-value $M(f) = \lim_{x \to \infty} x^{-1} \sum_{n \leq x} f(n)$, then g has a mean-value, too.
- If f has Fourier-coefficients $\hat{f}(\gamma) = M(f \cdot e_{-\gamma})$, then the Fourier-coefficients of g exist.
- If the RAMANUJAN-Fourier-coefficients $a_r(f) = (\varphi(r))^{-1} \cdot M(f \cdot c_r)$ do exist, the same is also true for the function g.
- If the RAMANUJAN-expansion $\sum_{1 \leq r < \infty} a_r(f) \cdot c_r(n)$ of f is [pointwise, absolutely, uniformly, ...] convergent, the same is true for g.
- If the series $\sum_{1 \leq n < \infty} n^{-1} \cdot f(n)$ is convergent [or summable by some method of summation], then the same result is valid for $\sum_{1 \leq n < \infty} n^{-1} \cdot g(n)$.
- If the function f is in the class $\mathcal{B}^1$ of functions which are arbitrarily near with respect to the semi-norm $\|f\|_1 = \limsup_{x \to \infty} x^{-1} \cdot \sum_{n \leq x} |f(n)|$ to finite linear combinations of RAMANUJAN sums, then g is in $\mathcal{B}^1$.

pletely multiplicative functions).

The set $\mathcal{M}$ of multiplicative arithmetical functions (not identically zero, and so $f(1) = 1$) is a commutative group $(\mathcal{M},*)$ under convolution

$$f*g:\ n \mapsto \sum_{d|n} f(d)\cdot g\left(\frac{n}{d}\right),$$

as we have seen in I.2. The unit element is given by the function ε, with values $\varepsilon(1) = 1$ and $\varepsilon(n) = 0$, if $n > 1$; the *convolution-inverse* of f is denoted by $f^{-1(*)}$, and it is possible to determine the values of $f^{-1(*)}$ recursively from the relation $f * f^{-1(*)} = \varepsilon$.

The "near-ness" of multiplicative arithmetical functions at the primes is measured using the following definition.

Definition. *Two functions* f, g *in* $\mathcal{M}$ *are called* <u>*related*</u>, *if the sum*

$$\sum_p p^{-1}\cdot |\, f(p) - g(p)\,| < \infty \tag{1.1}$$

is absolutely convergent.

The sum $\sum_p p^{-1}$ is divergent. Therefore, for related functions f, g, the differences $|\, f(p) - g(p)\,|$ have to be be rather small, at least in the mean, to ensure that f and g are related. In spite of the rarity of the higher prime-powers we need some restrictions on the size of the functions under consideration. We define the subset $\mathcal{G}$ of $\mathcal{M}$ as the set of multiplicative arithmetical functions satisfying

$$\sum_p \sum_{k\ge 2} p^{-k}\cdot |f(p^k)| < \infty, \text{ and } \sum_p p^{-2}\cdot |f(p)|^2 < \infty:$$

$$\mathcal{G} = \left\{ f \in \mathcal{M},\ \sum_p p^{-2}\cdot |f(p)|^2 < \infty \text{ and } \sum_p \sum_{k\ge 2} p^{-k}\cdot |f(p^k)| < \infty \right\}. \tag{1.2}$$

The following **notation** will frequently be used in this chapter: the set

$$\mathcal{G}^* = \left\{ f \in \mathcal{G},\ \varphi_f(p,s) \ne 0 \textit{ in } \operatorname{Re} s \ge 1 \textit{ for every prime } p \right\} \tag{1.3}$$

is the set of those multiplicative functions in $\mathcal{G}$ for which $\varphi_f(p,s)$ does not vanish in the complex half-plane $\operatorname{Re} s \ge 1$, where $\varphi_f(p,s)$ denotes the p-th factor in the EULER product for the generating DIRICHLET series $\mathcal{D}(f,s) = \sum_{n\ge 1} f(n)\cdot n^{-s}$ of the arithmetical function f, and so

(1.4) $$\varphi_f(p,s) = \left(1 + f(p)\cdot p^{-s} + f(p^2)\cdot p^{-2s} + \dots \right).$$

The set of multiplicative arithmetical functions f with absolutely convergent series $\sum_{n\geq 1} n^{-1}\cdot|f(n)|$ is denoted by

(1.5) $$\mathcal{ACM} = \left\{ f \in \mathcal{M} ;\ \sum_{n\geq 1} n^{-1}\cdot |f(n)| < \infty \right\}$$

(absolutely convergent, multiplicative); and, finally, we put

(1.6) $$\mathcal{R}\varepsilon = \left\{ f \in \mathcal{G};\ f \ \textit{related to}\ \varepsilon \right\} = \left\{ f \in \mathcal{G};\ \sum_p p^{-1}\cdot |f(p)| < \infty \right\}.$$

Examples.

(1) If f is multiplicative, then the functions f and $\mu^2 \cdot f$ are related.

(2) The functions $n \mapsto n^{-1}$, more generally $n \mapsto n^{-\alpha}$, where $\alpha > 0$, are in $\mathcal{R}\varepsilon$ and, at the same time, in $\mathcal{ACM}$.

(3) The functions $n \mapsto n^{-1}\cdot\varphi(n)$, $n \mapsto \mu^2(n)$ and $n \mapsto \mu(n)$ are in $\mathcal{G}^*$. Every bounded multiplicative function is in $\mathcal{G}$.

(4) If f is a multiplicative function in $\mathcal{G}$, then the multiplicative functions f_{cm} and f_{sm}, defined by

$$f_{cm}(p^k) = f(p)^k \text{ for every prime p and every } k \geq 1,$$

$$f_{sm}(p^k) = f(p) \text{ for every prime p and every } k \geq 1,$$

are completely multiplicative (resp. strongly multiplicative) are in $\mathcal{G}$ and are related to f if $|f(p)| < p$ in the first case.

(5) Another important **construction** ("multiplicative truncation") is the following: given a multiplicative function f in $\mathcal{G}$, then, with a fixed squarefree integer K, f may be written as a convolution product

$$f = f_1^{(K)} * f_2^{(K)},$$

where $f_1^{(K)}$ and $f_2^{(K)}$ are multiplicative, and

$$\begin{cases} f_1^{(K)}(p^k) = f(p^k), \text{ if } p|K, \text{ and } = 0 \text{ otherwise, and} \\ f_2^{(K)}(p^k) = 0, \text{ if } p|K, \text{ and } = f(p^k) \text{ otherwise.} \end{cases}$$

The function $f_2^{(K)}$ is in $\mathcal{G}$ again, and it is related to f; the function $f_1^{(K)}$ is in $\mathcal{R}\varepsilon$ as well as in $\mathcal{ACM}$. An important special case of this construction is the choice $K = 2\cdot 3\cdot 5\cdot \ldots \cdot P$ with some fixed prime P.

III.2. MAIN RESULTS

Using the notation introduced in III.1, the main result of this chapter is the following theorem.

Theorem 2.1. *Assume that* f *and* g *are multiplicative arithmetical functions which are related. If* f *is in* $\mathcal{G}^*$ *and if* g *is in* $\mathcal{G}$, *then the multiplicative function* h, *defined by* $g = f * h$, *has the properties*

(a) $\sum_{n\geq 1} n^{-1} \cdot |h(n)| < \infty$, *so* $h \in \mathcal{ACM}$.

(b) $\sum_{n\geq 1} n^{-s} \cdot h(n) = \prod_p \varphi_g(p,s) \cdot (\varphi_f(p,s))^{-1}$ *in* Re $s \geq 1$.

Remark. The formulation of Theorem 2.1 is not symmetrical in f and g; of course, $h = g * f^{-1(*)}$ is multiplicative, but in order to be able to show the absolute convergence of $\sum n^{-1} \cdot h(n)$, one must derive some additional properties of f: the non-vanishing of the factors of the EULER product implies nice properties of the convolution inverse $f^{-1(*)}$.

Example. In the formulation of Theorem 2.1, the condition $f \in \mathcal{G}^*$ cannot be replaced by $f \in \mathcal{G}$. The function $g = \mu$ (MÖBIUS function) is in $\mathcal{G}$, and the multiplicative function f, defined by

$$f(p^k) = \begin{cases} -2, & \text{if } p = 2, \text{ and } k = 1, \\ -1, & \text{if } p > 2, \text{ and } k = 1, \\ 0, & \text{if } k \geq 2, \end{cases}$$

is related to g, and is in $\mathcal{G}$ but not in $\mathcal{G}^*$. An easy computation shows that $f^{-1(*)}(p^k) = 1$, if $p \neq 2$ and k is arbitrary. For $p = 2$, however, we obtain $f^{-1(*)}(2^k) = 2^k$; therefore

$$h(p^k) = f^{-1(*)}\left(p^k\right) - f^{-1(*)}\left(p^{k-1}\right) = \begin{cases} 0, & \text{if } p \neq 2 \\ 2^{k-1}, & \text{if } p = 2, \end{cases}$$

and so $\sum n^{-1} \cdot |h(n)|$ is divergent.

However, in spite of this example, a condition weaker than $f \in \mathcal{G}^*$ is sometimes sufficient for applications.

Theorem 2.2. *Assume that* f *and* g *are related, and that both are in* $\mathcal{G}$. *Assume, furthermore, that the factors* $\varphi_f(p,s)$ *of the* EULER *product of the* DIRICHLET *series* $\mathcal{D}(f,s)$ *are* **not** *zero in the closed half-plane* $\text{Re } s \geq 1$ *for every prime* p *outside some finite exceptional set* $\mathcal{EP}$. *Then there exists a multiplicative function* h *in* $\mathcal{ACM}$ *(see* (1.5)*) satisfying* $g = f * h$, ***provided*** $f(p^k) = g(p^k)$ *for every exceptional prime* $p \in \mathcal{EP}$ *and every* $k = 1,2, \ldots$.

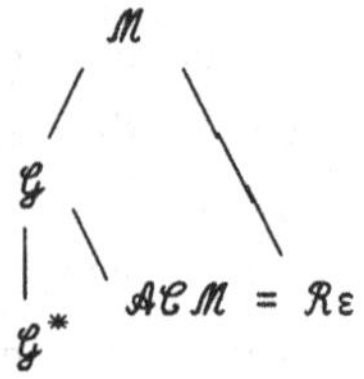

Figure III.1

Theorems 2.1 and 2.2 are deduced from the following theorem.

Theorem 2.3.

(1) *The set* $\mathcal{G}$ *is closed with respect to convolution:* $f*g$ *is in* $\mathcal{G}$, *if* f *and* g *both are both in* $\mathcal{G}$.

(2) *The set* $\mathcal{G}^*$ *is closed with respect to convolution.*

(3) *If* f *is in* $\mathcal{G}^*$, *then the convolution inverse* $f^{-1(*)}$ *is in* $\mathcal{G}^*$.

(4) *The set* $\mathcal{ACM}$ *of functions* f *with absolutely convergent series* $\sum n^{-1}\cdot|f(n)|$ *and the set* $\mathcal{R}\varepsilon$ *of functions in* $\mathcal{G}$, *related to the unit element* ε, *are identical.*

An extension of Theorem 2.1 to functions that are related in some (apparently) more general sense is easily possible.

Given β, $0 < \beta \leq 1$, the multiplicative functions f and g are called *β-related*, if $\sum_p p^{-\beta}\cdot| f(p) - g(p) |$ is convergent. In analogy with notation given earlier we use the abbreviations

(2.1) $$\mathcal{ACM}_\beta = \{ f \in \mathcal{M};\ \sum_n n^{-\beta}\cdot|f(n)| < \infty \},$$

(2.2) $$\mathcal{R}\varepsilon_\beta = \{ f \in \mathcal{G}_\beta;\ f\ \beta\text{-related to } \varepsilon \},$$

(2.3) $$\mathcal{G}_\beta = \Big\{ f \in \mathcal{M},\ \sum p^{-2\beta}\cdot|f(p)|^2 < \infty \text{ and } \sum\sum p^{-k\beta}\cdot |f(p^k)| < \infty \Big\},$$

and

(2.4) $$\mathcal{G}_\beta^* = \Big\{ f \in \mathcal{G}_\beta;\ \varphi_f(p,s) \neq 0 \text{ in Re } s \geq \beta \text{ for every prime } p \Big\}.$$

Theorem 2.4.

(1) *The set* $\mathcal{G}_\beta$ *is closed with respect to convolution.*

(2) *The set* $\mathcal{G}_\beta^*$ *is closed with respect to convolution.*

(3) *If* f *is in* $\mathcal{G}_\beta^*$, *then* $f^{-1(*)}$ *is in* $\mathcal{G}_\beta^*$.

(4) *The sets* $\mathcal{ACM}_\beta$ *and* $\mathcal{R}\varepsilon_\beta$ *are identical.*

Corollary 2.5. *If* $f \in \mathcal{G}_\beta^*$ *and* $g \in \mathcal{G}_\beta$ *are β-related, then the multiplicative function* $h = g * f^{-1(*)}$ *is in* $\mathcal{ACM}_\beta$.

Theorem 2.4 and its corollary are reduced to Theorem 2.1 and 2.3 by applying these results to the functions

$$f^{\#}: n \mapsto n^{(1-\beta)} \cdot f(n), \quad g^{\#}: n \mapsto n^{(1-\beta)} \cdot g(n). \qquad \square$$

An important generalization is due to L. LUCHT (preprint [1991]). The use of weights enables him to deal, for example, with more general remainder terms.

Proof of Theorem 2.1. Since f is in $\mathcal{G}^*$, by Theorem 2.3 (3) (this theorem will be proved in the next section) the inverse $f^{-1(*)}$ is in $\mathcal{G}^*$, and so, using Theorem 2.3 (1),

$$h = f^{-1(*)} * g \in \mathcal{G}.$$

If p is prime, then $h(p) = g(p) - f(p)$; f and g are related, and therefore

$$\sum_p p^{-1} \cdot |h(p)| = \sum_p p^{-1} \cdot |g(p) - f(p)| < \infty,$$

and so h is related to ε, and h is in $\mathcal{R}\varepsilon$. But, according to Theorem 2.3 (4), the sets $\mathcal{R}\varepsilon$ and $\mathcal{ACM}$ are identical; therefore

$$\sum n^{-1} \cdot |h(n)| < \infty.$$

(b) follows from I.Lemma 5.1 and $\varphi_g(p,s) = \varphi_f(p,s) \cdot \varphi_h(p,s)$. $\square$

Proof of Theorem 2.2. Assume that the multiplicative functions f and g satisfy the conditions of Theorem 2.2. Then, split

$$f = f_1 * f_2, \quad g = g_1 * g_2,$$

where, using the abbreviation $K = \prod_{p \in \mathcal{EP}} p$, the primes running through the set $\mathcal{EP}$ of exceptional primes, f_1 and f_2 are multiplicative and

$$f_1(p^k) = f(p^k), \text{ if } p|K, \text{ and } = 0 \text{ otherwise,}$$
$$f_2(p^k) = 0, \quad \text{ if } p|K, \text{ and } = f(p^k) \text{ otherwise,}$$

and similarly for g. Then f_2 is in $\mathcal{G}^*$, and $h_2 = f_2^{-1(*)} * g_2$ is in $\mathcal{ACM}$ by

Theorem 2.1.

But, by the assumptions on the values of f and g at "exceptional prime powers" p^k, where $p \in \mathscr{EP}$, $f_1 = g_1$, therefore $f_1^{-1(*)} = g_1^{-1(*)}$, and

$$h = f^{-1(*)} * g = f_1^{-1(*)} * g_1 * f_2^{-1(*)} * g_2 = h_2$$

is in $\mathscr{ACM}$, and Theorem 2.2 is proved as soon as Theorem 2.3 is proved. □

III.3. LEMMATA, PROOF OF THEOREM 2.3

A) WIENER-type-lemmata

NORBERT WIENER showed that the inverse of a non-vanishing, 2π-periodic function with an absolutely convergent FOURIER series again has an absolutely convergent FOURIER series. An elegant proof of this result may be given via GELFAND's theory of commutative BANACH algebras (see, for example, W. RUDIN [1966], [1973], or L. LOOMIS [1953]). The main part of the proof consists of the determination of the so-called maximal ideal space of the BANACH algebra of all functions $F(e^{i\vartheta}) = \sum c_n \cdot e^{in\vartheta}$, defined on the interval $[0, 2\pi]$ using the norm $\|F\| = \sum |c_n|$. The maximal ideal space is the set of algebra-homomorphisms of this BANACH algebra into $\mathbb{C}$, and this space is, in WIENER's case, built up precisely from the evaluation homomorphisms.

The same approach leads to the following

Lemma 3.1. *Denote by* $\mathscr{A}$ *the* BANACH *algebra of power series*

$$F(z) = \sum_{n=1}^{\infty} a_n \cdot z^n,$$

absolutely convergent in $|z| \le 1$, *with finite norm*

$$\|F\| = \sum_{n=1}^{\infty} |a_n|.$$

If F *is in* $\mathscr{A}$, *and if* $F(z) \neq 0$ *in the unit-disk* $\{z;\ |z| \le 1\}$, *then the power-series for the function* $z \mapsto 1/F(z)$

is in $\mathcal{A}$ again.

For a proof see, for example, W. RUDIN [1966], L. LOOMIS [1953].

The corresponding theorem for DIRICHLET series is more difficult to prove, the reason being that the maximal ideal space of the corresponding BANACH algebra contains many more functions.

Lemma 3.2. *The DIRICHLET series*

$$\sum_{n=1}^{\infty} a_n \cdot n^{-s}, \text{ where } \sum_{n=1}^{\infty} |a_n| < \infty,$$

has an inverse

$$\sum_{n=1}^{\infty} b_n \cdot n^{-s}, \text{ where } \sum_{n=1}^{\infty} |b_n| < \infty,$$

if and only if there is some positive lower bound $\delta > 0$ for $|\sum_{n=1}^{\infty} a_n \cdot n^{-s}|$ in the half-plane Re $s \geq 0$:

$$\left|\sum_{n=1}^{\infty} a_n \cdot n^{-s}\right| \geq \delta \quad \text{in Re } s \geq 0. \tag{3.1}$$

A proof for this result, using GELFAND's theory, is given in E. HEWITT and J. H. WILLIAMSON [1957]; according to HEWITT and WILLIAMSON this result can also be deduced from a paper by R. S. PHILLIPS [1951].

B) Splitting of functions $f \in \mathcal{G}$ into a convolution product

Assume that a function $f \in \mathcal{G}$ is given; by the definition of the set $\mathcal{G}$ it is possible to choose a constant P_0 with the properties

$$\left\{\begin{array}{l} |f(p)| < \frac{1}{6} \cdot p, \text{ if } p > P_0, \\ \text{and} \\ \sum_{p>P_0} \sum_{k\geq 2} p^{-k} \cdot |f(p^k)| < \frac{1}{3}. \end{array}\right. \tag{3.2}$$

Define multiplicative functions f_0, f_1', f_1'', f_2 by prescribing their values at prime-powers in the following way.

$$f_0(p^k) = \begin{cases} (f(p))^k, & \text{if } p > P_0, \\ 0, & \text{if } p \leq P_0. \end{cases} \tag{3.3}$$

The function f_0 is completely multiplicative. The function defined next, f_1', is 2-multiplicative and inverse to f with respect to convolution:

(3.4) $$f_1'(p^k) = \begin{cases} -f(p), & \text{if } k = 1 \text{ and } p > P_0, \\ 0 & \text{otherwise.} \end{cases}$$

The "tail" of f is defined as follows:

(3.5) $$f_1''(p^k) = \begin{cases} f(p^k), & \text{if } p > P_0, \\ 0, & \text{if } p \le P_0. \end{cases}$$

Finally, the "head" of f is

(3.6) $$f_2(p^k) = \begin{cases} 0, & \text{if } p > P_0, \\ f(p^k), & \text{if } p \le P_0. \end{cases}$$

We define $f_1 = f_1' * f_1''$. Looking at the generating DIRICHLET series, it is obvious that

$$f = f_0 * f_1' * f_1'' * f_2, \text{ and } f_0^{-1(*)} = f_1' ;$$

the second assertion can also be seen from the relation $h^{-1(*)} = \mu \cdot h$, which is true for completely multiplicative functions (see Exercise I.8).

C) The Main Lemma

Lemma 3.3. *If* f, g $\in \mathcal{G}$, *then the following assertions are true:*

(a) $f * g \in \mathcal{G}$.

(a*) *If* f *and* g *are in* $\mathcal{G}^*$, *the same is true for* $f * g$.

(b) $f_0 \in \mathcal{G}^*$, *and* $f_0^{-1(*)} \in \mathcal{G}^*$.

(c) f_1', f_1'' *are in* $\mathcal{G}$, f_1 *is in* $\mathcal{ACM} \cap \mathcal{G}^*$, *where* $f_1 = f_1' * f_1''$. [*In fact* f_1', f_1'' *are in* $\mathcal{G}^*$.]

(d) $f_1^{-1(*)} \in \mathcal{ACM}$, *where* $f_1 = f_1' * f_1''$.

(e) $f_2 \in \mathcal{ACM}$.

(f) *If, for every prime* $p \le P_0$, $\varphi_f(p,s) \ne 0$ *in the half-plane* $\mathrm{Re}\, s \ge 1$, *then* $f^{-1(*)}$ *is in* $\mathcal{ACM}$.

Proof. Recall that $\mathcal{G}$, $\mathcal{G}^*$, $\varphi_f(p,s)$ and $\mathcal{ACM}$ are defined in (1.2) to (1.5).

(a) For the moment we write $w = f * g$. Then $w(p) = f(p) + g(p)$, and so

$$\sum_p p^{-2} \cdot |w(p)|^2 \le 2 \sum_p p^{-2} \cdot \{ |f(p)|^2 + |g(p)|^2 \} < \infty$$

because f and g are in $\mathcal{G}$. If $k \ge 2$, the definition of convolution yields

$$w(p^k) = \sum\sum_{r,t\ge 0,\ r+t=k} f(p^r) \cdot g(p^t).$$

Thus the second sum in the definition of the set $\mathcal{G}$ may be estimated by

$$\sum_p \sum_{k\geq 2} p^{-k} \cdot |w(p^k)| \leq \sum_r \sum_t \sum_p \underset{r+t \geq 2}{} p^{-r} \ |f(p^r)| \cdot p^{-t} \ |g(p^t)|$$

$$\leq \ \Sigma_1 + \ \ldots \ + \Sigma_4,$$

with <u>conditions of summation</u>

p prime, $r \geq 2$, $t \geq 2$	in Σ_1,
p prime, $r = 0$ or $r = 1$, $t \geq 2$	in Σ_2,
p prime, $t = 0$ or 1, $r \geq 2$	in Σ_3,
p prime, $r = t = 1$	in Σ_4.

The convergence of $\Sigma_p \ \Sigma_{t\geq 2} \ p^{-t}|g(p^t)|$ implies the boundedness of $\Sigma_{t\geq 2} \ p^{-t} \ |g(p^t)|$, and by the boundedness of $\Sigma_p \ \Sigma_{t\geq 2} \ p^{-t} \ |f(p^t)|$ the sum Σ_1 is convergent. Since $\Sigma \ p^{-2} \ |f(p)|^2$ is convergent, we obtain $f(p) = \mathcal{O}(p)$, and $g \in \mathcal{G}$ implies the boundedness of Σ_2; similarly, Σ_3 is bounded. Finally, the convergence of Σ_4 comes from the CAUCHY-SCHWARZ inequality.

(a*) The equality $w = f * g$ implies the relation

$$\mathcal{D}(w,s) = \mathcal{D}(f,s) \cdot \mathcal{D}(g,s)$$

for the generating DIRICHLET series, and

$$\varphi_w(p,s) = \varphi_f(p,s) \cdot \varphi_g(p,s)$$

for the factors of the EULER products, and thus **(a*)** is clear.

(b) The factors $\varphi_{f_0}(p,s)$ of the EULER product are given by $(1 - p^{-s} \cdot f_0(p))^{-1}$, resp. 1 for $p > P_0$ resp. $p \leq P_0$, and therefore these are $\neq 0$ in $\mathrm{Re}\ s \geq 1$ by the choice of P_0. Thus $f_0 \in \mathcal{G}^*$. It is easy to check that $f_1' \in \mathcal{G}^*$, and so $f_0^{-1(*)} = f_1' \in \mathcal{G}^*$.

(c) $f \in \mathcal{G}$ obviously implies that $f_1' \in \mathcal{G}$ and $f_1'' \in \mathcal{G}$, and thus $f_1 = f_1' * f_1''$ is in $\mathcal{G}$, by **(a)**. The values of f_1 at prime-powers are

$$f_1(p^k) = \begin{cases} f(p^k) - f(p^{k-1}) \cdot f(p), & \text{if } p > P_0 \text{ and } k \geq 2, \\ 0 & \text{otherwise.} \end{cases}$$

Therefore, we obtain

$$(3.7)\quad \sum_{n\le N} n^{-1}\cdot|f_1(n)| \le \prod_{P_0<p\le N}\Big\{1+\sum_{k\ge 2} p^{-k}\cdot(|f(p^k)|+|f(p^{k-1})\cdot f(p)|)\Big\}.$$

The assumption $f \in \mathcal{G}$ immediately implies the convergence of the product on the right-hand side of (3.7). Therefore, $f_1 \in \mathcal{ACM}$. Finally, the sum

$$S_p = \sum_{k\ge 2} p^{-k}\cdot(|f(p^k)| + |f(p^{k-1})\cdot f(p)|)$$

is (if $p > P_0$)

$$S_p \le p^{-2}\cdot|f(p)|^2 + (1+p^{-1}\cdot|f(p)|)\cdot\sum_{k\ge 2} p^{-k}\cdot|f(p^k)| \le \tfrac{1}{2}$$

by our assumptions on P_0. Therefore, in the half-plane $\mathrm{Re}\ s \ge 1$

$$|\varphi_{f_1}(p,s)| \ge 1 - S_p \ge \tfrac{1}{2},$$

and so $f_1 \in \mathcal{G}^*$.

(e) The generating DIRICHLET series $\mathcal{D}(f_2,s)$ for f_2 is a product of finitely many, in $|z| \le p^{-1}$ absolutely convergent power series,

$$\sum_{k\ge 0} f(p^k)\cdot z^k,\ \text{where } z = p^{-s},\ \mathrm{Re}\ s \ge 1, \text{ and } p \le P_0.$$

Therefore, $\mathcal{D}(f_2,s)$ is absolutely convergent in $\mathrm{Re}\ s \ge 1$, particularly at the point $s = 1$, and so f_2 is in $\mathcal{ACM}$.

(f) Under the additional assumption $\varphi_f(p,s) \neq 0$ in $\mathrm{Re}\ s \ge 1$, for every $p \le P_0$, all the power series $\sum_{k\ge 0} f(p^k)\cdot z^k$ in **(e)** are invertible with absolutely convergent power series expansion by Lemma 3.1. Therefore, the DIRICHLET series for $1/\mathcal{D}(f_2,s)$ is absolutely convergent in $\mathrm{Re}\ s \ge 1$, and $f_2^{-1(*)}$ is in $\mathcal{ACM}$.

(d) This part is based on the rather difficult Lemma 3.2 and thus may be considered as the most difficult assertion of Lemma 3.3. The function $1 - x - \exp(-2x)$ is 0 for $x = 0$, is $\tfrac{1}{2} - e^{-1} > 0$ at $x = \tfrac{1}{2}$, and has a unique local maximum at $\tfrac{1}{2}\cdot\ln 2 = 0.346\ldots$; therefore (see Figure III.2)

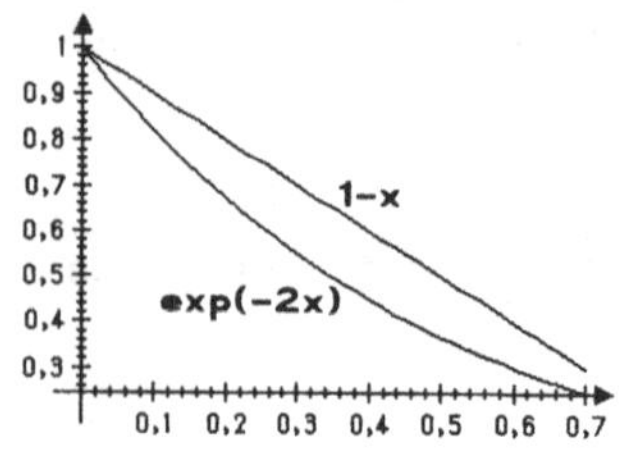

Figure III.2

$$1 - x \ge \exp(-2x) \text{ in } 0 \le x \le \tfrac{1}{2}.$$

Using this inequality, the multiplicativity of f_1 and the values $f_1(p^k)$ given in **(c)**, we arrive at the lower estimate

$$\left|\sum_{n=1}^{\infty} n^{-s}\cdot f_1(n)\right| \geq \prod_{p>P_0} \left\{1 - \sum_{k\geq 2} p^{-k}\cdot(|f(p^k)| + |f(p^{k-1})\cdot f(p)|)\right\}$$

$$\geq \exp\left\{-2\sum_{p>P_0} S_p\right\}.$$

for all s in Re $s \geq 1$. Making use of the fact that f is in $\mathcal{G}$ (and using the assumption $|f(p)|/p < 1/6$ as long as $p > P_0$), we obtain

$$\sum_{p>P_0} S_p \leq \sum_{p>P_0} p^{-2}\cdot|f(p)|^2 + \frac{7}{6}\cdot\sum_{p>P_0}\sum_{k\geq 2} p^{-k}\cdot|f(p^k)| \leq \gamma_1$$

with some constant γ_1, and so

$$\left|\sum_{n=1}^{\infty} n^{-s}\cdot f_1(n)\right| \geq \delta = \exp(-2\gamma_1) \text{ in Re } s \geq 1.$$

An application of Lemma 3.2 now gives the desired conclusion $f_1^{-1(*)} \in \mathcal{ACM}$. □

Proof of Theorem 2.3. Assertions (1) and (2) are already proved (see Lemma 3.3, **(a)** and **(a*)**). For (4), the assumption $h \in \mathcal{ACM}$ implies

$$\sum_p p^{-1}\,|h(p)| \leq \sum_{n=1}^{\infty} n^{-1}\cdot|h(n)| < \infty,$$

therefore $h(p) = \mathcal{O}(p)$ and

$$\sum_p p^{-2}\cdot|h(p)|^2 = \mathcal{O}\left\{\sum_p p^{-1}\,|h(p)|\right\} < \infty.$$

Moreover,

$$\sum_p\sum_{k\geq 2} p^{-k}\cdot|h(p^k)| \leq \sum_{n=1}^{\infty} n^{-1}\cdot|h(n)| < \infty,$$

and so $h \in \mathcal{G}$. The relation

$$\sum_p p^{-1}\cdot|h(p) - \varepsilon(p)| = \sum_p p^{-1}\cdot|h(p)| < \infty$$

shows that h and ε are related, therefore $h \in \mathcal{R}\varepsilon$ and $\mathcal{ACM} \subset \mathcal{R}\varepsilon$.

On the other hand, if h is in $\mathcal{R}\varepsilon$, then

$$\sum_p p^{-1}\cdot|h(p)| < \infty, \text{ and } \sum_p\sum_{k\geq 2} p^{-k}\cdot|h(p^k)| < \infty;$$

these relations imply

$$\sum_{n\leq N} n^{-1}\cdot|h(n)| \leq \prod_{p\leq N}\left\{1 + p^{-1}\cdot|h(p)| + \sum_{k\geq 2} p^{-k}\cdot|h(p^k)|\right\} = \mathcal{O}(1),$$

therefore $\mathcal{R}\varepsilon \subset \mathcal{ACM}$.

Finally, the proof of (3) is based on some of the assertions of Lemma

3.3. The multiplicative decomposition $f = f_o * f_1 * f_2$ gives

$$f^{-1(*)} = f_o^{-1(*)} * f_1^{-1(*)} * f_2^{-1(*)}.$$

According to Lemma 3.3 (d) and (f) the convolution inverses $f_1^{-1(*)}$ and $f_2^{-1(*)}$ are in $\mathcal{ACM}$; this set is equal to $\mathcal{R}\varepsilon$ by the, already proven, assertion (4) of Theorem 2.3, and $\mathcal{R}\varepsilon \subset \mathcal{G}$. On behalf of (b) of Lemma 3.3 the function $f_o^{-1(*)}$ is in $\mathcal{G}^* \subset \mathcal{G}$, and from (a) we deduce

$$f^{-1(*)} \in \mathcal{G}.$$

Using (c) and (e), we see that the two DIRICHLET series

$$\mathcal{D}(f_1,s) = \sum_{n=1}^{\infty} n^{-s} \cdot f_1(n) \text{ and } \mathcal{D}(f_2,s)$$

are absolutely convergent in Re s ≥ 1; therefore the inverse DIRICHLET series $\mathcal{D}(f_1^{-1(*)},s)$ and $\mathcal{D}(f_2^{-1(*)},s)$ cannot have any zeros in Re s ≥ 1, and so $f_1^{-1(*)}$ and $f_2^{-1(*)}$ are in $\mathcal{G}^*$. The set $\mathcal{G}^*$ is closed with respect to convolution. Therefore

$$f^{-1(*)} = f_o^{-1(*)} * f_1^{-1(*)} * f_2^{-1(*)} \text{ is in } \mathcal{G}^*. \qquad \square$$

III.4. APPLICATIONS

Some applications of the main theorem (Theorem 2.1) are given in this section which specify the vague remarks in the footnote at the beginning of this chapter. The methods used are well-known (see, for example, the proof of A. WINTNER's theorem in II.2) and are based on the representation

$$g = f * h,$$

with a rather "small" function h. The method is thus a linear method and does not apply aptly to problems such as the existence of higher moments.

For the definitions of the sets $\mathcal{G}$, $\mathcal{G}^*$, $\mathcal{G}_\beta$, $\mathcal{G}_\beta^*$ and of $\varphi_f(p,s)$ we refer to (1.2), (1.3), (2.3), (2.4), and (1.4).

Theorem 4.1. *Let* $0 < \beta \le 1$, $\beta \le \gamma \le 1$; $f \in \mathcal{G}_\beta$ *and* $g \in \mathcal{G}_\beta$ *denote multiplicative arithmetical functions which are* β*-related, so that*

$$(4.1) \qquad \sum_p p^{-\beta} \cdot | f(p) - g(p) | < \infty .$$

Assume that, as $x \to \infty$, *the following asymptotic formula holds:*

$$(4.2) \qquad \sum_{n \le x} f(n) = A \cdot x^\gamma + R(x), \ x \to \infty,$$

with the remainder term

$$R(x) = \mathcal{O}(x^\beta) \ \left[\textit{resp. } R(x) = o(x^\beta) \right].$$

Then, for $x \to \infty$,

$$(4.3) \qquad \sum_{n \le x} g(n) = A^* \cdot x^\gamma + R^*(x),$$

where

$$R^*(x) = \mathcal{O}(x^\beta) \ \left[\textit{resp. } R^*(x) = o(x^\beta) \right],$$

and where the constant A^* *is given by*

$$(4.4) \qquad A^* = A \cdot \sum_n \frac{1}{n} \cdot (f^{-1(*)} * g)(n) = A \cdot \prod_p \frac{\varphi_g(p,\gamma)}{\varphi_f(p,\gamma)}$$

$$= A \cdot \prod_p \left(1 + \frac{f(p)}{p^\gamma} + \frac{f(p^2)}{p^{2\gamma}} + \ldots \right)^{-1} \times \left(1 + \frac{g(p)}{p^\gamma} + \frac{g(p^2)}{p^{2\gamma}} + \ldots \right).$$

Remark. Using Theorem 2.2, the assertions of Theorem 4.1 remain true if the assumption "$f \in \mathcal{G}^*$" is weakened to

> "$f \in \mathcal{G}$, and $\varphi_f(p,s) \ne 0$ in $\mathrm{Re}\ s \ge 1$ outside some finite exceptional set $\mathcal{EP}$ of primes, and $f(p^k) = g(p^k)$ for every prime in $\mathcal{EP}$ and every k."

The following result from H. Delange [1961a] is a simple corollary; the crucial assumption $f \in \mathcal{G}^*$ is easily obtained from the conditions on the values of f at powers of the prime 2.

Corollary 4.2. *Let* f, g *be multiplicative functions satisfying* $|f| \le 1$, $|g| \le 1$, *which are related (that is,* $\sum_p p^{-1} \cdot |f(p) - g(p)| < \infty$*). If the mean-value* M(f) *exists, then the mean-value* M(g) *also exists,* ***provided*** *that* $g(2^k) = f(2^k)$ *for every* k *in the exceptional case, that* $|f(2)| = 1$, *and that* $f(2^k) = (-1)^{k-1} \cdot \{ f(2) \}^k$ *for every* $k \ge 2$.

Proof of Theorem 4.1. According to Corollary 2.5, the multiplicative function $h = f^{-1(*)} * g$ is in $\mathcal{ACM}_\beta$; we obtain

$$\sum_{n\le x} g(n) = \sum_{n\le x} \sum_{d|n} h(d) \cdot f\left(\frac{n}{d}\right) = \sum_{d\le x} h(d) \cdot \sum_{m\le x/d} f(m)$$

in the usual way. Inserting the asymptotic formula (4.2), and then extending the summation in the first sum to $1 \le d < \infty$, one obtains $\left[\text{using } R(x) = \mathcal{O}(x^\beta)\right]$

$$\sum_{n\le x} g(n) = A\cdot x^\gamma \cdot \sum_{d\ge 1} d^{-\gamma}\cdot h(d) + \mathcal{O}\left(x^\gamma \cdot \sum_{d>x} d^{-\gamma}\cdot|h(d)|\right) + \mathcal{O}\left(x^\beta \cdot \sum_{d\le x} d^{-\beta}\cdot|h(d)|\right).$$

Enlarging the first remainder term by multiplying every summand with the factor $({}^d/_x)^{\gamma-\beta}$ (which is greater than 1), we obtain the assertion of Theorem 4.1 in the case where $R(x) = \mathcal{O}(x^\beta)$.

Similar calculations allow to derive the result in the second case $R(x) = o(x^\beta)$. □

Example. If $f \in \mathcal{G}^*$ has a mean-value, then the function $f = \mu^2 \cdot f$ is in $\mathcal{G}$ and related to f. Thus it has the mean-value (see (4.4))

$$M(\mu^2 f) = M(f)\cdot \prod_p \left(1 + p^{-1}\cdot f(p)\right)\cdot\left(\sum_{k\ge 0} p^{-k}\cdot f(p^k)\right)^{-1}.$$

Corollary 4.3. *Let $r \ge 1$ be an integer, and f a multiplicative function, uniformly bounded at the prime powers, and let g be a multiplicative function such that the series*

$$\sum_p \sum_{k\ge 2} p^{-k}\cdot |g(p^k)|^r \tag{4.5}$$

is absolutely convergent. Finally, assume that f and g are related, and that

$$\sum_p p^{-1}\cdot| f(p) - g(p) |^r \tag{4.6}$$

is finite. If, for every prime p, $\varphi_{f^r}(p,s) \ne 0$ in the half-plane $\mathrm{Re}\ s \ge 1$, then the existence of the mean-value $M(f^r)$ implies the existence of $M(g^r)$.

Proof. The boundedness of f at prime-powers and the condition on $\varphi_{f^r}(p,s)$ show that the power f^r is in $\mathcal{G}^*$. Next, having shown (from (4.6) and (4.5)) that g^r is in $\mathcal{G}$, and that f^r and g^r are related, then Theorem 4.1 gives the assertion. Only the proofs of

$$\sum_p p^{-2}\cdot|g(p)|^{2r} < \infty \tag{4.7}$$

and of

$$\sum_p p^{-1} \cdot |g^r(p) - f^r(p)| < \infty \tag{4.8}$$

are not quite so obvious. Firstly, $g(p) = \mathcal{O}(p^{1/r})$, by (4.6), and, since $|f(p)| \le K$ for all primes, using $|g(p)| \le 2 \cdot |g(p)-f(p)|$, if $|g(p)| \ge 2 \cdot K$, (4.6) implies that the sum

$$\sum\nolimits_{p, |g(p)| \ge 2K} p^{-1} \cdot |g(p)|^r < \infty$$

is convergent. So, splitting the sum in (4.7) into two sums, according to $|g(p)| < 2K$, and $|g(p)| \ge 2K$, and using $p^{-1} \cdot |g(p)|^r = \mathcal{O}(1)$ in the second sum, the convergence of (4.7) is easily shown. The relation

$$\begin{aligned} \left|g^r(p) - f^r(p)\right| &= \left|g(p) - f(p)\right| \cdot \left|g(p)^{r-1} + \ldots + f(p)^{r-1}\right| \\ &\ll \left|g(p) - f(p)\right| \cdot \left\{\left|g(p)^{r-1}\right| + \left|f(p)^{r-1}\right|\right\} \end{aligned}$$

reduces (4.8) to a proof of the convergence of

$$\sum_p p^{-1} \cdot \left|g(p) - f(p)\right| \cdot \left|g(p)^{r-1}\right|,$$

which is achieved by using the same splitting process as for (4.7). □

Theorem 4.4. *Assume that the arithmetical functions* $f \in \mathcal{G}^*$ *and* $g \in \mathcal{G}$ *are related. If*

$$\sum\nolimits_{n=1}^{\infty} n^{-1} \cdot f(n)$$

is convergent, then the series

$$\sum\nolimits_{n=1}^{\infty} n^{-1} \cdot g(n)$$

is convergent, and

$$\sum\nolimits_{n=1}^{\infty} n^{-1} \cdot g(n) = \sum\nolimits_{n=1}^{\infty} n^{-1} \cdot h(n) \sum\nolimits_{n=1}^{\infty} n^{-1} \cdot f(n),$$

where $h = g * f^{-1(*)}$.

The proof, which is similar to that of Theorem 4.1, is omitted (Exercise 2).

Theorem 4.5. *Let* $A: \mathbb{N} \to \mathbb{Z}$ *be an integer-valued* *additive* *function with the property*

$$\sum_{p,\, A(p) \neq 0} p^{-1} < \infty .$$

Then, for any $q \in \mathbb{Z}$, *the densities*

$$\delta_q = \lim_{x \to \infty} x^{-1} \cdot \sum_{n \le x,\ A(n) = q} 1$$

exist.

This theorem may be found, for example, in J. KUBILIUS [1964], p.93. It can easily be deduced from the Relationship Theorem 2.1. Consider the multiplicative function

$$f_\alpha : n \mapsto \exp(2\pi i\, \alpha\, A(n)), \text{ where } \alpha \text{ is real.}$$

This function is bounded, and thus it is in $\mathcal{G}$, and it is related to $1 \in \mathcal{G}^*$, for

$$\sum_p p^{-1} \cdot \left| \exp(2\pi i\, \alpha\, A(p)) - 1 \right| \le 2 \sum_{p,\, A(p) \ne 0} p^{-1} < \infty.$$

Therefore, there exists a multiplicative function h_α [$= \mu * f_\alpha$] with the properties

$$f_\alpha = h_\alpha * 1, \text{ and } \sum_{n=1}^{\infty} n^{-1} \mid h_\alpha(n) \mid < \infty.$$

According to Theorem 4.1 the mean-value $M(f_\alpha)$ exists for every real α. For the calculation of δ_q we consider

$$x^{-1} \cdot \sum_{n \le x,\, A(n) = q} 1 = x^{-1} \cdot \sum_{n \le x} \int_0^1 \exp(2\pi i\alpha\, (A(n)-q))\, d\alpha$$

$$= \int_0^1 \exp(-2\pi i\alpha\, q) \cdot x^{-1} \cdot \sum_{n \le x} e^{2\pi i\alpha A(n)} d\alpha.$$

The function $\alpha \mapsto M(f_\alpha)$ is bounded and LEBESGUE-measurable; using LEBESGUE's Dominated Convergence Theorem we obtain the existence of

$$\delta_q = \lim_{x \to \infty} \int_0^1 \exp(-2\pi i\alpha\, q) \cdot x^{-1} \cdot \sum_{n \le x} e^{2\pi i\alpha A(n)} d\alpha$$

$$= \int_0^1 \exp(-2\pi i\alpha\, q) \cdot \lim_{x \to \infty} x^{-1} \cdot \sum_{n \le x} e^{2\pi i\alpha A(n)}\, d\alpha,$$

and thus the densities δ_q do exist. □

The calculation of these densities is laborious. We perform this calculation in the special case of a strongly additive function A(.); in this case $A(p^k) = A(p)$ does not depend on k. Firstly, by Theorem 4.1, the mean-value of the function $f_\alpha : n \mapsto \exp(2\pi i\alpha\, A(n))$ is

$$M(f_\alpha) = \prod_p \left(1 + p^{-1} \cdot (e^{2\pi i\alpha A(p)} - 1) \right) = \sum_{n=1}^{\infty} n^{-1} \cdot (\mu * f_\alpha(n)).$$

The density δ_q is then

$$\begin{aligned}\delta_q &= \int_0^1 \exp(-2\pi i\alpha\, q)\cdot M(f_\alpha)\, d\alpha \\ &= \int_0^1 \exp(-2\pi i\alpha\, q\,)\cdot \sum_{p|n \Rightarrow A(p)\neq 0} \Big(n^{-1}\cdot\mu(n)\cdot\prod_{p|n}(1-e^{2\pi i\alpha A(p)})\Big)\, d\alpha \\ &= \int_0^1 \exp(-2\pi i\alpha\, q\,)\cdot \sum_{p|n \Rightarrow A(p)\neq 0} \Big(n^{-1}\cdot\mu(n)\cdot\sum_{d|n}\mu(d)\cdot e^{2\pi i\alpha A(d)}\Big)\, d\alpha .\end{aligned}$$

Interchanging the order of summation and integration (this is possible by the dominated convergence theorem) we obtain

$$\begin{aligned}\delta_q &= \sum_{p|n\Rightarrow A(p)\neq 0} n^{-1}\cdot\mu(n)\cdot \sum_{d|n,\ A(d)=q} \mu(d) \\ &= \sum_{\substack{d,\ A(d)=q,\\ p|d\Rightarrow A(p)\neq 0}} d^{-1}\cdot\mu^2(d)\cdot \sum_{\substack{m,(m,d)=1\\ p|m\Rightarrow A(p)\neq 0}} m^{-1}\cdot\mu(m) \\ &= \prod_{A(p)\neq 0}(1-p^{-1})\cdot \sum_{\substack{d,\ A(d)=q,\\ p|d\Rightarrow A(p)\neq 0}} d^{-1}\,\mu^2(d)\cdot\prod_{p|d}(1-p^{-1})^{-1}.\end{aligned}$$

III.5. ON A THEOREM OF L. LUCHT

Given $q \geq 1$, denote by $\mathcal{L}^q$ the $\mathbb{C}$-vector-space of arithmetical functions $f: \mathbb{N} \to \mathbb{C}$ with finite semi-norm

(5.1) $$\|f\|_q = \Big\{ \limsup_{x\to\infty} x^{-1}\cdot\sum_{n\leq x}|f(n)|^q \Big\}^{1/q}.$$

Note that $\mathcal{L}^q$ is **not** closed with respect to convolution. For example, the constant function $1 : n \mapsto 1$ is in $\mathcal{M}^q = \mathcal{L}^q \cap \mathcal{M}$, where $\mathcal{M}$ is the set of multiplicative functions. But, for any $q > 1$, the divisor function $\tau = 1 * 1$ is not in $\mathcal{M}^q$. By contrast, $(\mathcal{G},*)$ is a semi-group with identity element ε, and $(\mathcal{G}^*,*)$ is a group.

L. Lucht [1978] proved the following theorem, which may be considered an important step towards Theorem 2.1.

Theorem 5.1. *Assume* $q > 1$, *and let the multiplicative functions* f *and* g *in* $\mathcal{L}^q$ *be related. Assume, further, that, for every prime* p, *the factors* $\varphi_f(p,s)$ *of the generating* DIRICHLET *series for* f *are non-zero in the half-plane* Re $s \geq 1$. *Put* $h = f^{-1(*)} * g$; *then the series*

$$\sum_{n=1}^{\infty} n^{-1} \cdot h(n)$$

is absolutely convergent. In Re $s \geq 1$, *it has the product representation*

$$\sum_{n=1}^{\infty} n^{-s} \cdot h(n) = \prod_p \varphi_g(p,s) \cdot \{ \varphi_f(p,s)\}^{-1}$$

Theorem 5.1 easily follows from Theorem 2.1 for the following lemma.

Lemma 5.2. *Denote by* $\mathcal{M}$ *the set of multiplicative functions, and by* $\mathcal{M}^q$ *the intersection* $\mathcal{M}^q = \mathcal{L}^q \cap \mathcal{M}$. *Then, for any* $q > 1$,

$$\mathcal{M}^q \subset \mathcal{G}.$$

Remark. The assertion is wrong if $q = 1$ (see Exercise 8).

Proof of Lemma 5.2. Choose a real number $\varepsilon > 0$ so small that $(1+2\varepsilon)/q$ is less than 1; this choice is possible since $q > 1$. Denote the conjugate index by q', $q' = q/(q-1)$. HÖLDER's inequality gives

$$\sum_p \sum_{k\geq 2} p^{-k} \cdot |f(p^k)| = \sum_p \sum_{k\geq 2} |f(p^k)|\ p^{-\frac{1+\varepsilon}{q}\cdot k} \cdot p^{-k\cdot(1-\frac{1+\varepsilon}{q})}$$

$$\leq \Big\{\sum_p \sum_{k\geq 2} |f(p^k)|^q\ p^{-(1+\varepsilon)k}\Big\}^{1/q} \times \Big\{\sum_p \sum_{k\geq 2} p^{-k\cdot(1-\frac{1+\varepsilon}{q})q'}\Big\}^{1/q'}.$$

Using the assumption $f \in \mathcal{L}^q$ and partial summation, resp. the inequality $(1 - \frac{1+\varepsilon}{q})\cdot q' > \frac{1}{2}$, both of the double series on the right-hand-side of the above formula are convergent.

The assumption $f \in \mathcal{L}^q$ implies $|f(n)| \leq \gamma \cdot n^{1/q}$ with some constant γ. Assuming $q < 2$ without loss of generality, we obtain

$$\sum_p p^{-2}\ |f(p)|^2 \leq \gamma \sum_p p^{-2+1/q}\ |f(p)|$$

$$\leq \gamma\ \Big\{\sum_p p^{-(1+\varepsilon)}\ |f(p)|^q\Big\}^{1/q} \cdot \Big\{ \sum_p p^{-(2-(2+\varepsilon)/q)\cdot q'}\Big\}^{1/q'},$$

and because $f \in \mathcal{L}^q$, $\varepsilon > 0$, and $(2-(2+\varepsilon)/q)\cdot q' > 1$, the expression on the right-hand side is finite, and so $f \in \mathcal{G}$. □

III.6. THE THEOREM OF SAFFARI AND DABOUSSI, II

The theorem mentioned in the title was proved in Chapter II, 8. We prove it again easily in the special case of *multiplicative* functions.

Theorem 6.1. *Assume that* A *and* B *are subsets of* $\mathbb{N}$ *with the property*

$$\mathbb{N} = A \times B \text{ (direct product);}$$

so every $n \in \mathbb{N}$ *is representable as a product* $n = a \cdot b$, $a \in A$, $b \in B$, *in a unique way.*

(1) *If* $\sum_{b \in B} b^{-1}$ *is convergent, then the density*

$$\delta(A) = \lim_{x \to \infty} x^{-1} \cdot \sum_{a \in A,\ a \le x} 1 = \left\{ \sum_{b \in B} b^{-1} \right\}^{-1}$$

exists.

(2) *If* $\sum_{b \in B} b^{-1}$ *is divergent, then the density* $\delta(A)$ *is equal to zero.*

The representability of the semi-group $(\mathbb{N}, \cdot)$ as a direct product $\mathbb{N} = A \times B$ is equivalent to the possibility of decomposing the constant function 1 as a convolution product $1 = \alpha * \beta$ of the characteristic functions α, β of the sets A, resp. B.

In the **special case** where α and β are multiplicative, it is easy to deduce Theorem 6.1 (1) from Theorem 2.1. First $\alpha = 1 * \beta^{-1(*)}$; next the constant function 1 is in $\mathcal{G}^*$, α is in $\mathcal{G}$, and α and 1 are related: $1 = 1(p) = \alpha(p) + \beta(p)$ for primes p implies

$$\sum_p p^{-1} \cdot | 1 - \alpha(p) | = \sum_p p^{-1} \cdot (1 - \alpha(p)) = \sum_p p^{-1} \cdot \beta(p) < \infty,$$

and so $\alpha = 1 * \beta^{-1(*)}$, where $\sum n^{-1} \cdot |\beta^{-1(*)}(n)| < \infty$. Theorem 4.1 and the relation $\varphi_\alpha(p,1) \cdot \varphi_\beta(p,1) = \varphi_1(p,1)$ give the assertion. □

These remarks may be considered as a hint that there might be a more general version of Theorem 2.1 in which the assumption of multiplicativity can be weakened.

III.7. APPLICATION TO ALMOST-PERIODIC FUNCTIONS

Denote by $\mathcal{A}$ [resp. $\mathcal{D}$] the $\mathbb{C}$-vector-space of linear combinations of exponential sums $n \mapsto e_\alpha(n) = \exp(2\pi i \alpha n)$, α real [resp. $\alpha \in \mathbb{Q}$], and denote by $\mathcal{B}$ the $\mathbb{C}$-vector-space of linear combinations of RAMANUJAN sums

$$c_r : n \mapsto \sum_{d|(n,r)} d \cdot \mu(r/d) = \sum_{a \bmod r,\ (a,r)=1} e_{a/r}(n).$$

Using the semi-norm $\|f\|_q = \left\{\limsup_{x \to \infty} x^{-1} \cdot \sum_{n \le x} |f(n)|^q\right\}^{1/q}$, the spaces

$$\mathcal{A}^q = \|.\|_q \text{ - closure of } \mathcal{A} \text{ [q-almost-periodic functions]},$$

$$\mathcal{D}^q = \|.\|_q \text{ - closure of } \mathcal{D} \text{ [q-limit-periodic functions]},$$

$$\mathcal{B}^q = \|.\|_q \text{ - closure of } \mathcal{B} \text{ [q-almost-even functions]}$$

may be constructed. These spaces will be studied in Chapters VI and VII in more detail. In this section we are going to prove the following result.

Theorem 7.1. *Assume that the multiplicative arithmetical functions* f *and* g *are related, that* $f \in \mathcal{G}^*$ *and that* $g \in \mathcal{G}$:

(i) *if* $f \in \mathcal{A}^1$, *then* $g \in \mathcal{A}^1$;
(ii) *if* $f \in \mathcal{D}^1$, *then* $g \in \mathcal{D}^1$;
(iii) *if* $f \in \mathcal{B}^1$, *then* $g \in \mathcal{B}^1$;
(iv) *if* $\|f\|_1 < \infty$, *then* $\|g\|_1 < \infty$.

Remark. These assertions follow from the fact that $g = f * h$ with a "small" function h. So the existence of such a function is also a sufficient condition for Theorem 7.1.

Proof of Theorem 7.1. (i) The assumptions imply $g = f * h$, where $\sum_{n=1}^{\infty} n^{-1} \cdot |h(n)| < \infty$ (see Theorem 2.1). Given $\varepsilon > 0$, put $\delta = \varepsilon \cdot (1 + \|f\|_1)^{-1}$ and choose N so large that

$$\sum_{n \ge N} n^{-1} \cdot |h(n)| < \delta.$$

Select a finite linear combination of exponentials near f; more exactly, choose $t = \sum_\alpha a_\alpha \cdot e_\alpha$ with the property $\| f - t \|_1 < \varepsilon'$, where $\varepsilon' = \varepsilon \cdot \left(\sum_{n=1}^{\infty} n^{-1} \cdot |h(n)| \right)^{-1}$. Define the function H by

$$H(n) = \begin{cases} h(n), & \text{if } n \le N, \\ 0, & \text{if } n > N. \end{cases}$$

The convolution $e_\alpha * H$ is in $\mathcal{A}$:

$$(e_\alpha * H)(n) = \sum_{d|n,\ d\le N} h(d) \cdot e_\alpha(n/d) = \sum_{d\le N} h(d) \cdot \psi_{d,\alpha}(n),$$

with the function

$$\psi_{d,\alpha}(n) = \begin{cases} 0, & \text{if } d \nmid n, \\ e_\alpha(n/d), & \text{if } d|n. \end{cases} \tag{7.1}$$

The relation $\sum_{1\le m\le d} \exp\left(2\pi i \cdot \frac{m}{d} \cdot n \right) = d$, if $d|n$, and $= 0$ otherwise, implies

$$\psi_{d,\alpha}(n) = d^{-1} \cdot \sum_{1\le m\le d} \exp\left(2\pi i \left(\frac{m}{d} + \frac{\alpha}{d} \right) \cdot n \right), \tag{7.2}$$

and so $\psi_{d,\alpha}$ is a linear combination of exponentials, in fact $\psi_{d,\alpha} \in \mathcal{D} \subset \mathcal{A}$, and $e_\alpha * H$ is in $\mathcal{A}$. Using the inequality

$$\| F * G \|_1 \le \| G \|_1 \cdot \sum_{n=1}^{\infty} n^{-1} \cdot |F(n)| \tag{7.3}$$

(see Exercise 6), we find

$$\begin{aligned} \| g - t*H \|_1 &\le \| (f-t)*H \|_1 + \| f*(h-H) \|_1 \\ &\le \| f - t \|_1 \cdot \sum_{n=1}^{\infty} n^{-1} \cdot |H(n)| + \| f \|_1 \cdot \sum_{n>N} n^{-1} \cdot |h(n)| \\ &\le \varepsilon' \cdot \sum_{n=1}^{\infty} n^{-1} \cdot |h(n)| + \| f \|_1 \cdot \delta < 2\varepsilon. \end{aligned}$$

Any of the (finitely many) functions $e_\alpha * H$ is in $\mathcal{A}$; therefore $t * H = \sum_\alpha a_\alpha \cdot (e_\alpha * H)$ is in $\mathcal{A}$, and g is in $\mathcal{A}^1$.

(ii) If $\alpha \in \mathbb{Q}$, then $\psi_{d,\alpha}$ is in $\mathcal{D}$, as shown above, and so $e_\alpha * H$ is in $\mathcal{D}$ if α is rational. These remarks are sufficient to obtain a proof of (ii) by repeating the proof of (i) almost verbatim.

(iii). The convolution $c_r * H : n \mapsto \sum_{d\le N} h(d)\cdot\psi_d(n)$ is in $\mathcal{B}$; the reason for this being that the function $\psi_d(n) = c_r(n/d)$ if $d|n$, and $= 0$ otherwise, is even modulo $d\cdot r$ and so is a linear combination of RAMANUJAN sums. Using this observation the proof of (iii) is performed as before.

(iv) is left as Exercise 9. □

Example. Assume that $f \in \mathcal{G}$ is multiplicative and that the series

$$\sum_p p^{-1}\cdot(\,f(p) - 1\,) \tag{7.4}$$

is **absolutely** convergent. Then f is in $\mathcal{B}^1$. A special case of this example is the function $n \mapsto \mu^2(n)$. Other examples are $n \mapsto n^{-1}\cdot\varphi(n)$, and $n \mapsto n^{-1}\cdot\sigma(n)$.

This follows from Theorem 7.1: condition (7.4) states that f is related to the constant function $1 \in \mathcal{G}^*$, and this function is obviously in $\mathcal{B} \subset \mathcal{B}^1$.

Results for membership to $\mathcal{B}^2$ are not as smooth as the results of Theorem 7.1. We have to use a norm-estimate similar to estimate (7.3) used in the proof of Theorem 7.1.

Lemma 7.2. *Let* F *and* G *be arithmetical functions, where* F *has finite* $\|.\|_2$*-norm (see (5.1)). Assume that* $G = F*h$, *where* h *satisfies the condition*

$$\sum_{n=1}^{\infty} n^{-\frac12}\cdot|h(n)| < \infty. \tag{7.5}$$

Then $\|G\|_2 < \infty$. *More precisely,*

$$\|G\|_2 \le \Big(\sum_{n=1}^{\infty} n^{-\frac12}\cdot|h(n)|\Big)\cdot\|F\|_2. \tag{7.6}$$

Proof.

$$\|F*h\|_2^2 = \limsup_{N\to\infty} N^{-1}\cdot\sum_{n\le N} |h*F(n)|^2$$

$$\le \limsup_{N\to\infty} N^{-1}\cdot\sum_{n\le N}\sum_{d|n} |h(d)\,F(n/d)| \sum_{t|n} |h(t)\,F(n/t)|$$

$$\le \limsup_{N\to\infty} N^{-1}\cdot\sum_{d\le N}\sum_{t\le N} |h(d)\cdot h(t)|\cdot \sum_{\substack{n\le N,\\ n\equiv 0 \bmod [d,t]}} |F(n/d)\cdot F(n/t)|.$$

Using the CAUCHY-SCHWARZ inequality, we obtain

$$\| F*h \|_2^2 \le \limsup_{N\to\infty} \sum_{d\le N} \sum_{t\le N} \frac{|h(d)\cdot h(t)|}{\sqrt{d\cdot t}} \cdot \Big((N/d)^{-1}\cdot \sum_{\substack{n\le N\\ n\equiv 0(d)}} | F(n/d)|^2$$

$$\times (N/t)^{-1}\cdot \sum_{\substack{n\le N\\ n\equiv 0\ (t)}} | F(n/t)|^2 \Big)^{1/2},$$

and (7.6) is proven. □

Now, using the same ideas as for the proof of Theorem 7.1, we immediately obtain the following theorem.

Theorem 7.3. *Suppose that* f *is an arithmetical function in* $\mathcal{B}^2$, *and* $h\colon \mathbb{N} \to \mathbb{C}$ *satisfies condition* (7.5). *Then, again, the function* $g = f*h$ *is in* $\mathcal{B}^2$.

III.8. EXERCISES

1) Let f be a multiplicative function in $\mathcal{L}^*$. For a fixed integer m define

$g(n) = f(n)$ if g.c.d.$(n,m) = 1$, and $g(n) = 0$ otherwise.

Prove: if f has a mean-value M(f), then the mean-value M(g) exists and is equal to

$$M(g) = M(f) \cdot \prod_{p|m} \Big(\sum_{k=0}^{\infty} p^{-k} \cdot f(p^k) \Big)^{-1}.$$

2) Assume that f is a *strongly multiplicative* arithmetical function, for which the mean-value M(f) exists and is non-zero. Let g be the arithmetical function defined in Exercise 1). Prove: the mean-value M(g) exists and

$$M(g) = M(f) \cdot \prod_{p|m} \frac{p-1}{p} \cdot \prod_{p|m} \Big(1 + \frac{f(p)-1}{p}\Big)^{-1}.$$

3) Consider the additive function $A = \Omega - \omega$. For the densities δ_q defined in section 4, use Theorem 4.5 to obtain the formulae

$$\delta_0 = \prod_p (1 - p^{-2}) = 6\pi^{-2},$$

$$\delta_q = \int_0^1 \exp(-2\pi i\, \alpha\, q) \cdot \prod_p (1 - p^{-1}) \cdot (1 + (p - e^{2\pi i\alpha})^{-1})\, d\alpha.$$

4) Denote by $r(n)$ the number of pairs $(x,y) \in \mathbb{Z} \times \mathbb{Z}$ with the property that $n = x^2 + y^2$. Thus $r(n)$ counts the number of representations of n as a sum of two squares. Then $f(n) = \frac{1}{4}\, r(n)$ is multiplicative and

$$f(p^k) = \begin{cases} k+1, & \text{if } p \equiv 1 \bmod 4, \\ 1, & \text{if } p \equiv 3 \bmod 4 \text{ and } k \text{ even, or if } p = 2, \\ 0, & \text{if } p \equiv 3 \bmod 4 \text{ and } k \text{ is odd.} \end{cases}$$

Using $\sum_{n \le x} f(n) = \frac{1}{4}\pi \cdot x + \mathcal{O}(x^\beta)$, where β is some constant less than 1, give an asymptotic formula for $\sum_{n \le x} g(n)$, where $g = \mu^2 \cdot f$.

5) Show the existence of $M(f)$ for the function $f\colon n \mapsto \tau(n) \cdot 2^{-\omega(n)}$, and give a formula for this mean-value.

6) Prove inequality (7.3). *Hint:* without loss of generality, $\sum n^{-1} \cdot |F(n)|$ is convergent. Use $y^{-1} \cdot \sum_{m \le y} |G(m)| \le \|G\|_1 + \varepsilon$, if y is large.

7) Calculate the coefficients a_k in $\psi_d = \sum_{k|dr} a_k \cdot c_k$. The function ψ_d is defined in section 7).

8) Define an infinite set P of primes $p_1, p_2, \ldots$ by $p_1 = 2, \ldots, p_{n+1} \ge 2 \cdot p_1 p_2 \cdot \ldots \cdot p_n$, and define a multiplicative function by

$$f(p^k) = \begin{cases} p & \text{if } p \in P,\ k = 1, \\ 0 & \text{otherwise.} \end{cases}$$

Prove: $f \in \mathcal{M}^1$, $f \notin \mathcal{G}$. — Hints: $x^{-1} \sum_{n \le x} f(n) \le 2$, but $\sum_p (p^{-1} \cdot f(p))^2 = \infty$.

9) Let $f \in \mathcal{G}^*$ and $g \in \mathcal{G}$ be related functions, $g = f * h$, where $\sum_{n=1}^\infty n^{-1} \cdot |h(n)| < \infty$. Prove:

a) if $\|f\|_1 < \infty$, then

$$\|g\|_1 = \|f\|_1 \cdot \sum_{n=1}^\infty n^{-1} \cdot |h(n)| = \|f\|_1 \cdot \prod_p \sum_{k \ge 0} p^{-k} \cdot |h(p^k)|;$$

b) if $\alpha \in \mathbb{R}$ and if the FOURIER-coefficient $\hat{f}(n\alpha)$ exists for every $n \in \mathbb{N}$, then $\hat{g}(\alpha) = \sum_{n=1}^\infty n^{-1} \cdot h(n) \cdot \hat{f}(n\alpha)$.

10) Let f be multiplicative and $q \ge 1$. If the series $\sum_p p^{-1} \cdot \big| |f(p)|^q - 1 \big|$, $\sum_p |p^{-1} \cdot f(p)|^2$, $\sum_p \sum_{k \ge 2} p^{-k} \cdot |f(p^k)|^q$ are convergent, then $\|f\|_q < \infty$.

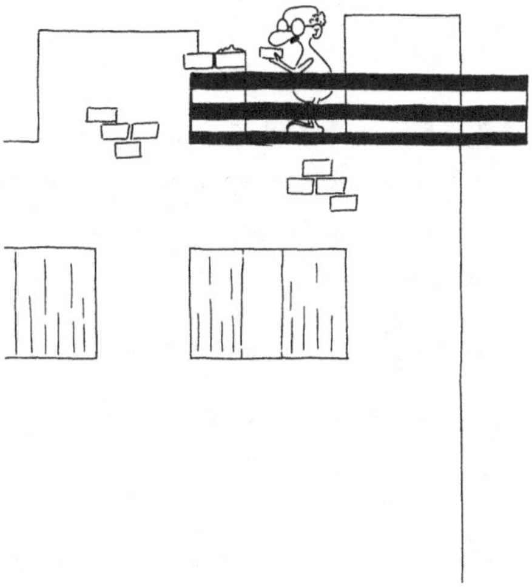

Chapter IV

Uniformly Almost-Periodic Arithmetical Functions

Abstract. *This chapter deals with completions* $\mathcal{B}^u \subset \mathcal{D}^u \subset \mathcal{A}^u$ *of the vector-spaces* $\mathcal{B} \subset \mathcal{D} \subset \mathcal{A}$ *of even and periodic functions* [*and of linear combinations of exponential functions* $n \mapsto \exp(2\pi i\alpha \cdot n)$] *with respect to the supremum-norm* $\|f\|_u = \sup_{n\in\mathbb{N}} |f(n)|$. *The fundamental properties of these spaces are proved, and additive and multiplicative arithmetical functions in the space* $\mathcal{B}^u$ *are characterized. Then some topics from probability theory are discussed and applied in order to study limit distributions of real-valued functions in* $\mathcal{B}^u$. *The maximal ideal spaces from* GELFAND*'s theory of commutative* BANACH *spaces,* $\Delta_{\mathcal{B}}$ *and* $\Delta_{\mathcal{D}}$, *are given for* $\mathcal{B}^u$ *and for* $\mathcal{D}^u$ *which are isomorphic to spaces of continuous functions defined on the compact sets* $\Delta_{\mathcal{B}}$ *and* $\Delta_{\mathcal{D}}$. *Some properties of arithmetical functions in* $\mathcal{B}^u$ *and* $\mathcal{D}^u$ *can then be derived from this knowledge and standard theorems of analysis (for example, the* STONE-WEIERSTRASS *or the* TIETZE *theorem). Finally, a theory of integration is developed in these spaces and applied to the calculation of some mean-values.*

IV.1. EVEN AND PERIODIC ARITHMETICAL FUNCTIONS

The $\mathbb{C}$-vector-spaces $\mathcal{B}_r$ of r-*even* and $\mathcal{D}_r$ of r-*periodic* arithmetical functions are defined in I. 3. The pointwise product converts these vector-spaces into $\mathbb{C}$-algebras. The $\mathbb{C}$-vector-spaces of all *even*, resp. all *periodic, functions* will be denoted by

(1.1) $$\mathcal{B} = \bigcup_{r=1}^{\infty} \mathcal{B}_r, \text{ resp. } \mathcal{D} = \bigcup_{r=1}^{\infty} \mathcal{D}_r.$$

$\mathcal{B}$ and $\mathcal{D}$ are $\mathbb{C}$-algebras. If f is even mod k then it is even mod r·k also. Thus

$$\mathcal{B}_{\gcd[r_1,r_2]} \subset \mathcal{B}_{r_1} \cup \mathcal{B}_{r_2} \subset \mathcal{B}_{\mathrm{lcm}\{r_1,r_2\}}.$$

The spaces $\mathcal{B}_r$, resp. $\mathcal{D}_r$, may be described as vector-spaces with some natural bases. In fact,

(1.2) $$\mathcal{D}_r = \mathrm{Lin}_{\mathbb{C}} \left[\exp\left(2\pi i \tfrac{a}{r}\cdot n\right), \ a = 1, 2, \ldots, r\right]$$

is the *vector-space of linear combinations of exponential-functions*

(1.3) $$e_{a/r}: \ n \mapsto \exp\left(2\pi i \tfrac{a}{r}\cdot n\right), \text{ where a runs from 1 to r.}$$

All these functions are r-periodic and thus in $\mathcal{D}_r$, and every r-periodic function is a linear combination of the functions (1.3), as may be seen from the linear independence of the functions (1.3) [which may be deduced from the orthogonality relations] and the fact that the dimension of $\mathcal{D}_r$ is exactly r. From (1.2), again, it is clear that $\mathcal{D}_r$ is an algebra (a pointwise product of r-periodic functions is an r-periodic function).

The space of r-even functions is a vector-space of linear combinations of RAMANUJAN sums,

(1.4) $$\mathcal{B}_r = \mathrm{Lin}_{\mathbb{C}} \, [c_d, \ d|r].$$

Obviously, since c_d is even mod d and thus even mod r, all these linear combinations are in $\mathcal{B}_r$. Next, the dimension of $\mathcal{B}_r$ over $\mathbb{C}$ is $\tau(r)$ [for example, the functions g_d, $d|r$, $g_d(n) = 1$ if $\gcd(n,r) = d$, and $g_d(n) = 0$ otherwise, are a basis for $\mathcal{B}_r$]. But the $\tau(r)$ RAMANUJAN sums c_d, where $d|r$, are linearly independent, as follows from the orthogonality relations (see I.3), and (1.4) is proved. If f is an r-even arithmetical function, then it is representable by the basis $[\, c_d,\ d|r\,]$ in the form

$$(1.5) \qquad f = \sum_{d|r} a_d(f) \cdot c_d$$

with the "RAMANUJAN-[FOURIER]-coefficients"

$$(1.6) \qquad \begin{cases} a_d(f) = (\ \varphi(d)\ )^{-1} \cdot M(f \cdot c_d) \\ \quad = (\ \varphi(d)\ )^{-1} \cdot r^{-1} \cdot \sum_{1 \le \rho \le r} f(\rho) \cdot c_d(\rho)\ . \end{cases}$$

Formula (1.6) easily follows from the orthogonality relations for RAMANUJAN sums, already proved in Chapter I (3.3).

The vector-space $\mathcal{B}$ of even functions is actually an algebra: the product of an r-even and a t-even function is an r·t-even function.

Another proof of this result follows from the fact that $\mathcal{B}_r$ is the space of linear combinations of the functions c_d, $d|r$. This result is combined with the multiplicativity of the RAMANUJAN sums, considered as a function of the index (see I, Theorem 3.1), and with the relations

$$(1.7') \qquad c_{p^k} \cdot c_{p^\ell} = \varphi(p^k) \cdot c_{p^\ell}, \text{ if } \ell > k\ ,$$

and

$$(1.7'') \quad c_{p^k} \cdot c_{p^k} = \varphi(p^k) \cdot \left(c_1 + c_p + \dots + c_{p^{k-1}} \right) + (p^k - 2\, p^{k-1}) \cdot c_{p^k}.$$

These relations may be checked pointwise, using I (3.4). Summarizing, we obtain the following theorem.

Theorem 1.1. *The vector-space $\mathcal{B}$ of all even functions is equal to the space of all linear-combinations of RAMANUJAN sums. Thus*

$$(1.8) \qquad \mathcal{B} = \mathrm{Lin}_{\mathbb{C}}\,[\, c_r,\ r = 1, 2, \dots\,],$$

and the vector-space $\mathcal{D}$ of all periodic arithmetical functions is equal to

$$(1.9) \qquad \mathcal{D} = \mathrm{Lin}_{\mathbb{C}}\,[\, e_{a/r},\ r = 1, 2, \dots, 1 \le a \le r,\ \gcd(a,r) = 1\,].$$

We mention that another proof for Theorem 1.1 , which is applicable in much more general situations, is given in SCHWARZ & SPILKER [1971], the main idea being to use the WEIERSTRASS - STONE Approximation Theorem.

Finally, we define the $\mathbb{C}$-vector-space (in fact, again, $\mathcal{A}$ is an algebra)

(1.10) $$\mathcal{A} = \mathrm{Lin}_{\mathbb{C}} [\, e_\alpha,\ \alpha \in \mathbb{R} \bmod \mathbb{Z} \,]$$

of complex linear combinations of the functions $e_\alpha : n \mapsto \exp(2\pi i\alpha \cdot n)$.

The *mean-value* $M(f)$ for functions in $\mathcal{A}$ exists, and $f \in \mathcal{A}$ has the FOURIER expansion

(1.11) $$\sum_{\alpha \in \mathbb{R}/\mathbb{Z}} M(f \cdot e_{-\alpha}) \cdot e_\alpha .$$

In general, arithmetical functions are neither even nor periodic. So the spaces defined up to now are (by far) too small. In order to enlarge the spaces $\mathcal{B} \subset \mathcal{D} \subset \mathcal{A}$ we use the supremum norm

(1.12) $$\|f\|_u = \sup_{n \in \mathbb{N}} |f(n)|,$$

where f is a (bounded) arithmetical function. Obviously, $\|.\|_u$ ***is a norm*** with the usual properties

(i) $\|f\|_u \geq 0$, and $\|f\|_u = 0$ if and only if $f = 0$,

(ii) $\|\lambda \cdot f\|_u = |\lambda| \cdot \|f\|_u$ for every complex λ,

(iii) $$\|f + g\|_u \leq \|f\|_u + \|g\|_u .$$

Moreover, in addition,

(iv) $$\|f \cdot g\|_u \leq \|f\|_u \cdot \|g\|_u ,$$

(v) $$\|f^2\|_u = \|f\|_u^{\,2} ,$$

and there is an *involution* on $\mathcal{B}$, $\mathcal{D}$, $\mathcal{A}$, namely the complex conjugation, with the properties

(*i*) $\overline{(f+g)} = \overline{f} + \overline{g}$,

(*ii*) $\overline{(\lambda \cdot f)} = \overline{\lambda} \cdot \overline{f}$,

(*iii*) $\overline{(f \cdot g)} = \overline{g} \cdot \overline{f}$,

(*iv*) $\overline{(\overline{f})} = f$.

In addition, this involution satisfies

(υ) $$\| f\cdot\bar{f} \| = \|f\|^2,$$

and so (after proving that $\mathcal{B}$, $\mathcal{D}$, $\mathcal{A}$ are BANACH algebras: we shall do this in the next section), we know that $\mathcal{B}$, $\mathcal{D}$, $\mathcal{A}$ are B^*-algebras (for the definition see RUDIN [1973], or the Appendix A.6).

Working in the vector-space of all $\|.\|_u$-bounded arithmetical functions, we construct the desired enlargements of our spaces by forming the closures of $\mathcal{B}$, $\mathcal{D}$, and $\mathcal{A}$ with respect to the $\|.\|_u$-norm,

(1.13i) $$\mathcal{B}^u = \|.\|_u\text{-closure of } \mathcal{B}$$

(the vector-space of *uniformly almost-even functions*),

(1.13ii) $$\mathcal{D}^u = \|.\|_u\text{-closure of } \mathcal{D}$$

(the vector-space of *uniformly limit-periodic functions*), and

(1.13iii) $$\mathcal{A}^u = \|.\|_u\text{-closure of } \mathcal{A}$$

(the vector-space of *uniformly almost-periodic functions*).

Since $\mathcal{A}$, $\mathcal{B}$, $\mathcal{D}$ are $\mathbb{C}$-algebras, the spaces $\mathcal{A}^u$, $\mathcal{B}^u$, $\mathcal{D}^u$ are again $\mathbb{C}$-algebras. Figure IV.1 shows inclusion relations between the spaces defined up to now.

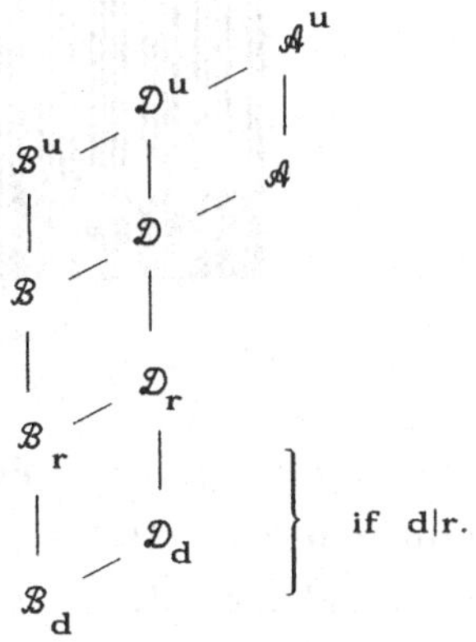

Figure IV.1.

Moreover, $\mathcal{B}^u \neq \mathcal{D}^u \neq \mathcal{A}^u$ ($\mathcal{B} \neq \mathcal{D} \neq \mathcal{A}$ is obvious). *The non-principal character* χ mod 4, $\chi(n) = 1$ if $n \equiv 1 \bmod 4$, $\chi(n) = -1$ if $n \equiv 3 \bmod 4$ and $\chi(n) = 0$ otherwise, *is periodic and so is in* $\mathcal{D}$, *but not in* $\mathcal{B}^u$: assume, that X is an M-even function near χ; there exist primes $p \equiv 3 \bmod 4$ not dividing M, so $\gcd(M,p) = 1$, $X(p) = X(1)$, but $X(p)$ is near -1 and $X(1)$ near 1, a contradiction.

If α is irrational, the function e_α is in $\mathcal{A}$, but not in $\mathcal{D}^u$. Otherwise, choose $g = \sum_{a \bmod r} \beta(a/r)\cdot e_{a/r}$ near e_α with respect to $\|.\|_u$ (where r is sufficiently large). If $n = r!\cdot m$, $m = 1,2,\ldots$, then

$$g(n) = \sum_{a \bmod r} \beta(a/r)$$

is constant, but the values of $e_\alpha(n)$ are dense on the unit circle (the sequence $\left\{\{\alpha\cdot r!\cdot m\},\ m = 1, 2, \ldots\right\}$, is uniformly distributed mod 1; see II.7).

As illustration of the rather regular behaviour of functions from the spaces defined above, Figures IV.2, IV.3 and IV.4 show *ad hoc* constructed functions in $\mathcal{B}^u$ (one multiplicative, one additive) and in $\mathcal{D}^u$ (multiplicative) in the range between 1 and 198 [resp. 298]. The first function is strongly multiplicative, with values $f(2) = 0.3$, $f(5) = 0$, and $f(p) = 1 - \chi(p)\cdot p^{-3/2}$ for primes $p \neq 2, 5$; the character χ mod 5 is defined by $\chi(2) = \chi(3) = -1$, $\chi(1) = \chi(4) = 1$, $\chi(5) = 0$.

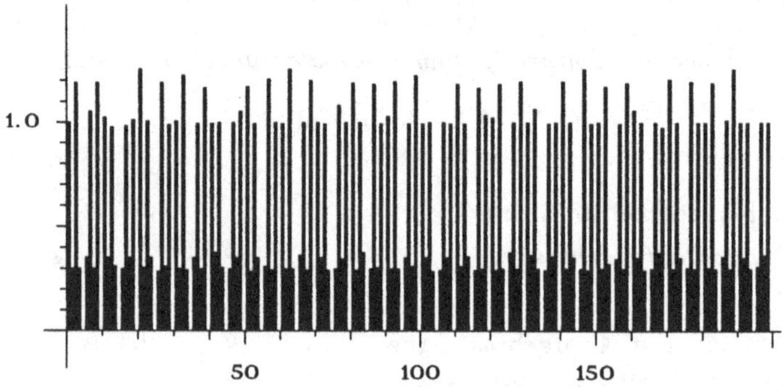

F i g u r e IV.2

The function given in Figure IV.3 is strongly additive. Its values at the primes $p > 2$ are $f(p) = \chi(p)\cdot p^{-3/2}$, with the same character as in Figure IV.2, and the value at $p = 2$ is $f(2) = 1$.

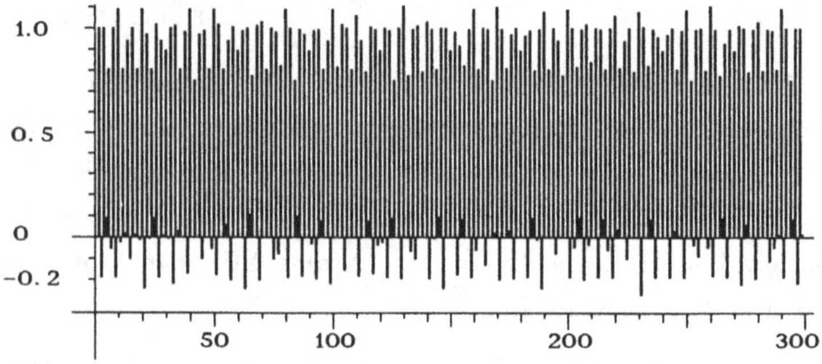

Figure IV.3

The final example, given graphically in Figure IV.4, is a strongly multiplicative function in $\mathcal{D}^u$, defined at the primes by $f(p) = \chi(p)\cdot(1 - p^{-3/2})$ if $p > 2$, and $f(2) = -\frac{1}{2}$.

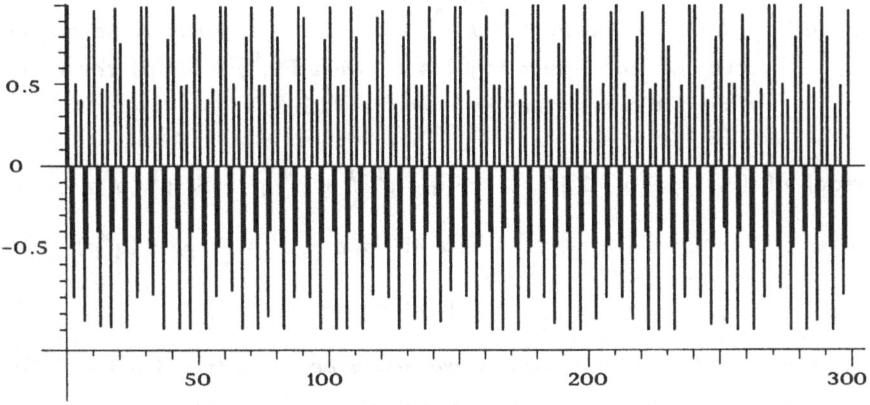

Figure IV.4

There is a fascinating interplay between evenness and periodicity, on the one hand, and additivity or multiplicativity, on the other. We give some simple examples below.

Proposition 1.2. *If* F *is additive and* q*-periodic, then the following hold:*

(i) $F(\ell) = 0$, *if* $\gcd(\ell,q) = 1$.

(ii) $F(p^k) \neq 0$ *is possible at most for primes* p *dividing* q.

(iii) *Put* $q = \prod_{1\le t\le T} p_t^{\alpha_t}$. *If* $\beta \ge \alpha_t$, *then* $F(p_t^{\beta}) = F(p_t^{\alpha_t})$.

(iv) F *is* q*-even.*

Proof. (i) follows from $F(q) = F(q\cdot\ell) = F(q) + F(\ell)$; (i) implies (ii); (iii) follows from

$$\sum_{1\le t\le T} F(p_t^{\alpha_t}) = F(q) = F(p_1^{\beta-\alpha_1}\cdot q) = \sum_{2\le t\le T} F(p_t^{\alpha_t}) + F(p_1^{\beta}).$$

(iv) If $n = \prod p^{\beta_p}$, then, using (ii) and (iii),

$$F(n) = \sum_{p|q} F(p^{\beta_p}) = \sum_{p|q} F(p^{\min(\beta_p,\alpha_p)}) = F(\gcd(n,q)). \qquad \square$$

Generalization. *If* F *is* q*-periodic and* ε*-nearly-additive (this means: if* $\gcd(n,m) = 1$, *then* $|F(n\cdot m) - F(n) - F(m)| < \varepsilon$ *), then*

(i') $|F(\ell)| < \varepsilon$, *if* $\gcd(\ell,q) = 1$,

(ii') If $\beta \ge \alpha_t$, *then* $|F(p_t^{\beta}) - F(p_t^{\alpha_t})| < 2\varepsilon$.

Proposition 1.3. *If* F *is* q*-periodic and multiplicative, and if* $F(q) \neq 0$, *then*

(i) $F(\ell) = 1$, *if* $\gcd(\ell,q) = 1$.

(ii) $F(p^\beta) = F(p^\alpha)$, *if* $p^\alpha \parallel q$ *and* $\beta \geq \alpha$.

(iii) *If the condition* $F(q) \neq 0$ *is weakened to: there is some prime-power* q^α *for which* $F(q^\alpha) \neq 0$, *then* $F(p^k) = F^k(p)$ *for every prime* $p \neq q$ *not dividing* q, *and* $k = 1,2,\ldots$.

Proof of (iii). $F(q^\alpha)\cdot F(p^{k+1}) = F(q^\alpha\cdot p^{k+1} + p^k\cdot q) = F(p^k\cdot(q^\alpha p + q))$

$$= F(p^k)\cdot F(q^\alpha p + q) = F(p^k)\cdot F(q^\alpha p)$$

$$= F(p^k)\cdot F(q^\alpha)\cdot F(p). \qquad \square$$

Proposition 1.4. *If* f *is multiplicative and* q*-periodic, then* $f(p^k) = 0$ *for some* k *is possible for at most finitely many primes.*

Proof. Assume there are infinitely many prime-powers $p_r^{k_r}$, for which $f(p_r^{k_r}) = 0$. Without loss of generality, we may assume that these prime-powers are coprime with q. Denote by a_r the residue-class of $p_r^{k_r}$ mod q; these residue-classes are in $(\mathbb{Z}/q\mathbb{Z})^\times$. At least two of the residue-classes $\prod_{1\leq r\leq R} a_r$ are equal, so there are integers R and S, for which

$$\prod_{R<r\leq S} a_r \equiv 1 \bmod q.$$

Then

$$1 = f(1) = f\Big(\prod_{R<r\leq S} a_r\Big) = f\Big(\prod_{R<r\leq S} p_r^{k_r}\Big) = 0,$$

a contradiction. $\square$

In the case of completely multiplicative functions the same argument gives:

If f *is completely multiplicative and* q*-periodic, then*

(i) $f(p) = 0$ *is possible only if* $p|q$.

(ii) $|f(n)| \in \{0,1\}$ *for any integer* n, $f(n)$ *is zero or a root of unity,*

(iii) *if* $q > 0$ *is the least period of* f, *then* $f(p) = 0$ *for any* $p|q$.

Therefore f *is a* <u>*DIRICHLET character*</u> mod q, *where* q *is the least period of* f.

Proof of (iii). If $q = q_1\cdot q_2$, where $f(q_1) \neq 0$, then

$$f(a)\cdot f(q_1) = f(aq_1+q_2\cdot q_1) = f(q_1) \cdot f(a+q_2),$$

and so q_2 is again a period of f. □

The next result is due to D. LEITMANN and D. WOLKE [1976]. The proof given is from SCHWARZ (Monatshefte Math. 1978). An arithmetical function is called p-multiplicative if the relation $f(p_1 \cdot p_2) = f(p_1)\cdot f(p_2)$ holds for all primes $p_1 \neq p_2$.

Proposition 1.5. *Let* $F\colon \mathbb{R} \to \mathbb{C}$ *be a periodic function with irrational period* $\alpha > 0$. *If the set*

$$\mathcal{S} = \{ t \in \mathbb{R},\ F \textit{ continuous at } t \}$$

is non-empty, and if the restriction f *of* F *to* $\mathbb{N}$ *is* p*-multiplicative, then* $F(\mathcal{S}) = \{0\}$ *or* $F(\mathcal{S}) = \{1\}$. *In particular, if* F *is continuous, then* $F = 0$ *or* $F = 1$ *is constant.*

Remark. The example $F(t) = \cos(2\pi t)$ shows that the condition "α is irrational" is necessary.

Proof. The results needed from the theory of uniform distribution may be found, for example, in KUIPERS-NIEDERREITER [1974], Theorem 6.3, example 6.1 and p.22. Assume there is some $t \in \mathcal{S}$ for which $F(t) \neq 0$. Fix ε, $0 < \varepsilon < \frac{1}{4}|F(t)|$. Let t' be an element of $\mathcal{S}$. The continuity of F at t, t' implies the existence of a $\delta > 0$ so that

$$(1.14) \qquad |F(t) - F(x)| < \varepsilon,\ |F(t') - F(y)| < \varepsilon, \text{ if } |t-x|,\ |t' - y| < 2\delta.$$

Choose a real t^*, $|t^*-t| < \delta$, such that 1, α^{-1} and $\alpha^{-1}\cdot t^*$ are $\mathbb{Q}$-linearly independent. Then the 2-dimensional sequence of points

$$\left(\{\alpha^{-1}q\},\ \{\alpha^{-1}t^*\cdot q\}\right),\ q \text{ prime},$$

is uniformly distributed modulo 1 in the unit-square in $\mathbb{R}^2$. Therefore, we find a prime q and integers m_1, m_2 satisfying

$$|\ \alpha^{-1}q + m_1 - \alpha^{-1}t'\ | < \alpha^{-1}\cdot \delta,$$

and

$$|\ \alpha^{-1}\, t^*\cdot q + m_2 - \alpha^{-1}\, t\ | < \alpha^{-1} \cdot \delta.$$

Having fixed q, there exists a prime $p \neq q$ and an integer m for which

$$| \alpha^{-1} \cdot p + m - \alpha^{-1} \cdot t^* | < \alpha^{-1} \cdot q^{-1} \cdot \delta,$$

and so

$$| p \cdot q + (mq + m_2) \cdot \alpha - t | < 2\delta.$$

Continuity at t and α-periodicity, resp. continuity at t', give

$$| F(pq) - F(t) | < \varepsilon , |F(q) - F(t')| < \varepsilon.$$

And

$$|F(p) - F(t)| < \varepsilon,$$

because $|p + m\alpha - t| < 2\delta$, for $|t-t^*| < \delta$ and (1.14). Using p-multiplicativity, we deduce

$$| F(q) - 1 | < 2\varepsilon \cdot |F(p)|^{-1} < 4\varepsilon \cdot |F(t)|^{-1},$$

and so

$$|F(t') - 1| < \left(1 + 4 |F(t)|^{-1} \right) \cdot \varepsilon.$$

Thus $F(\mathcal{P}) = \{1\}$. □

Finally, we mention that a characterization of all multiplicative, periodic arithmetical functions was given by N. G. DE BRUIJN [1943] and also by D. LEITMANN and D. WOLKE [1976]. We do not reproduce the proof here, but simply quote the result.

Theorem 1.6. *A multiplicative arithmetical function* f *is periodic if and only if there exists an integer* N *and a* DIRICHLET *character* χ mod N *with the following properties:*

(i) *If* $p|N$ *and* $k \in \mathbb{N}$, *then* $f(p^k) = 0$.

(ii) *If* $p \nmid N$, *then the function* $k \mapsto \overline{\chi}(p^k) \cdot f(p^k)$ *is constant and* $\neq 0$.

(iii) *There are at most* <u>*finitely many*</u> *primes* p *for which* $\overline{\chi}(p^k) \cdot f(p^k) \neq 1$ *for* <u>*some*</u> *exponent k.*

IV.2. SIMPLE PROPERTIES

First we prove the following theorem.

Theorem 2.1. *The algebras* $\mathcal{B}^u$, $\mathcal{D}^u$, $\mathcal{A}^u$ *are* BANACH *algebras (and so are complete with respect to* $\|\,.\,\|_u$*), and the supremum-norm has the properties* (i) - (v) *and* (υ) *of section* IV.1.

Proof. Let us prove, for example, (iii) for $\mathcal{B}^u$: given f, g $\in \mathcal{B}^u$ and some $\varepsilon > 0$, there are functions F, G in $\mathcal{B}$ satisfying $\| f - F \|_u < \varepsilon$, $\| g - G \|_u < \varepsilon$. Then $\| f+g - (F+G) \|_u < \varepsilon$, and so $(f+g) \in \mathcal{B}^u$. Next, $\mathcal{B}^u$ is an algebra: given f, g in $\mathcal{B}^u$, and $\varepsilon > 0$, there are functions F, G in $\mathcal{B}$ satisfying $\|f - F\|_u < \varepsilon$, $\|g - G\|_u < \varepsilon$. Then $\|F\cdot G - f\cdot g\|_u \le \|f-F\|_u \cdot \|G\|_u + \|f\|_u \cdot \|g-G\|_u < C\cdot\varepsilon$, since $\|g\|_u$ and $\|f\|_u$ are bounded and $\|G\|_u$ is near $\|g\|_u$; property (iv) of the norm is used. F·G is in $\mathcal{B}$, and so $f\cdot g \in \mathcal{B}^u$.

Concerning the completeness of, say, $\mathcal{B}^u$, we assume that $\{F_k\}$, k = 1, 2, ... is a $\|\,.\,\|_u$- CAUCHY-sequence in $\mathcal{B}^u$. Then the values $F_k(n)$ are a CAUCHY- sequence in $(\mathbb{C}, |.|)$, and are therefore convergent to some complex number F(n). The function $F : n \mapsto F(n)$ satisfies $\|F - F_k\|_u \le \varepsilon$ if $k \ge k_0(\varepsilon)$, so F is the $\|.\|_u$ - limit of the sequence F_k. Finally F is in $\mathcal{B}^u$ because F is near F_k if k is large, and F_k is near some f_k in $\mathcal{B}$. □

Theorem 2.2. *Assume* f, g $\in \mathcal{B}^u$ [*resp.* $\in \mathcal{D}^u$, *resp.* $\in \mathcal{A}^u$]. *Then the functions*

$$\mathrm{Re}\, f,\ \mathrm{Im}\, f,\ |f|,$$

are again in $\mathcal{B}^u$ [*resp.* $\mathcal{D}^u$, *resp.* $\mathcal{A}^u$]. *If* f, g *are real-valued, then*

$$f^+ = \max(0, f) \text{ and } f^- = -\min(0, f),$$

and, more generally

$$\max(f,g) \text{ and } \min(f,g)$$

are again in $\mathcal{B}^u$ [*resp.* $\mathcal{D}^u$, *resp.* $\mathcal{A}^u$]. *The shifted functions (with positive integers* a, b)

$$f_a : n \mapsto f(n+a), \text{ and } f_{b;a} : n \mapsto f(bn + a)$$

are in $\mathcal{D}^u$ if f *is in $\mathcal{B}^u$ or $\mathcal{D}^u$, and in $\mathcal{A}^u$ if* f *is in $\mathcal{A}^u$.*

Proof. The result for Re f and Im f is obvious. If f is near $\varphi \in \mathcal{B}$ [resp. $\mathcal{D}$], then $\big| |f| - |\varphi| \big| \le | f-\varphi |$, so $|f|$ is near $|\varphi|$ and $|\varphi|$ is even [resp. periodic] and so again is in $\mathcal{B}$ [resp. $\mathcal{D}$].

If f is in $\mathcal{A}^u$ and near φ in $\mathcal{A}$, then there seems to be no easily accessible structural property[1)] which is obviously true for $|\varphi|$. But the WEIERSTRASS Theorem (see Appendix, Theorem A.1.1) shows that $|\varphi|$ is in $\mathcal{A}^u$: for $|\varphi|$ is bounded, say $|\varphi| \le M$. Given $\varepsilon > 0$, there is, by the WEIERSTRASS Theorem, a polynomial P(X) with real coefficients, satisfying $\big|P(x) - |x|\big| < \varepsilon$ in $-M \le x \le M$. $\mathcal{A}$ being an algebra, the function $P(\varphi)$ is in $\mathcal{A}$, and $\| P(\varphi) - |\varphi| \|_u < \varepsilon$, and so $|\varphi| \in \mathcal{A}^u$. Therefore, $|f|$ is in $\mathcal{A}^u$.

The formulae $\max(f,g) = \frac{1}{2}(f+g) + \frac{1}{2}\cdot|f-g|$, $\min(f,g) = \frac{1}{2}(f+g) - \frac{1}{2}\cdot|f-g|$ show the assertions concerning $\max(f,g)$ and $\min(f,g)$.

If φ is in $\mathcal{D}$, resp. $\mathcal{A}$, then the shifted function φ_a is clearly in $\mathcal{D}$, resp. $\mathcal{A}$ (similarly, $\varphi_{b;a}$ is in $\mathcal{D}$, resp. $\mathcal{A}$); and φ_a is near f_a if φ is near f. □

Theorem 2.3. *If f is in $\mathcal{A}^u$ then the mean-value* M(f) *exists. Moreover, the* FOURIER *coefficients*

$$\hat{f}(\alpha) = M(f\cdot e_{-\alpha})$$

and the RAMANUJAN *coefficients*

$$a_r(f) = \{\varphi(r)\}^{-1} \cdot M(f\cdot c_r)$$

exist.

Proof. Without loss of generality, let $f \in \mathcal{A}^u$ be real-valued. Given $\varepsilon > 0$, there exists a function $F \in \mathcal{A}$ with the property

$$F(n) - \varepsilon < f(n) < F(n) + \varepsilon$$

for every $n \in \mathbb{N}$. The mean-value M(F) exists, therefore the difference of the upper and lower mean-value of f, is

---

1) Of course, φ is (see, for example, CORDUNEANU [1968]) almost-periodic, and so there are ε-translation numbers for φ; these are also ε-translation numbers for $|\varphi|$, and so $|\varphi|$ is in $\mathcal{A}^u$.

$$|M_{-}(f) - M^{-}(f)| \le \varepsilon,$$

and so M(f) exists. If $f \in \mathcal{A}^u$, then $f \cdot c_r$ and $f \cdot e_{-\alpha}$ are also in $\mathcal{A}^u$, and thus the assertions about the FOURIER and RAMANUJAN coefficients are clear. □

Theorem 2.4. *Let* $f \in \mathcal{A}^u$, *and let* $\mathcal{K} \subset \mathbb{C}$ *be a compact set with the following property: there is some* $\delta > 0$ *such that* $\bigcup_{n \in \mathbb{N}} B(f(n), \delta) \subset \mathcal{K}$. $B(f(n), \delta)$ *denotes the ball with radius* δ *around* $f(n)$. *Assume that*

$$\psi: \mathcal{K} \to \mathbb{C} \text{ is LIPSCHITZ-continuous;}$$

so there is a constant L *with the property*

$$|\psi(z) - \psi(z')| \le L \cdot | z - z' |, \text{ if } z, z' \in \mathcal{K}.$$

Then the composed function

$$\psi \circ f : \mathbb{N} \to \mathbb{C}$$

is again in $\mathcal{A}^u$. *The same result is valid in* $\mathcal{B}^u$.

Proof. Let ε be less than δ. If F in $\mathcal{A}$ is near f, $\| f - F \|_u < \varepsilon$, then the values of f and F are in $\mathcal{K}$; by the LIPSCHITZ-continuity, $\| \psi \circ f - \psi \circ F \|_u \le L \cdot \varepsilon$. We have to show that $\psi \circ F$ is in $\mathcal{A}^u$. According to the complex version of the WEIERSTRASS Approximation Theorem, there is a polynomial $P(z, \overline{z})$ with complex coefficients, so that

$$| \psi(z) - P(z, \overline{z}) | < \varepsilon, \text{ if } z \in \mathcal{K}.$$

Thus $| \psi(F(n)) - P(F(n), \overline{F(n)}) | < \varepsilon$ for any $n \in \mathbb{N}$; the function $n \mapsto P(F(n), \overline{F(n)})$ is in $\mathcal{A}$, and so $\psi \circ f$ is in $\mathcal{A}^u$. □

Corollary 2.5.

(1) *If* $f \in \mathcal{A}^u$, *then* $e^{i\lambda \cdot f} \in \mathcal{A}^u$ *for every complex constant* λ.

(2) *If* $f \in \mathcal{A}^u$ *and* $|f| \ge \delta$, *where* $\delta > 0$, *then* $1/f$ *is in* $\mathcal{A}^u$.

(3) *If* f *is in* $\mathcal{A}^u$, $|f| \ge \delta$, *where* $\delta > 0$, *and if there is an angle* $\{z \in \mathbb{C}, |\arg(z) - \alpha| \ge \delta\}$ *free of values of* f, *then* $\log(f)$ *is in* $\mathcal{A}^u$.

Theorem 2.4 is a special case of the next, more general, theorem.

Theorem 2.6. *Let* $f \in \mathcal{A}^u$ *(resp.* $f \in \mathcal{B}^u$*), and, for* $\gamma > 0$,

$$K_\gamma = \{ z \in \mathbb{C}: \exists n \in \mathbb{N} \text{ with the property } |f(n) - z| < \gamma \}.$$

Then, for every continuous function $\psi: K_\gamma \to \mathbb{C}$, *the composed function*

$$\psi \circ f : \mathbb{N} \to \mathbb{C}$$

is again in $\mathcal{A}^u$ *(resp. in* $\mathcal{B}^u$*).*

Proof. The function f is bounded, therefore the closure $\overline{K_{\gamma/2}}$ is compact and ψ, restricted to $\overline{K_{\gamma/2}}$, is uniformly continuous. Given $\varepsilon > 0$, there is a δ, $0 < \delta < \frac{1}{2}\gamma$ such that

$$|\psi(z) - \psi(z')| < \varepsilon \text{ for all } z, z' \in K_{\gamma/2}, \ |z-z'| < \delta.$$

Choose a function F in $\mathcal{A}$ (resp. in $\mathcal{B}$) near f, $\|f - F\|_u < \delta$. Then

$$\|\psi\circ f - \psi\circ F\|_u \le \varepsilon.$$

If $f \in \mathcal{B}^u$, $F \in \mathcal{B}$, then $\psi\circ F \in \mathcal{B}$, and $\psi\circ f \in \mathcal{B}^u$. If $f \in \mathcal{A}^u$, $F \in \mathcal{A}$, then $\psi\circ F \in \mathcal{A}^u$ by the WEIERSTRASS Approximation Theorem (as in the proof of Theorem 2.4). Therefore, $\psi\circ f$ is in $\mathcal{A}^u$. □

The next result contains a characterization of the additive functions of to $\mathcal{B}^u$.

Theorem 2.7. (1) *If* f *is in* $\mathcal{A}^u$ *and is additive, then*

(i) $$\sum_p \sup_k | f(p^k) | < \infty.$$

If f *is in* $\mathcal{B}^u$, *then*

(ii) $$\lim_{k \to \infty} f(p^k) \text{ exists for every prime.}$$

(2) *If* f *is additive and if relations* (i) *and* (ii) *are true, then* f *is in* $\mathcal{B}^u$.

(3) *If* f *is in* $\mathcal{D}^u$ *and is additive, then* (ii) *is true.*

(4) *Therefore, the intersection of the vector-space of additive functions with* $\mathcal{D}^u$ *is equal to the intersection of this space with* $\mathcal{B}^u$.

Proof. (1.i) Without loss of generality, f is real-valued; f is uniformly bounded, and so $\left|\sum f(p^k)\right| \le \|f\|_u$, summed over any finite set of prime-powers for which $f(p^k) \ge 0$ (and the same is true for every finite set of prime-powers for which $f(p^k) < 0$). These remarks imply

$$\sum_p \sup_k \left|f(p^k)\right| \le 2\cdot \|f\|_u + 1.$$

(1.ii) The values $f(p^k)$ are bounded, so there is a subsequence $k_1 < k_2 < \ldots$, for which $f(p^{k_r})$ is convergent, $| L - f(p^{k_r})| < n^{-1}$, if $k_r \ge K_1(n)$. Choose $F_n \in \mathcal{B}$ near f, $\|F_n - f\|_u < n^{-1}$; if $k \ge K_2(n)$ is large, then the values $F_n(p^k)$ are constant, and thus

$$| L - f(p^k)| \le | L - f(p^{k_r}) | + | f(p^{k_r}) - F_n(p^{k_r})| + | F_n(p^k) - f(p^k)| < 2\cdot n^{-1}$$

if $k, k_r \ge \max (K_1(n), K_2(n))$, and (ii) is proved.

(2) Assume f is additive and satisfies (i) and (ii), $\lim_{k\to\infty} f(p^k) = g(p)$. Choose $\varepsilon > 0$. There are constants P_0 and k_0 {depending on ε}, so that

$$\sum_{p>P_0} \sup_k |f(p^k)| < \varepsilon, \text{ and } |f(p^k) - g(p)| < P_0^{-1}\cdot\varepsilon \text{ for every } p \le P_0,$$

if $k \ge k_0$. Put $K = \prod_{p\le P_0} p^{k_0}$ and define a K-even function F by

$$F(n) = f(\gcd(n,K)).$$

We aim at $\| f - F \|_u < 2\cdot\varepsilon$. Write $n = \prod p^{\nu_p(n)} = n'\cdot n''$, where n' contains those prime-factors of n which are $\le P_0$, and n" contains the "large" prime-factors $p > P_0$. Then $|f(n'')| < \varepsilon$ by the choice of P_0 (and by additivity). Decompose $n' = n_1'\cdot n_2'$, where n_1' contains the primes p with $\nu_p(n) \le k_0$ and n_2'' contains the others. Then

$$F(n') = f(n_1') + \sum_{p|n_2'} f(p^{k_0}),$$

and so

$$|F(n) - f(n)| \le |f(n'')| + \sum_{p|n_2'} |f(p^{\nu_p(n)}) - f(p^{k_0})| < \varepsilon + \omega(n_2')\cdot\varepsilon\cdot P_0^{-1} < 2\varepsilon.$$

(3) Let p be a prime, $\varepsilon > 0$. Choose a function $F \in \mathcal{D}$, r-periodic, $\| f - F \|_u < \varepsilon$. If $p \nmid r$, then

$$|F(p^k)| \le |F(p^k)-f(p^k)| + |F(r) - f(r)| + |F(p^k\cdot r) - f(p^k\cdot r)| < 3\varepsilon,$$

therefore $|f(p^k)| < 4\varepsilon$ for every k if $p\nmid r$. If $p|r$, say $p^k \| r$, then, similarly,

$$|F(p^k) - F(p^{k_0})| < 4\varepsilon,$$

therefore

$$|f(p^k) - f(p^{k_0})| < 6\varepsilon \text{ for every } k \ge k_0,$$

$$|f(p^k) - f(p^\ell)| < 12\varepsilon \text{ if } k, \ell \geq k_0.$$

So $k \mapsto f(p^k)$ is a CAUCHY sequence, which proves (ii).

(4) follows from (1) - (3). □

Theorem 2.8. *Assume that f is in* $\mathcal{D}^u$.

(1) *If* $P \in \mathbb{C}[X]$ *is a polynomial with complex coefficients, then*

$$P \circ f \in \mathcal{D}^u.$$

(2) *If* $P \in \mathbb{Z}[X]$ *is a polynomial with integer coefficients and* $P > 0$, *then*

$$f \circ P \in \mathcal{D}^u.$$

Proof. $\mathcal{D}^u$ and $\mathcal{B}^u$ are algebras, and so (1) is clear for $\mathcal{D}^u$ and $\mathcal{B}^u$. Approximating f by a finite linear combination of functions $e_{a/r}$, it is easy to reduce assertion (2) to the problem of showing that $n \mapsto e_{a/r}(\, P(n)\,)$ is in $\mathcal{D}^u$; but, due to $P(n+r) \equiv P(n) \bmod r$ this function is periodic and so it is in $\mathcal{D}$. □

Finally, we give the following uniqueness theorem.

Theorem 2.9. *Assume that* $q \geq 1$, $f \in \mathcal{D}^u$ *and* $\|f\|_q = 0$, *where*

$$\|f\|_q = \Big\{ \limsup_{x \to \infty} x^{-1} \cdot \sum_{n \leq x} |f(n)|^q \Big\}^{1/q}.$$

Then $f = 0$.

Proof. Assume, on the contrary, that there is some n_0, for which $|f(n_0)| = \delta > 0$. Choose $\varepsilon = \frac{1}{4}\cdot\delta$; there is a function F in $\mathcal{D}$ near f, so that $\|f-F\|_u < \varepsilon$. F is periodic with some period K. Therefore (for any m in $\mathbb{N}$)

$$|f(n_0+m\cdot K)| \geq |F(n_0+m\cdot K)| - \varepsilon = |F(n_0)| - \varepsilon \geq |f(n_0)| - 2\varepsilon = \tfrac{1}{2}\delta,$$

and

$$\|f\|_q \geq \Big\{ \limsup_{x \to \infty} x^{-1} \cdot \sum_{n \leq x,\, n \equiv n_0 \bmod K} |f(n)|^q \Big\}^{1/q}$$
$$\geq \tfrac{1}{2} K^{-1/q} \cdot (\, \tfrac{1}{2}\delta\,) > 0,$$

a contradiction. □

Remark. Other uniqueness theorems are proved in VI, Theorems 1.5, 1.6.

IV.3. LIMIT DISTRIBUTIONS

First we have to repeat some definitions and notation from probability theory; see, for example, RÉNYI [1970], and the Appendix.

A function $F: \mathbb{R} \to \mathbb{R}$ is called a **distribution function** if

(i) F is monotonically non-decreasing,

(ii) F is right-continuous, so that for every $x \in \mathbb{R}$

$$\lim_{\xi \to 0, \xi > 0} F(x+\xi) = F(x),$$

(iii) $F(-\infty) = \lim_{x \to -\infty} F(x) = 0,$
$F(+\infty) = \lim_{x \to \infty} F(x) = 1.$

So, in fact, $F: \mathbb{R} \to [0,1]$. Note that the set of discontinuities of a distribution function is at most denumerable.

Examples of distribution functions are

- the function $x \mapsto \varepsilon(x) = \begin{cases} 0, & \text{if } x < 0, \\ 1, & \text{if } x \geq 0. \end{cases}$

- the GAUSSian normal distribution

$$x \mapsto F(x) = \int_{-\infty}^{x} \left\{\sigma \cdot \sqrt{2\pi}\right\}^{-1} \cdot \exp\left\{-\tfrac{1}{2} \cdot \sigma^{-2} \cdot (\xi-\mu)^2\right\} d\mu.$$

- if (Ω, P) is a probability space, $X: \Omega \to \mathbb{R}$ a real-valued random variable, then

(3.1) $$F(u) = P(t: X(t) < u)$$

is the distribution function associated with X.

A sequence F_N of distribution functions is said to *converge weakly* [in the sense of probability theory] to F if

$$\lim_{N \to \infty} F_N(x) = F(x)$$

is true for all points of continuity of F. The limit-function F need not be a distribution function, but it is non-decreasing and bounded.

In the theory of distributions the FOURIER-STIELTJES transform of a distribution function F is called its *characteristic function* f,

$$f(t) = \int_{-\infty}^{\infty} e^{itx} \, dF(x). \tag{3.2}$$

There is an inversion formula: if a and (a+h) are points of continuity for F, then

$$F(a+h) - F(a) = \lim_{T\to\infty} \int_{-T}^{T} (it)^{-1}\cdot(1-e^{-ih}) \cdot e^{ita}\cdot f(t)\, dt. \tag{3.3}$$

Further properties of the *characteristic function* f of a distribution function are listed below:

(1) $f(0) = 1$.

(2) $|f(t)| \le 1$.

(3) $f(-t) = \overline{f(t)}$.

(4) f is uniformly continuous on $\mathbb{R}$.

(5) If a_k are non-negative real numbers, $\sum_{1\le k\le K} a_k = 1$, and if f_k are characteristic functions, then $\sum_{1\le k\le K} a_k \cdot f_k$ is a characteristic function again; in particular, Re f is a characteristic function.

(6) If f is k times differentiable at 0, then the **moments**

$$\alpha_k = \int_{-\infty}^{\infty} x^k \cdot dF(x)$$

of F do exist up to the order k if k is even, and up to the order (k-1) if k is odd.

Continuity Theorem. *If* $\{F_n\}$ *is a sequence of distribution functions with characteristic functions* $t \mapsto f_n(t)$, *then* $\{F_n\}$ *converges weakly to a* **distribution function** F *if and only if*

(i) $\lim_{n\to\infty} f_n(t) = f(t)$ *exists for every* $t \in \mathbb{R}$,

(ii) *the limit* f *is continuous at* 0.

Then f *is the characteristic function of* F.

For a proof see, for example, LUKACS [1970].

The **application to arithmetical functions** rests on the fact that

$$x \mapsto F_N(x) = N^{-1} \cdot \sum_{n\le N,\ g(n)\le x} 1$$

is a distribution function, if g is a real-valued arithmetical function; $x \mapsto F_N(x)$ is non-decreasing, $F_N(-\infty) = 0$, $F_N(\infty) = 1$, and is right-continuous. Its characteristic function is

$$f_N(t) = \int_{-\infty}^{\infty} e^{itx}\, dF_N(x) = N^{-1} \int_{-\infty}^{\infty} e^{itx}\, d\Big(\sum_{n\le N,\ g(n)\le x} 1\Big)$$

$$= N^{-1} \cdot \sum_{n\le N} e^{it\cdot g(n)}.$$

Theorem 3.1. *Let* $g \in \mathcal{A}^u$ *be real-valued. Then there is a limit-distribution*

$$F(x) = \lim_{N\to\infty} F_N(x) \quad (\textit{if } x \textit{ is a point of continuity for } F\,)$$

for g.

The **proof** is a direct application of the continuity theorem (note the fact that it is not assumed that g is additive). If t is an arbitrary real number, then the function $n \mapsto \exp(\, itg(n)\,)$ is in $\mathcal{A}^u$ according to Corollary 2.5. Thus the sequence of characteristic functions

$$f_N(t) = N^{-1} \cdot \sum_{n\le N} \exp(\, itg(n)\,)$$

converges for $n \to \infty$ to the mean-value

$$M(\, n \mapsto \exp(itg(n))\,) =: f(t).$$

The inequality

$$|\, e^{iu} - 1 \,| = \Big|\, i\cdot\int_0^u e^{i\xi}\, d\xi \,\Big| \le \Big| \int_0^u |\, e^{i\xi} |\, d\xi \Big| \le K\cdot|u|,$$

if u is in the disc B(0,R), with some constant $K = K(R)$ [if u is real or, more generally, if $\mathrm{Im}(u) \ge 0$, then it is possible to take $K = 1$], gives

$$\Big|\, N^{-1} \cdot \sum_{n\le N} \Big(\exp(\, itg(n)\,) - 1\Big) \Big| \le N^{-1} \cdot \sum_{n\le N} t\cdot|g(n)|,$$

and so, as N tends to infinity,

$$|f(t) - f(0)| \le t\cdot M(|g|),$$

so that f is continuous at $t = 0$. An application of the continuity theorem for characteristic functions gives the assertion. □

The theorem given above may be extended to classes of arithmetical functions that are much larger than $\mathcal{A}^u$; this will be done in Chapter VI, 8 A.

IV.4. GELFAND's THEORY: MAXIMAL IDEAL SPACES

Some notions and definitions from functional analysis are used in this section. We refer to the Appendix, A.6.

The algebras $\mathcal{B}^u \subset \mathcal{D}^u \subset \mathcal{A}^u$ are commutative BANACH algebras with identity element $e = 1$, and there is the "standard" involution $f \mapsto \overline{f}$ (complex conjugation) satisfying $\| f \cdot \overline{f} \|_u = \| f \|_u^2$. So these spaces are B^*-algebras, and, according to GELFAND and NAIMARK's Theorem, these algebras are essentially algebras of continuous functions on the [compact] maximal ideal space Δ. The GELFAND transform

$$(4.1) \quad \begin{cases} \hat{}: f \mapsto \left(\hat{f} : \Delta \to \mathbb{C},\ \hat{f}(h) = h(f) \right), \\ \hat{}: \mathcal{B}^u \to \mathcal{C}(\Delta_{\mathcal{B}}) \text{ resp. } \hat{}: \mathcal{D}^u \to \mathcal{C}(\Delta_{\mathcal{D}}) \text{ resp. } \hat{}: \mathcal{A}^u \to \mathcal{C}(\Delta_{\mathcal{A}}) \end{cases}$$

is an isometric isomorphism in each case.

IV.4.A. The maximal ideal space $\Delta_{\mathcal{B}}$ of $\mathcal{B}^u$.

a) Construction of some algebra-homomorphisms.

Clearly, for any integer $n \in \mathbb{N}$, the *evaluations* $h_n : f \mapsto f(n)$ are elements of $\Delta_{\mathcal{B}}$. Next, for any prime p, and for $f \in \mathcal{B}^u$, the limit

$$f(p^\infty) = \lim_{k \to \infty} f(p^k)$$

exists, as shown in Theorem 2.7, and so the functions $h_{p^\infty} : f \mapsto f(p^\infty)$ are elements of $\Delta_{\mathcal{B}}$. More generally, given exponents k_p, $0 \le k_p \le \infty$, a (complex) value $f(\mathcal{K})$ can be defined for the vector

$$\mathcal{K} = (k_p)_{p \text{ prime}}$$

in the following manner[2)]: consider the increasing sequence n_r of positive integers

---

2) We think of the sequence of primes being ordered according to size. An integer n may be described as a special vector $\mathcal{K}$, where at most finitely many of the k_p are non-zero and none is infinity.

$$n_r = \prod_{1\le\rho\le r} p_\rho^{\min(r,k_{p_\rho})}, \quad r = 1,2, \dots,$$

with the property $n_r | n_{r+1}$. Then

$$f(\mathcal{K}) = \lim_{r\to\infty} f(n_r)$$

exists[3], and

$$h_{\mathcal{K}}: f \mapsto f(\mathcal{K})$$

is an element of $\Delta_{\mathcal{B}}$. All these functions $h_{\mathcal{K}}$ are different, as can be seen by evaluating $h_{\mathcal{K}}$ on suitable RAMANUJAN sums c_{q^ℓ}, where q is prime.

Our goal is to prove that we obtained all the elements of $\Delta_{\mathcal{B}}$. Before doing this, we calculate the values of $h_{\mathcal{K}}$ at RAMANUJAN sums c_{q^ℓ} for prime powers q^ℓ. Obviously (giving the greatest common divisor on the right-hand-side a natural interpretation),

$$h_{\mathcal{K}}(c_{q^\ell}) = c_{q^\ell}\Big(\gcd\big(\prod_p p^{k_p}, q^\ell\big)\Big),$$

and this equals

$$(4.2) \qquad \begin{cases} = c_{q^\ell}(q^\ell) = \varphi(q^\ell), & \text{if } k_q \ge \ell, \\ = c_{q^\ell}(q^{\ell-1}) = -q^{\ell-1}, & \text{if } k_q = \ell - 1, \\ = 0, & \text{if } k_q < \ell - 1. \end{cases}$$

b) Determination of $\Delta_{\mathcal{B}}$.

Following the paper by T. MAXSEIN, W. SCHWARZ and P. SMITH [1991] rather closely, we are going to prove Theorem 4.1.

Theorem 4.1. *The maximal ideal space $\Delta_{\mathcal{B}}$ consists precisely of the functions $h_{\mathcal{K}}$, where $\mathcal{K}$ runs through the set of vectors $(k_p)_{p \text{ prime}}$, $0 \le k_p \le \infty$.*

Proof. Assume $h \in \Delta_{\mathcal{B}}$; h being continuous, it is sufficient to know the

---

[3] Given $\varepsilon > 0$, choose $F \in \mathcal{B}$ satisfying $\|f-F\|_u < \varepsilon$. The function F is even, and so $F(n_r) = \beta$ is constant for $r \ge r_0(\varepsilon)$. Thus the sequence $r \mapsto f(n_r)$ is a Cauchy sequence.

values of h on the subalgebra $\mathcal{B}$ of $\mathcal{B}^u$. The RAMANUJAN sums c_r, considered as functions of the index r, are multiplicative. Therefore, it is sufficient to know the values $h(c_{q^\ell})$ for prime-powers q^ℓ.

Since $h(f) \in \text{spec}(f)$, and $\text{spec}(c_{q^\ell})$ is $\{ \varphi(q^\ell), -q^{\ell-1}, 0 \}$ if $\ell > 1$, $\{\varphi(q), -1\}$ if $\ell = 1$, and $\{1\}$ if $\ell = 0$, there are at most three possibilities for choosing the value $h(c_{q^\ell})$. However, not every choice is admissible. The relations

$$(4.3') \qquad c_{p^m} \cdot c_{p^\ell} = \varphi(p^\ell) \cdot c_{p^m}, \text{ if } m > \ell,$$

and

$$(4.3'') \qquad c_{p^\ell} \cdot c_{p^\ell} = \varphi(p^\ell) \cdot (c_1 + c_p + \ldots + c_{p^{\ell-1}}) + (p^\ell - 2p^{\ell-1}) \cdot c_{p^\ell}$$

imply (using the fact that h is an algebra-homomorphism; q denotes a prime)

(a) $h(c_{q^m}) = 0$, if $h(c_{q^\ell}) = 0$ and $m > \ell$,

(b) $h(c_{q^m}) = \varphi(q^m)$, if $h(c_{q^\ell}) \neq 0$ and $0 \le m < \ell$,

(c) $h(c_{q^\ell}) < 0$ is possible for at most *one* ℓ (q fixed),

(d) if $h(c_{q^{\ell+1}}) = 0$ but $h(c_{q^\ell}) \neq 0$, then $h(c_{q^\ell}) = - q^{\ell-1} < 0$.

Therefore, either $h(c_{q^m}) = \varphi(q^m)$ for any $m \ge 0$ (define $k_q = \infty$ in that case), or there exists an exponent k_q such that

$$(4.4) \qquad h(c_{q^\ell}) = \begin{cases} \varphi(q^\ell), & \text{if } \ell \le k_q, \\ -q^{\ell-1}, & \text{if } \ell = k_q + 1, \\ 0, & \text{if } \ell > k_q + 1. \end{cases}$$

Then, for the vector $\mathcal{K} = (k_q)_{q \text{ prime}}$, we obtain $h = h_{\mathcal{K}}$, and so $\Delta_{\mathcal{B}}$ is completely determined. □

c) Topology.

The GELFAND topology of $\Delta_{\mathcal{B}}$ is the weakest topology that makes every GELFAND transform (4.1)

$$\hat{f}: \Delta_{\mathcal{B}} \to \mathbb{C}, \ \hat{f}(h) = h(f)$$

continuous. So, for any prime power q^ℓ and any open set $\mathcal{O}$ in $\mathbb{C}$, the sets

$$\hat{c}_{q^\ell}^{-1}(\mathcal{O}) = \{ h \in \Delta; \ h(c_{q^\ell}) \in \mathcal{O} \}$$

are open. Therefore, using (4.4), the sets

$$\{ h_{\mathcal{X}} \ ; \ k_p \text{ arbitrary for } p \neq q, \ k_q \geq \ell \},$$

where $\ell \in \mathbb{N}$, and

$$\{ h_{\mathcal{X}} \ ; \ k_p \text{ arbitrary for } p \neq q, \ k_q = \ell-1 \}$$

are open. Choosing these sets as a subbasis for the topology, we see that every $\hat{f}$ is continuous. For:

Given $\varepsilon > 0$ and f, choose $g = \sum_{1\leq r\leq R} \gamma_r \cdot c_r$ satisfying $\|f-g\|_u < \frac{1}{2}\varepsilon$. Assume that $h \in \Delta_{\mathcal{B}}$, $h = h_{\mathcal{X}}$, $\mathcal{X} = (k_p(h))$, is given. An open neighbourhood U(h) of h is defined by the condition

$$h^* \in U(h) \text{ iff } h^* = h_{\mathcal{X}^*}, \text{ and } k_p(h^*) = k_p(h) \text{ for any } p\leq R.$$

Then $h(g) = h^*(g)$ for any h^* in U(h), and so

$$| \hat{f}(h) - \hat{f}(h^*)| = | h(f) - h^*(f) | \leq | h(f) - h(g) | + | h^*(f) - h^*(g)|$$

$$\leq \| f - g \|_u + \| f - g \|_u < \varepsilon.$$

Thus, $\hat{f}$ is continuous and so the topology of $\Delta_{\mathcal{B}}$ is completely determined. It coincides with the product topology on the space

$$\mathbb{N}^* = \prod_p \left\{1, p, p^2, \ldots, p^\infty\right\}, \tag{4.5}$$

where each factor is the ALEXANDROFF-one-point-compactification of the discrete (and locally compact) space $\{1, p, p^2, \ldots\}$.

d) Main result.

For functions f in $\mathcal{B}^u$, obviously $\|f^2\|_u = \|f\|_u^2$, and so we obtain from 11.12 in W. RUDIN [1966] a result already mentioned at the beginning of this section.

Theorem 4.2. *The Banach-algebra $\mathcal{B}^u$ is semi-simple, and the GELFAND transform $f \mapsto \hat{f}$ is an isometric algebra-isomorphism from $\mathcal{B}^u$ onto $\mathcal{C}(\Delta_{\mathcal{B}})$.*

Note that semi-simplicity immediately also follows from the fact that the evaluation homomorphisms $h_n : f \mapsto f(n)$ are in $\Delta_{\mathcal{B}}$, and so the assumption $f \in \text{radical}(\mathcal{B}^u) = \bigcap_{h\in \Delta_{\mathcal{B}}} \textit{kernel}(h)$ implies $f = 0$.

Next, RUDIN [1966], section 11.20, implies the following corollary.

Corollary 4.3. *If* $f \in \mathcal{B}^u$ *is real-valued and if* $\inf_{n\in\mathbb{N}} f(n) > 0$, *then there exists a [real-valued] square-root* g *of* f *in* $\mathcal{B}^u$.

e) Applications.

The following result is well-known and can also be derived from the WEIERSTRASS approximation theorem (see Corollary 2.5); we deduce it from our knowledge of $\Delta_{\mathcal{B}}$.

Corollary 4.4. *Assume that* $f \in \mathcal{B}^u$. *Then* $1/f \in \mathcal{B}^u$ *if and only if* $\inf_{n\in\mathbb{N}}|f(n)|$ *is positive.*

Proof. If $1/f \in \mathcal{B}^u$, then this function is bounded and so $|f|$ is bounded from below. On the other hand, according to GELFAND's Theory (see RUDIN [1966], 18.17), $1/f \in \mathcal{B}^u$ if for any $h \in \Delta_{\mathcal{B}}$ the value h(f) is not zero. The values h(f) are given as certain limits in section 2, and the condition $|f| \geq \delta$ obviously implies that all these limits are non-zero, and corollary 4.4 is proved. □

This corollary may be extended considerably.

Theorem 4.5. *Let* $f \in \mathcal{B}^u$ *be given. If the function* F *is holomorphic in some region of* $\mathbb{C}$, *including the range* $\hat{f}(\Delta_{\mathcal{B}})$ *of* $\hat{f}$, *then the composed function* $F\circ\hat{f}$ *is in* $\mathcal{C}(\Delta_{\mathcal{B}})$ *and thus is equal to some* $\hat{g}$, $g \in \mathcal{B}^u$. *Therefore,* $F\circ f$ *is in* $\mathcal{B}^u$ *again.*

Except for the last sentence, this is a specialization of L. H. LOOMIS [1953], **24 D**. Next, $\hat{g} = F\circ\hat{f}$ implies $h(g) = F(h(f))$ for any h in $\Delta_{\mathcal{B}}$, and so the assertion is true if F is a polynomial [then $F(h(f)) = h(F(f))$]. The general case follows from this. □

In the case of multiplicative functions, the following results are true.

Theorem 4.6a. *Let* $f \in \mathcal{B}^u$ *be given. If* f *is multiplicative, then* $f(p^k) = 0$ *is possible for at most finitely many primes* p,

and the same argument gives the following stronger version of Theorem 4.6a.

Theorem 4.6b. *Let* $f \in \mathcal{B}^u$ *be given. If* $\delta > 0$ *and* f *is multiplicative, then there are at most finitely many primes with the property*

$|f(p^k) - 1| > \delta$ *for some* k.

Proof. $\hat{f}(h_{\mathcal{K}_0}) = 1$ where $\mathcal{K}_0 = (k_p)$, $k_p = 0$ for any p. Given $\varepsilon = \frac{1}{2}\delta$, then there is some neighbourhood $\mathcal{U}$ of h with the property $|\hat{f}(h) - 1| < \varepsilon$ for h in $\mathcal{U}_0$. But this neighbourhood contains all $h_{\mathcal{K}}$ with k_p arbitrary except for finitely many primes; for these exceptional primes $k_p = 0$ may be taken. Next, f being multiplicative,

$$\hat{f}(h) = \lim_{L\to\infty} \prod_{p\le L} f(p^{\min(k_p,L)}), \tag{4.6}$$

and this implies, by a suitable choice of the k_p, and noting $|\hat{f}(h) - 1| < \varepsilon$, that $|f(p^k) - 1| > \varepsilon$ is impossible for any "non-exceptional" prime and any k. □

IV.4.B. The maximal ideal space $\Delta_{\mathcal{D}}$ of $\mathcal{D}^u$

a) Embedding of $\Delta_{\mathcal{D}}$ in $\prod_{r\in\mathbb{N}} \mathbb{Z}/r\cdot\mathbb{Z}$

Define, using the abbreviation $\omega_r = \exp(2\pi i/r)$, an element $f_r \in \mathcal{D}$ by $f_r(n) = \omega_r^n$. The set of functions

$$\left\{ f_r^{\ell},\ 1 \le \ell \le r,\ \gcd(\ell,r) = 1,\ r = 1,2,\ldots \right\} \tag{4.7}$$

is a basis of $\mathcal{D}$. A function f in $\mathcal{D}$ is r-periodic for some r, and so 1/f is again r-periodic and in $\mathcal{D} \subset \mathcal{D}^u$, if f does not assume the value zero. Therefore,

$$\text{spec}(f_r) = \{ \omega_r^j,\ 1 \le j \le r \}.$$

If $h \in \Delta_{\mathcal{D}}$, then

$$h(f_r) = \omega_r^{j(r,h)}, \tag{4.8}$$

where j(r,h) is some uniquely determined integer modulo r, depending on h. Thus we obtain a map

$$\varphi:\ \Delta_{\mathcal{D}} \to \prod_{r\in\mathbb{N}} \mathbb{Z}/r\cdot\mathbb{Z}, \tag{4.9}$$

defined by $\varphi(h) = (\ j(r,h)\ )_{r=1,2,\ldots}$, where h und j are related by (4.8). Obviously, φ is *injective.*

Examples. (1) If f is a periodic function with period M, and if H is a homomorphism in $\Delta_{\mathcal{D}}$, then $H(f) = f(j(M,H))$.

Proof. $H(e_{1/r}) = e_r(\,j(r,H))$. The FOURIER expansion $f = \sum_{\mu} a_\mu \cdot e_{\mu/M}$

implies the result.

(2) If g is in $\mathcal{D}^u$, and G is M-periodic, $\|g - G\|_u < \varepsilon$, then $|H(g) - g(j(M,H))| < 2\varepsilon$ for every H in $\Delta_{\mathcal{D}}$. (This depends on the fact that $|H(f)| \le \|f\|_u$.)

(3) If h_r is the evaluation homomorphism $f \mapsto f(r)$, then

$$j(k,h_r) \equiv r \bmod k \text{ for } k = 1, 2, \dots .$$

b) The Prüfer Ring $\hat{\mathbb{Z}}$

For any $n \in \mathbb{N}$ consider the residue class ring $\mathbb{Z}/n\cdot\mathbb{Z}$ with discrete topology. If $m|n$, then there is a continuous projection

$$\pi_{m,n} : \mathbb{Z}/_{n\cdot\mathbb{Z}} \to \mathbb{Z}/_{m\cdot\mathbb{Z}} \text{ , (a mod n) } \mapsto \text{(a mod m).} \tag{4.10}$$

The set $X = \prod_{r\in\mathbb{N}} \mathbb{Z}/_{r\cdot\mathbb{Z}}$ with the product topology is a compact HAUSDORFF space, and the set

$$\hat{\mathbb{Z}} = \{ (\alpha_n) \in X \text{ , } \alpha_n \in \mathbb{Z}/_{n\cdot\mathbb{Z}} \text{ and } \pi_{m,n}(\alpha_n) = \alpha_m, \text{ if } m|n \} \tag{4.11}$$

is a closed subspace of X and therefore is again compact (and HAUSDORFF). Note that $\mathbb{N}$ is dense in $\hat{\mathbb{Z}}$; the reason is that, given an element $(\alpha_r)_r$ in $\hat{\mathbb{Z}}$, and given positive integers $r_1, \dots, r_N$, there exists an integer $m \in \mathbb{N}$ satisfying $m \equiv \alpha_{r_i} \bmod r_i$ for $1 \le i \le N$.

Since $f^s_{r\cdot s} = f_r$ it follows that $j(r\cdot s,h) \equiv j(r,h) \bmod r$ for any $h \in \Delta_{\mathcal{D}}$. Therefore, the image of the map φ is contained in $\hat{\mathbb{Z}}$.

c) Surjectivity of φ: $\Delta_{\mathcal{D}} \to \hat{\mathbb{Z}}$

Let some element $(\alpha_r)_r$ in $\hat{\mathbb{Z}}$ be given. Our aim is to construct an algebra-homomorphism $h \in \Delta_{\mathcal{D}}$ satisfying $\varphi(h) = (\alpha_r)_r$. Define a linear map $h: \mathcal{D} \to \mathbb{C}$ on the elements of the basis of $\mathcal{D}$ by

$$h(f_r^k) = \omega_r^{k\cdot\alpha_r},\ 1 \le k \le r,\ \gcd(k,r) = 1,\ r = 1,2,\dots,$$

and extend h linearly to $\mathcal{D}$. Then h is multiplicative on $\mathcal{D}$: assume first that $\gcd(r,s) = 1$; then the relation

$$s\cdot k\cdot\alpha_r + r\cdot\ell\cdot\alpha_s \equiv (\, s\cdot k + r\cdot\ell \,)\cdot\alpha_{r\cdot s} \bmod r\cdot s$$

implies

$$h(f_r^k \cdot f_s^\ell) = h(f_r^k)\cdot h(f_s^\ell).$$

This is also true if $\gcd(r,s) \neq 1$; without loss of generality, r and s may be assumed to be powers of the same prime, and then the assertion is easily checked.

h *is continuous on* $\mathcal{D}$: given an element $\psi \in \mathcal{D}$, $\psi = \sum_{1\le\nu\le N} a_\nu \cdot f_{r_\nu}^{k_\nu}$, there exists an $m \in \mathbb{N}$, for which $m \equiv \alpha_{r_\nu} \bmod r_\nu$ for $1 \le \nu \le N$. Since $h(\psi) = \psi(m)$, we obtain

$$|h(\psi)| \le |\psi(m)| \le \|\psi\|_u,$$

and so h is continuous on $\mathcal{D}$. This space being dense in $\mathcal{D}^u$, h may be extended continuously to an algebra-homomorphism of $\mathcal{D}^u$, and $\varphi(h) = (\alpha_r)_{r=1,2,\dots}$.

d) Continuity of $\varphi : \Delta_{\mathcal{D}} \to \hat{\mathbb{Z}}$.

Fix $\alpha_k \in \mathbb{Z}/k\cdot\mathbb{Z}$, $1 \le k \le N$, with the property $\alpha_n \equiv \alpha_m \bmod m$ if $m|n$. Then

$$V(\alpha_1, \dots, \alpha_N) = \{ (\beta_n) \in \hat{\mathbb{Z}} \,,\; \beta_k = \alpha_k \text{ for } 1 \le k \le N \}$$

is a typical basis element of the [product] topology of $\hat{\mathbb{Z}}$. Moreover, $h \in \varphi^{-1}(V(\alpha_1, \dots, \alpha_N))$ if and only if $h(f_k) = \omega_k^{\alpha_k}$ for any k in $1 \le k \le N$. This is equivalent to $\hat{f}_k(h) = \omega_k^{\alpha_k}$, $1 \le k \le N$, where $\hat{f}_k$ is the GELFAND transform of f_k.

If U_k is a neighbourhood of $\omega_k^{\alpha_k}$ not containing any other k^{th} root of unity, then it follows that

$$\varphi^{-1}(V(\alpha_1, \dots, \alpha_N)) = \bigcap_{k=1}^{N} \hat{f}_k^{-1}(U_k)$$

is an open set in the GELFAND topology of $\Delta_{\mathcal{D}}$, and so φ is continuous. Since $\Delta_{\mathcal{D}}$ and $\hat{\mathbb{Z}}$ are compact Hausdorff spaces, φ is a homeomorphism. Thus we obtain the following theorem.

Theorem 4.7. *The maximal space $\Delta_{\mathcal{D}}$ is homeomorphic with the Prüfer Ring $\hat{\mathbb{Z}}$, defined in* (4.11).

Remark 1. The evaluation homomorphisms h_n are dense in $\Delta_{\mathcal{D}}$.

Proof of Remark 1. Given H in $\Delta_{\mathcal{D}}$, choose a neigbourhood U(H) "defined by R"; this means that $h \in U(H)$ iff $j(r,h) = j(r,H)$ for r in $1 \le r \le R$. Define the integer n as $j(R!, H)$. Then

$$n \equiv j(r,H) \bmod r \text{ for } r = 1,2,\dots,R, \tag{4.12}$$

and h_n is obviously in U(H). □

e) Arithmetical Applications

Next, we apply our knowledge of the maximal ideal space to the problem of the characterization of additive and multiplicative functions in $\mathcal{B}^u$. Some of the results have already been proved in section 2 using ad hoc elementary methods from number theory.

In [1943] N. G. DE BRUIJN characterized multiplicative, almost-periodic arithmetical functions. Additive, almost-periodic functions were characterized by E. R. VAN KAMPEN (1940). The results are as follows.

Theorem 4.8. *Assume* f *to be fibre-constant. Then* f *is in* $\mathcal{B}^u$ *if and only if* $\lim_{k\to\infty} f(p^k)$ *exists for every prime* p.

This result is not true for $\mathcal{D}^u$, as the example of a character χ satisfying $\chi(p) \neq 0, 1$ shows.

Remark 2. f is termed *fibre-constant* if there is a prime q such that $f(n) = f(\gcd(n,q^\infty))$ for any n. Obviously, $\lim_{k\to\infty} f(p^k)$ exists for any prime $p \neq q$ trivially.

Theorem 4.9. *An additive function is in* $\mathcal{B}^u$ *if and only if*

(4.13) $$\lim_{k\to\infty} f(p^k) \text{ exists for any prime}$$

and

(4.14) $$\sum_p \sup_k |f(p^k)| < \infty .$$

Theorem 4.10. *A multiplicative function is in* $\mathcal{B}^u$ *if and only if* (4.13) *holds and if*

(4.15) $$\sum_p \sup_k | f(p^k) - 1 | < \infty$$

is true.

Remark 3. If f is in $\mathcal{B}^u$ then the GELFAND transform $\hat{f}$ is continuous at $h_{\mathcal{K}}$, where $\mathcal{K} = (k_p)_p$, and $k_q = \infty$, $k_p = 0$, if $p \neq q$. All the functions $h_{\mathcal{K}'}$, where $k_p' = k_p = 0$ for $p \neq q$, and $k_q' = L$, L sufficiently large, are near $h_{\mathcal{K}}$, and thus the limit relation (4.13) is true.

The **proof** of Theorem 4.8 now follows from the preceding remark and the fact that for fibre-constant functions $\hat{f}(h)$ may be defined in an obvious manner using the limit relation (4.13) at q. The resulting function $\hat{f}$ is obviously continuous and so f is in $\mathcal{B}^u$. □

For additive functions in $\mathcal{D}^u$ we prove the following theorem.

Theorem 4.11. *If* f *is in* $\mathcal{D}^u$ *and additive, then* $\lim_{k\to\infty} f(p^k)$ *exists for every prime* p, *and relation* (4.14) *is true; therefore an additive function from* $\mathcal{D}^u$ *is in fact already in* $\mathcal{B}^u$.

Proof. Given $\varepsilon > 0$, choose an M-periodic function F in $\mathcal{D}$ satisfying $\| f - F \|_u < \frac{1}{3}\varepsilon$. Then F is ε - nearly additive, and so, according to section 1, $|F(p^\beta)| < \varepsilon$, if p does not divide M, and $|F(p^\beta) - F(p^\alpha)| < 2\varepsilon$ if $\beta \ge \alpha$ and $p^\alpha \| M$. This implies that $k \mapsto f(p^k)$ is a CAUCHY-sequence.

Concerning (4.14), without loss of generality, let f be real-valued. The function $\hat{f}$, continuous on the compact maximal ideal space, is bounded by $\|f\|_u$. Therefore, for any evaluation homomorphism h_n, $|\hat{f}(h_n)| \le \|f\|_u$. Now put

$$n_1 = \prod_p p^{k_p}, \quad n_2 = \prod_p p^{k_p},$$

where in the first [resp. second] product the product runs over all powers p^{k_p} for which $f(p^{k_p})$ is positive [resp. negative]. Then $f(n_1)$ and $|f(n_2)|$ are uniformly bounded and the theorem follows. □

We use the following *notation*: given any arithmetical function, define

(4.16) $$f_{(p)}(n) = f(\gcd(n, p^\infty)), \text{ if p is prime,}$$

and

(4.17) $$F_R(n) = f(\gcd(n, \prod_{p>R} p^\infty)).$$

The functions $f_{(p)}$ are fibre-constant.

Proof of Theorem 4.9.

(a) Assume that (4.13) and (4.14) hold. The function f being additive, we obtain

(4.18) $$f = \sum_{p\le R} f_{(p)} + F_R,$$

and the functions $f_{(p)}$ are in $\mathcal{B}^u$ by Theorem 4.8. Next,

$$|F_R(n)| = | f(n) - \sum_{p\le R} f_{(p)}(n) | \le \sum_{p>R} \sup_k |f(p^k)| < \varepsilon$$

if R is sufficiently large, and so $f \in \mathcal{B}^u$.

(b) If $\mathcal{K} = (0,0,\ldots)$, $\mathcal{K}' = (k_p)_p$, where k_p is arbitrary for $p > R$ and $k_p = 0$ if $p \le R$, then $h_{\mathcal{K}'}$ is near $h_{\mathcal{K}}$. Since f is additive, we obtain $\hat{f}(h_{\mathcal{K}}) = 0$; $\hat{f}$ is continuous, and so $|\hat{f}(h_{\mathcal{K}'})| < \varepsilon$ if R is sufficiently large. Therefore, evaluating $\hat{f}(h_{\mathcal{K}'})$, one obtains

$$| \sum_{R<p<R'} f(p^{k_p}) | < \varepsilon$$

for any system k_p of exponents [$k_p = \infty$ is admissible, $f(p^\infty) = \lim_k f(p^k)$], and so every subseries of $\sum_p f(p^{k_p})$ is convergent. Therefore this series is absolutely convergent for any choice of the exponents. This implies (4.14). □

Proof of Theorem 4.10.

(a) Assume that (4.13) and (4.15) hold. Being multiplicative,

$$f = \prod_{p\le R} f_{(p)} \cdot F_R,$$

where the fibre-constant functions $f_{(p)}$ are in $\mathcal{B}^u$. Next, using (4.15),

$$\left|\prod_{p\le R} f_{(p)}(n)\right| \le \exp\left\{ \sum_{p\le R}^{*}(| f_{(p)}(n)|- 1)\right\} \le C,$$

uniformly in R, where * means that summation is only over those primes, for which $| f_{(p)}(n)| \ge 1$. And

$$\left| f(n) - \prod_{p\le R} f_{(p)} (n)\right| \le C\cdot\left|F_R(n)-1\right| < C\cdot\varepsilon,$$

uniformly in n if R is large, again using (4.15). Therefore, f is in $\mathcal{B}^u$.

(b) If f is in $\mathcal{B}^u$ and multiplicative, then the proof is similar to the corresponding proof of Theorem 4.9. The details, a little more complicated than before, are omitted. It is helpful to use the fact that the absolute convergence of a product $\prod x_i$ is equivalent with the absolute convergence of the series $\sum \{x_i -1\}$. □

A second proof of the result for multiplicative functions is possible by reducing the assertion to the corresponding result for additive functions. First we prove the following lemma.

Lemma 4.12. *Assume that* f *is in* $\mathcal{B}^u$ *and* $f(1) = 1$. *Then*

(1) $\lim_{k \to \infty} f(p^k) = 0$ *is possible for at most finitely many primes.*

(2) *For any* $\varepsilon > 0$ *there is an integer* R *so that* $|f(n) - 1| < \varepsilon$ *if* $\gcd(n,R) = 1$.

(3) *If* $\delta > 0$, *the inequality* $|f(p^k) - 1| > \delta$ *for some* k *is possible at most for finitely many primes.*

(4) *If* f *is in* $\mathcal{D}^u$ *and multiplicative, then* $|f(p^k)| \le \frac{1}{2} \cdot \|f\|_u^{-1}$ *for some* k *is possible for at most finitely many primes* p.

Proof. (1) If (1) is not true, then an ascending, infinite sequence $p_1, p_2, \ldots$ exists with the property $\hat{f}(h_{p_1^\infty}) = \hat{f}(_{p_2^\infty}) = \ldots = 0$. But the sequence of evaluation homomorphisms $h_{p_k^\infty}$, k =1, 2, ..., is convergent to h_1 in $\Delta_{\mathcal{B}}$. Therefore,

$$0 = \lim_{k \to \infty} \hat{f}(h_{p_k^\infty}) = \hat{f}(h_1) = h_1(f) = f(1) = 1,$$

a contradiction.

(2) is proved using the same idea: if n is not divisible by the first r primes, r sufficiently large, then $\hat{f}(h_n)$ is near $\hat{f}(h_1) = 1$.

(3) is obvious now.

(4) Let $\varepsilon > 0$. For the evaluation homomorphism H_1 the integers $j(r,H_1)$ are $\equiv 1 \bmod r$. Choose R so large that for H R-near H_1 [this means that $j(r,H) = j(r,H_1)$ for r = 1, 2, ..., R, no condition for r > R] $|\hat{f}(H) - \hat{f}(H_1)| < \varepsilon$. Assume that n is coprime with R!; then there are integers x,y, so that

$$n \cdot y = 1 + R! \cdot x,$$

and so $f(n) \cdot f(y) = f(1 + R! \cdot x)$. The boundedness of f(y) implies

$$|f(n)| \ge \|f\|_u^{-1} \cdot |f(1 + R! \cdot x)|.$$

Choosing $H = H_{1+R!\cdot x}$, then $f(1+R!\cdot x)$ is near 1 and the result is proved.

In the proof of (4), **complete** multiplicativity was used. However, a variation of the proof also applies for the general case: Let $\varepsilon > 0$. For H_1, the evaluation homomorphism at 1, the integers $j(r, H_1)$ are $\equiv 1 \bmod r$. Choose R so large, that for any H "R-near" to H_1 [this means that $j(r,H) = j(r, H_1)$ for r = 1, 2, ..., R] $|\hat{f}(H) - \hat{f}(H_1)| < \frac{1}{2}$. Assume $\gcd(p,R!) = 1$; then, for every k in $\mathbb{N}$ there are integers x, y such that

$$p^k \cdot y = 1 + R! \cdot x.$$

We may assume that $p \nmid y$ [otherwise take the solution $x' = x + p^k$, $y' = y + R!$]. Then $f(p^k) \cdot f(y) = f(1 + R! \cdot x)$, hence $|f(p^k)| \geq |f(1 + R! \cdot x)| \cdot \|f\|_u^{-1}$. Choosing $H = H_{1+R!\cdot x}$, we obtain

$$|f(1 + R! \cdot x) - 1| = |\hat{f}(H) - \hat{f}(H_1)| < \tfrac{1}{2}.$$

Therefore $|f(p^k)| > \frac{1}{2} \|f\|_u^{-1}$ for every prime $p \nmid R!$, and for every $k \in \mathbb{N}$, and the result is proved. □

Lemma 4.13. **(i)** *Assume that* f *is multiplicative and in* $\mathcal{B}^u$. *If* p *is fixed, and* $\lim_{k\to\infty} a(p^k) = a$ *exists, then the multiplicative function* $f^{\#}$, *with values* $f^{\#}(q^k) = f(q^k)$ *if* q *is a prime* $\neq p$, *and* $f^{\#}(p^k) = a(p^k) \cdot f(p^k)$, *is again in* $\mathcal{B}^u$.

(ii) *With the same assumptions, the multiplicative function* $f_{\#}$ *with values* $f_{\#}(q^k) = f(q^k)$ *if* q *is a prime* $\neq p$, *and* $f_{\#}(p^k) = 1$ *for* $k = 1, 2, \ldots$, *is in* $\mathcal{B}^u$.

Proof. (i) is clear; f is multiplied by a fibre-constant function in $\mathcal{B}^u$. (ii) Choose F in $\mathcal{B}$, F R-even, ε-near f, so that $\|f - F\|_u < \varepsilon$. If $p^r \| R$, then write $R = R' \cdot p^r$, $p \nmid R'$. The function $G\colon n \mapsto F(\gcd(n, R'))$ is even; if $n = p^{\ell} \cdot n'$, $p \nmid n'$, then $|f_{\#}(n) - G(n)| = |f(n') - F(n')| < \varepsilon$. Therefore G is near $f_{\#}$, and so $f_{\#}$ is in $\mathcal{B}^u$. □

Now we give a second proof for one direction[1)] in Theorem 4.10. Let $f \in \mathcal{B}^u$ be multiplicative. We would like to look for $g = \log \circ f$, but in order to do this some preparations are necessary. According to Lemma 4.12 the relation $|f(n) - 1| < \frac{1}{3}$ is true for all integers which are coprime with some finite set $\mathcal{E}$ of exceptional primes. Change the function f into a multiplicative function $f_{\#}$ with values $f_{\#}(p^k) = 1$ at these finitely many exceptional primes. Then $f_{\#}$ is again in $\mathcal{B}^u$, for Lemma 4.13 (ii). Now the logarithm behaves nicely in the disc $B(1, \frac{1}{2})$, and $g = \log \circ f_{\#}$ is additive, and again in $\mathcal{B}^u$ by Corollary 2.5 (3). Then Theorem 4.9 shows

$$\sum_p \sup_k |g(p^k)| \leq K,$$

---

1) $f \in \mathcal{B}^u$ implies the convergence of (4.15). The other direction is simpler.

and, using the inequalities $\frac{1}{2}\cdot|z| \le |\log(1+z)| \le \frac{3}{2}\cdot|z|$ in $|z| \le \frac{1}{2}$, this implies

$$\sum_{p \text{ not in } \mathscr{E}} \sup_k | f(p^k) - 1 | \le 2 \cdot K, \tag{4.19}$$

where p runs through non-exceptional primes. The finiteness of the other primes finally gives

$$\sum_p \sup_k | f(p^k) - 1 | < \infty.$$

IV.5. APPLICATION OF TIETZE'S EXTENSION THEOREM

Using our knowledge of $\Delta_{\mathscr{B}}$, $\Delta_{\mathscr{D}}$ and the TIETZE Extension Theorem [see, for example, HEWITT-STROMBERG [1965], or the Appendix, Theorem A.1.3], we prove the following theorem.

Theorem 5.1. *Given a sequence* (n_j) *of (pairwise distinct) integers greater than one with the property*

(5.1) *the minimal prime-divisors* $p_{min}(n_j) = p_j$ *of* n_j *tend to* ∞ *as* $j \to \infty$,

and given complex numbers a_j *converging to* $a \in \mathbb{C}$, *then there exists a function f in* $\mathscr{B}^u$ *assuming the values* a_j *at* n_j.

Proof. Condition (5.1) implies that $\lim_{j\to\infty} h_{n_j} = h_1$ in $\Delta_{\mathscr{B}}$. The subset $\mathcal{K} = \{h_1\} \cup \{h_{n_j}\}$ of $\Delta_{\mathscr{B}}$ is closed and therefore compact. Define a complex-valued function F on $\mathcal{K}$ by

$$F(h_1) = a \text{ , and } F(h_{n_j}) = a_j .$$

F is continuous on $\mathcal{K}$, and TIETZE's Extension Theorem gives the existence of a continuous function F^* on $\Delta_{\mathscr{B}}$ extending F, which is the image of some f in $\mathscr{B}^u$ under the GELFAND transform, and

$$f(n_j) = \hat{f}(h_{n_j}) = F(h_{n_j}) = a_j. \qquad \square$$

Theorem 5.2. *Given a strictly increasing sequence* n_ℓ *of positive integers and given complex numbers* a_ℓ *with the properties*

(i) $\lim_{\ell \to \infty} a_\ell = a$ *exists, and*

(ii) *the evaluation maps* $h_{n_\ell} \in \Delta_{\mathcal{D}}$ *converge to some* H *in* $\Delta_{\mathcal{D}}$,

then there exists a function $f \in \mathcal{D}^u$ *assuming the values* a_ℓ *at* n_ℓ.

Proof. The subset $\mathcal{K} = \{H\} \cup \left(\cup \{h_{n_\ell}\}\right)$ is closed and therefore compact. Define a function F on $\mathcal{K}$ by $F(H) = a$, $F\left(h_{n_\ell}\right) = a_\ell$. Then F is continuous on $\mathcal{K}$, and by TIETZE's Theorem F is extendable to a continuous function F^* on $\Delta_{\mathcal{D}}$. This function is, under the GELFAND transform, the image of some $f \in \Delta_{\mathcal{D}}$. Then $f(n_\ell) = h_{n_\ell}(f) = F^*(h_{n_\ell}) = a_\ell$. □

The definition of the topology of $\mathbb{Z}^\wedge$ immediately gives the following example.

Example. Given a strictly increasing sequence n_ℓ of non-negative integers with the property

$$\text{(5.2)} \qquad \begin{cases} \text{given } R \in \mathbb{N}, \text{ there exists an } \ell_o \in \mathbb{N} \text{ such that for every } L \geq \ell \geq \ell_o \\ \qquad\qquad n_L \equiv n_\ell \mod r \text{ for } 1 \leq r \leq R, \end{cases}$$

then the evaluation homomorphisms h_{n_ℓ} are convergent and Theorem 5.2 is applicable. For example:

(a) If $n_\ell = \ell!$, then condition (5.2) is obviously true.

(b) If $n_{\ell+1} = n_\ell \cdot u_\ell$, and $u_\ell \equiv 1 \mod r$ for $1 \leq r \leq R(\ell)$, $R(\ell) \to \infty$, then the sequence h_{n_ℓ} is convergent.

IV.6. INTEGRATION OF UNIFORMLY ALMOST-EVEN FUNCTIONS

The GELFAND transform $^\wedge\colon \mathcal{B}^u \to \mathcal{C}(\Delta_{\mathcal{B}})$, defined by $\hat{f}(h) = h(f)$, is an isometric algebra-homomorphism. The inverse map is simply the restriction map

(6.1) $$\iota\colon \mathcal{C}(\Delta_{\mathcal{B}}) \to \mathcal{B}^u\ ,\ \iota(f^*)\colon n \mapsto f^*(h_n),$$

where h_n is the evaluation homomorphism at n, and where f^* is any function in $\mathcal{C}(\Delta_{\mathcal{B}})$, the space of continuous functions on $\Delta_{\mathcal{B}}$. Equation (6.1) is clear from

$$\iota(\hat{f})(n) = \hat{f}(h_n) = h_n(f) = f(n).$$

Examples. 1) Multiplicativity reads as follows: $f \in \mathcal{B}^u$ is multiplicative if and only if $\hat{f}(h_{nm}) = \hat{f}(h_n)\cdot \hat{f}(h_m)$ if $\gcd(n,m) = 1$.

This result may be extended by continuity of $\hat{f}$: Given H, H′ in $\Delta_{\mathcal{B}}$, represented by the vectors (k_p), resp. $(k_p{}')$, and assuming $\min(k_p, k_p{}') = 0$ for each prime p (so that H, H′ are "coprime"), define the product $H\cdot H'$ as that homomorphism belonging to the vector $(\max(k_p, k_p{}'))$; then f is multiplicative if and only if $\hat{f}(H\cdot H') = \hat{f}(H)\cdot \hat{f}(H')$ for all coprime homomorphisms H, H′. Similar remarks apply to additive functions.

2) We construct the image of the RAMANUJAN sum c_{p^k} under the GELFAND map ^. Let the homomorphism H in $\Delta_{\mathcal{B}}$ be described by its vector of exponents $\{k_p\}$. Then put

(6.2) $$C(p^k,H) = \begin{cases} 0, & \text{if } k_p < k-1,\\ -p^{k-1}, & \text{if } k_p = k-1,\\ \varphi(p^k), & \text{if } k_p > k-1.\end{cases}$$

Clearly, this function $C(p^k,H)$, defined on arguments H in $\Delta_{\mathcal{B}}$, is an extension of c_{p^k}, the values $C(p^k,h_n)$ being equal to $c_{p^k}(n)$. And the function $H \mapsto C(p^k,H)$ is continuous since the sets

$$\mathcal{O} = \{\ H \in \Delta_{\mathcal{B}},\ k_p < k-1 \text{ [resp. } = 1,\ \text{resp. } > \text{]}\ \}$$

are open in $\Delta_{\mathcal{B}}$. So $C(p^k, .\,)$ is the GELFAND transform of c_{p^k}. Using the multiplicativity of the RAMANUJAN sums with respect to the index, we obtain the transforms of all RAMANUJAN sums c_r.

The mean-value $M\colon \mathcal{B}^u \to \mathbb{C}$, $f \mapsto M(f)$ is a non-negative (that is, $f \ge 0$ implies $M(f) \ge 0$) linear functional on $\mathcal{B}^u$. Due to the obvious relation $|M(f)| \le \|f\|_u$ it is continuous. The map

(6.3) $$M^{\#}\colon F \mapsto M(\iota(F)),\ M^{\#}\colon \mathcal{C}(\Delta_{\mathcal{B}}) \to \mathbb{C},$$

is nothing more than an extension of the mean-value-functional M to $\mathcal{C}(\Delta_{\mathcal{B}})$, and so

$$M^{*}: \mathcal{C}(\Delta_{\mathcal{B}}) \to \mathbb{C}$$

is a non-negative linear functional; it is continuous ($|M^{*}(F)| \le \|F\|$). Then RIESZ's Theorem (see Appendix A.3) immediately gives the following result.

Theorem 6.1. *There exists a complete and regular probability measure* μ, *defined on a* σ*-algebra* $\mathcal{A}$, *containing the Borel sets of* $\Delta_{\mathcal{B}}$, *with the property*

$$\int_{\Delta_{\mathcal{B}}} F \, d\mu = M^{*}(F) = M(\iota(F)). \tag{6.4}$$

for every $F \in \mathcal{C}(\Delta_{\mathcal{B}})$.

So the mean-value $M(f) = \lim_{x\to\infty} x^{-1}\cdot\sum_{n\le x} f(n)$ of functions f in $\mathcal{B}^{u}$ can be represented as an integral,

$$M(f) = \int_{\Delta_{\mathcal{B}}} \hat{f} \, d\mu. \tag{6.5}$$

In fact, it will be proved that μ *is a product measure.* Write

$$\Delta_{\mathcal{B}} = \prod_{p} \{ 1, p, p^{2}, \ldots, p^{\infty} \} = \prod_{p} \mathbb{N}_{p}, \tag{6.6}$$

and define probability measures μ_{p} on the factors $\mathbb{N}_{p}$ by

$$\mu_{p}(p^{k}) = p^{-k}\cdot(1-p^{-1}), \quad \mu_{p}(p^{\infty}) = 0. \tag{6.7}$$

Then μ_{p} is defined on the Borel sets $\mathcal{B}(\mathbb{N}_{p})$ of $\mathbb{N}_{p}$ (these are all subsets of $\mathbb{N}_{p}$). The product measure

$$\tilde{\mu} = \prod_{p} \mu_{p} \tag{6.8}$$

is defined on the least σ-algebra $\mathcal{S} = \prod_{p} \mathcal{B}(\mathbb{N}_{p})$ with the property that all the projections $\pi_{p}: \Delta_{\mathcal{B}} \to \mathbb{N}_{p}$ are $\mathcal{S}$-$\mathcal{B}(\mathbb{N}_{p})$-measurable (this means that $\pi_{p}^{-1}(A_{p}) \in \mathcal{S}$ for any Borel set A_{p} in $\mathcal{B}(\mathbb{N}_{p})$).

Proposition 6.2. *The product* σ*-algebra* $\mathcal{S} = \prod_{p} \mathcal{B}(\mathbb{N}_{p})$ *is equal to the* σ*-algebra of Borel sets in* $\Delta_{\mathcal{B}}$.

Proof. Both the σ-algebras mentioned in the proposition are generated by the measurable rectangles $\prod \mathcal{R}_p$, where $\mathcal{R}_p \subset \mathbb{N}_p$ and $\mathcal{R}_p = \mathbb{N}_p$ for all but a finite number of primes p. This is true for $\mathcal{S}$ by definition of the product σ-algebra; and by definition of the topology of $\Delta_{\mathcal{B}}$ it is clear that all the measurable rectangles $\prod \mathcal{R}_p$ are Borel sets, and that all these rectangles belong to the Borel sets. □

Example 3. Denote by $\mathcal{P}$ a finite set of primes, and, with each $p \in \mathcal{P}$, associate an integer (including ∞) m(p), $0 \le m(p) \le \infty$. Characterize an element h in $\Delta_{\mathcal{B}}$ by the vector $\{ k_p(h)\}_p$ of "exponents". The set

$$Y = \{ h \in \Delta_{\mathcal{B}}: k_p(h) = m(p) \text{ for each } p \text{ in } \mathcal{P} \}$$

has measure

$$\mu(Y) = \int_{\Delta_{\mathcal{B}}} \chi_Y \, d\mu = \prod_{p \in \mathcal{P}} p^{-m(p)} \cdot \ell_p,$$

where $\ell_p = 1 - p^{-1}$, if $m(p) < \infty$, $\ell_p = 1$ otherwise. The expression $p^{-\infty}$ is to be interpreted as zero.

Proof. (a) Let $m(p) < \infty$ for each p in $\mathcal{P}$. Y is open and closed, so the characteristic function χ_Y is continuous, and

$$\mu(Y) = \lim_{N \to \infty} N^{-1} \cdot \sum_{n \le N} \chi_Y(n).$$

The relation $n \in Y$ is equivalent to $p^{m(p)} \| n$ for each p in $\mathcal{P}$. Therefore, using $\sum_{d|(a,b)} \mu(d) = 1$ iff $\gcd(a,b) = 1$, and writing $\mathcal{P} = \{p_1, \dots, p_r\}$, $m(p_\rho) = m_\rho$ and $P = p_1^{m_1} \cdot \ldots \cdot p_r^{m_r}$, we obtain

$$N^{-1} \cdot \sum_{n \le N} \chi_Y(n) = \sum_{d_1 | p_1} \mu(d_1) \ldots \sum_{d_r | p_r} \mu(d_r) \cdot N^{-1} \cdot \sum_{\substack{n \le N \\ n \equiv 0 \bmod (Pd_1 \dots d_r)}} 1.$$

For $N \to \infty$, this expression tends to

$$\prod_{p \in \mathcal{P}} p^{-m(p)} \cdot (1 - p^{-1}).$$

(b) In the case where $m(p) = \infty$ for at least one p in $\mathcal{P}$, the set Y is contained in every set

$$Z_m = \{ h \in \Delta_{\mathcal{B}}, m(p) \ge m \}$$

with measure

$$\mu(Z_m) = (1-p^{-1})\cdot(p^{-m} + p^{-m-1} + \dots) = p^{-m},$$

according to case (a), and thus $\mu(Y) = 0$. □

Example 4. *The set* $\mathbb{N}$ *of positive integers* (embedded in $\Delta_{\mathcal{B}}$) $\mathbb{N} \subset \Delta_{\mathcal{B}}$ *has measure zero.*

Enumerating the primes as $p_1 < p_2 < \dots$, the measure of the set

$$Y_{r,s} = \{ x \in \Delta_{\mathcal{B}},\ m_p(x) = 0 \text{ for } p_r \le p \le p_s \}$$

is $\prod_{r\le\rho\le s} (1 - p_\rho^{-1})$ (according to example 3). Therefore,

$$Y_r = \bigcap_{s\ge r} Y_{r,s}$$

has measure $\prod_{p\ge p_r} (1 - p^{-1}) = 0$ and the assertion follows from $\mathbb{N} \subset \bigcup_r Y_r$.

Theorem 6.3. *The measure space* $(\Delta_{\mathcal{B}}, \mathcal{B}(\Delta_{\mathcal{B}}), \mu)$ *coincides with the product measure space*

$$(\prod_p \mathbb{N}_p, \mathcal{S}, \tilde{\mu}).$$

Proof. According to HEWITT-STROMBERG [1965] a product measure is determined uniquely by the values of the measure on measurable rectangles. Without loss of generality, these may be taken as $\prod_p A_p$, where $A_p = \mathbb{N}_p$ for $p \ge p_0$, and $A_p = \{p^{m(p)} ;\ m(p) \in M(p)\}$, where $M(p) \subset \{0\} \cup \mathbb{N} \cup \{\infty\}$. Then

$$\tilde{\mu} (\prod_p A_p) = \prod_{p<p_0} \mu_p(A_p) = \prod_{p<p_0} \sum_{m(p)\in M(p)} p^{-m(p)}\cdot(1-p^{-1}).$$

On the other hand, the same expression is obtained for $\mu(\prod_p A_p)$ by example 3. □

Corollary 6.4. *If* $\mathcal{P}$ *is a finite non-void set of primes, and* $f^{(p)}: \mathbb{N}_p \to \mathbb{C}$ *is* μ_p*-integrable for each prime* $p \in \mathcal{P}$, *then the function*

$$f : \Delta_{\mathcal{B}} \to \mathbb{C},\ h \mapsto \prod_{p\in\mathcal{P}} f^{(p)}(\pi_p(h))$$

is μ*-measurable, and*

$$\int_{\Delta_{\mathcal{B}}} f \, d\mu = \prod_{p\in\mathcal{P}} \int_{\mathbb{N}_p} f^{(p)} \, d\mu_p .$$

Of course, μ_p is the projection of $\Delta_{\mathcal{B}}$ to its "p-th factor" $\mathbb{N}_p$.

Example 5. The continuous extension of the RAMANUJAN sums c_r to $\Delta_{\mathcal{B}}$ was given (see (6.2)) as $c_r^{\#}: x \mapsto \prod_{p^k \| r} c_{p^k}^{\#}(\pi_p(x))$. Therefore, the mean-value is

$$M(c_r) = \int_{\Delta_{\mathcal{B}}} c_r^{\#} \, d\mu = \prod_{p^k \| r} \int_{\mathbb{N}_p} c_{p^k}^{\#} \, d\mu_p$$

$$= \prod_{p^k \| r} \left\{ -p^{k-1} \cdot (1-p^{-1}) \cdot p^{-(k-1)} + \varphi(p^k) \cdot \Sigma_{m \ge k} \, p^{-m} \cdot (1-p^{-1}) \right\}$$

$$= 0 \text{ if } r \ge 2, \text{ and } 1 \text{ if } r = 1.$$

Similarly,

$$M(|c_r|) = \begin{cases} 2^{\omega(r)} \cdot \prod_{p|r} (1-p^{-1}), & \text{if } r \ge 2, \\ 1, & \text{if } r = 1, \end{cases}$$

and

$$\int_{\mathbb{N}_p} c_{p^k} \cdot c_{p^m} d\mu_p = p^{k-1} \cdot (p-1) \text{ if } k = m, \text{ otherwise } = 0,$$

and, therefore, the orthogonality relations $M(c_r \cdot c_s) = \varphi(r)$ if $r=s$, and $= 0$ otherwise, are proved again.

The final **example 6** gives a calculation of the RAMANUJAN coefficients for functions f in $\mathcal{B}^u$ which are finite products of fibre-constant functions $f^{(p)}$,

$$f = \prod_{p \in \mathcal{P}} f^{(p)}, \text{ where } \mathcal{P} \text{ is a finite set of primes.}$$

The mean-value $M(f \cdot c_r)$ equals $\int \hat{f} \cdot c_r^{\#} \, d\mu$, and, this being a product over simpler integrals,

$$M(f \cdot c_r) = \prod_{p \in \mathcal{P}, p \nmid r} (1-p^{-1})\left(f(1) + p^{-1} \cdot f(p) + p^{-2} f(p^2) + \ldots \right)$$

$$\times \prod_{p \in \mathcal{P}, p^k \| r} (1-p^{-1}) \cdot \left\{ -f(p^{k-1}) + \varphi(p^k) \cdot \Sigma_{m \ge k} p^{-m} \cdot f(p^m) \right\}$$

if all primes dividing r are in $\mathcal{P}$, and otherwise

$$M(f \cdot c_r) = 0.$$

An **extension of the integral** to the larger class (vector-space!)

$$\mathcal{L}^q(\Delta_{\mathcal{B}}, \mathcal{B}(\Delta_{\mathcal{B}}), \mu), \text{ where } q \ge 1,$$

of measurable functions F: $\Delta_{\mathcal{B}} \to \mathbb{C}$ with the property $\int_{\Delta_{\mathcal{B}}} |F|^q \, d\mu < \infty$ is possible. Identifying functions F, G with

$$\| F - G \|_q := = \left\{ \int_{\Delta_{\mathcal{B}}} |F-G|^q \, d\mu \right\}^{1/q} = 0,$$

the well-known L^q-spaces are obtained.

L^2 is a complex HILBERT-space with inner product

$$\langle F,G \rangle = \int_{\Delta_{\mathcal{B}}} F \cdot \overline{G} \, d\mu.$$

The set of functions $\{\varphi(r)\}^{-\frac{1}{2}} \cdot c_r^{*}$, r = 1,2, ... is an orthonormal basis in L^2. This follows from the fact that the continuous functions on Δ are dense in L^2 and the linear combinations of [extensions of] RAMANUJAN sums are dense in $\mathcal{C}(\Delta_{\mathcal{B}})$.

Finally, we note that a more powerful theory of integration of arithmetical functions was developed by E. V. NOVOSELOV about 1962-1964, and the most powerful theory of integration, due to J.-L. MAUCLAIRE, is presented in his monograph [1986].

IV.7. EXERCISES

1) The pointwise product of an r-even and a t-even function is { l.c.m[r, t]}-even. Prove this and a similar result for periodic functions.

2) Let $r \in \mathbb{N}$ and f the indicator-function $1_{r\mathbb{N}}$ of the set $r \cdot \mathbb{N}$. Calculate the RAMANUJAN coefficients $a_d(f)$ and the FOURIER series of f.

3) For given $r \in \mathbb{N}$, calculate the RAMANUJAN coefficients $a_d(f)$ for the function f defined by f(n) = 1 if gcd(n,r) = 1 and f(n) = 0, if gcd(n,r) > 1.

Solution:

$$a_d(f) = \begin{cases} \frac{\mu(d)}{\varphi(d)} \frac{\varphi(r)}{r}, & \text{if } d|r, \\ 0, & \text{if } d \nmid r. \end{cases}$$

4) Prove: the quotient space $\mathcal{D}/\mathcal{B}$ is of infinite dimension. (Hint: the residue-classes $e_{1/r} + \mathcal{B}$, r = 3, 4, ... are pairwise different.)

5) The quotient space $\mathcal{A}/\mathcal{D}$ is of infinite dimension.

6) Let k be a positive integer, and f an arithmetical function. Put $f_k(n) = f(\gcd(n,k))$. Prove the equivalence of the following three properties:

(1) $f \in \mathcal{B}^u$,

(2) for every $\varepsilon > 0$ there is a k in $\mathbb{N}$ so that $\| f-f_k \|_u < \varepsilon$,

(3) the set $\{ f_k, k \in \mathbb{N} \}$ is relatively compact in the set of bounded functions with the topology induced by $\|.\|_u$.

7) Prove: the assumptions $f \in \mathcal{A}$, $\inf_{n\in\mathbb{N}} |f(n)| \geq \frac{1}{2}$, do not imply $f^{-1} \in \mathcal{A}$. (Hint: $f(n) = 1 + \frac{1}{2} e_\alpha(n)$.)

8) Let $f \in \mathcal{A}^u$ have no zeros. If $|f|^{-1} \in \mathcal{A}^u$, prove that $f^{-1} \in \mathcal{A}^u$.

9) Give a formula for the GELFAND transform $\hat{c}_r \in \Delta_{\mathcal{B}}$ of the RAMANUJAN sum $c_r \in \mathcal{B}^u$.

10) Describe a countable base for the system of neighbourhoods of the evaluation homomorphism $h_1 \in \Delta_{\mathcal{B}}$.

11) Let $\{n_j\}$ be a sequence of positive integers with the property that the least prime-divisor p_j of n_j tends to infinity. Then the evaluation homomorphisms h_{n_j} converge in $\Delta_{\mathcal{B}}$ to h_1.

12) Let $\{n_j\}$ be an increasing sequence of positive integers with the following property: for every $R \in \mathbb{N}$ there exists an $i_0 \in \mathbb{N}$ so that $n_j \equiv n_i \bmod r$ for $1 \leq r \leq R$, $j \geq i \geq i_0$. Then, in $\Delta_{\mathcal{D}}$, the evaluation homomorphisms h_{n_j} are convergent. [Example: $n_j = j!$]

13) Prove: the evaluation homomorphisms h_n, n = 1, 2, ..., are a dense subset of $\Delta_{\mathcal{D}}$.

14) Show in detail that $\Delta_{\mathcal{B}}$ is homeomorphic to

$$\mathbb{N}^* = \prod_p \{1, p, p^2, \dots, p^\infty\},$$

where each factor is the ALEXANDROFF-one-point-compactification of the discrete space $\{1, p, p^2, \dots\}$.

15) Let $\chi_{r,s}$ be the characteristic function of the residue-class s mod r. Prove: $M(\chi_{r_1,s_1}, \chi_{r_2,s_2}) = 0$ if $(r_1,r_2) \nmid |s_1 - s_2|$, and $= \{\mathrm{lcm}[r_1,r_2]\}^{-1}$ otherwise.

16) (A. HILDEBRAND).

a) Prove, for all integers q_1, q_2, N, the asymptotic formula

$$N^{-1} \cdot \sum_{n \le N} c_{q_1}(n) \cdot c_{q_2}(n) = \delta_{q_1,q_2} \cdot \varphi(q_1) + O\left(N^{-1} \cdot \sigma(q_1) \cdot \sigma(q_2)\right).$$

b) There exists a positive constant c_1 such that the inequality

$$N^{-1} \cdot \sum_{n \le N} \left| \sum_{q \le Q} a_q \cdot c_q(n) \right|^2 \le c_1 \cdot \sum_{q \le Q} |a_q|^2 \cdot \varphi(q)$$

is true for all integers N, $Q \le N^{\frac{1}{2}}$ and all complex sequences $(a_1, a_2, \dots, a_Q)$.

c) Prove, by dualizing this inequality,

$$\sum_{q \le Q} \frac{1}{\varphi(q)} \cdot \left| \hat{f}^{(N)} * \mu(q) \right|^2 \le c_1 \cdot N^{-1} \cdot \sum_{n \le N} |f(n)|^2$$

for all $Q \le N^{\frac{1}{2}}$ and complex values f(1), f(2), ..., f(N), where $\hat{f}^{(N)}(d) = N^{-1} \cdot d \cdot \sum_{n \le N, n \equiv 0 \bmod d} f(n)$.

d) Prove, for all integers N and complex numbers f(1), f(2), ..., f(N) the inequality

$$\sum_{p^k \le N} p^{-k} \left| p^k \cdot N^{-1} \cdot \sum_{n \le N, p^k | n} f(n) - N^{-1} \cdot \sum_{n \le N} f(n) \right|^2 \le c_2 \cdot N^{-1} \cdot \sum_{n \le N} |f(n)|^2.$$

This is another dualized TURÁN-KUBILIUS inequality (see I, Thm. 4.3).

<u>Hints</u>: a) Use I, Theorem 3.1 (b).

b) Estimate the error term with CAUCHY-SCHWARZ's inequality and apply WINTNER's Theorem II, Corollary 2.3.

c) Use I, Theorem 4.2.

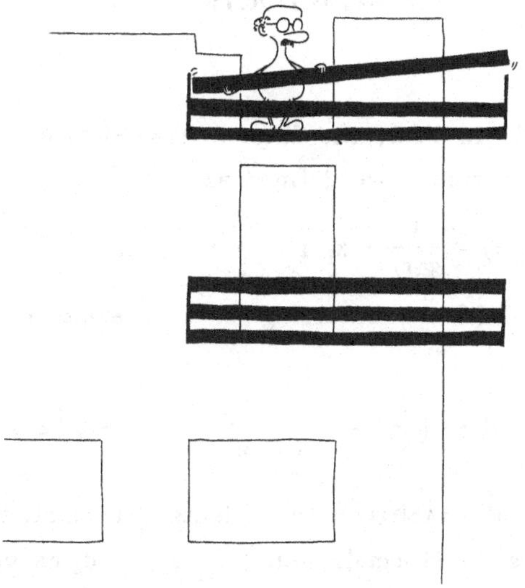

Chapter V

RAMANUJAN Expansions of Functions in $\mathcal{B}^u$

Abstract. *This chapter gives the main parts of A. HILDEBRAND's dissertation, written in Freiburg (1984), which deals with the pointwise convergence of RAMANUJAN expansions*

$$f(n) = \sum_{1\le r<\infty} a_r \cdot c_r(n)$$

of arithmetical functions $f \in \mathcal{B}^u$, *where the RAMANUJAN coefficients are the "natural ones" defined by*

$$a_r(f) = \{\varphi(r)\}^{-1} \cdot M(f \cdot c_r).$$

The assertion that the RAMANUJAN expansion for functions in $\mathcal{B}^u$ *is convergent is shown to be equivalent to some other assertions, in particular to the following, which is to be proved: the* $\|\,.\,\|_1$ *- norm of the kernel function* $S_{Q,k}(n) = \sum_{r\le Q} \{\varphi(r)\}^{-1} \cdot c_r(k) \cdot c_r(n)$ *is bounded. The proof uses estimates for incomplete sums containing the MÖBIUS function, for example for*

$$M(n,z) = \sum_{d|n,\, d\le z} \mu(d),\quad M_1(n,z) = \sum_{d|n,\, d\le z} \mu(d) \cdot \log(z/d).$$

V.1. INTRODUCTION

As mentioned already in IV.2, the RAMANUJAN-FOURIER coefficients of an arithmetical function f are defined as

(1.1) $$a_r(f) = \frac{1}{\varphi(r)} \cdot M(f \cdot c_r), \; r = 1, 2, \ldots,$$

if the mean-values $M(f \cdot c_r)$ occurring in (1.1) exist; c_r denotes the RAMANUJAN sum

$$c_r(n) = \sum_{d | \gcd(r,n)} d \cdot \mu\left(\frac{r}{d}\right) = \sum_{1 \le a \le r,\; \gcd(a,r)=1} \exp\left\{2\pi i \cdot \frac{a}{r} \cdot n\right\}.$$

In the case where the RAMANUJAN coefficients $a_r(f)$ exist, the RAMANUJAN **expansion** of f is the [formal] sum $\sum_{r=1}^{\infty} a_r(f) \cdot c_r$ associated with f:

(1.2) $$f \sim \sum_{r=1}^{\infty} a_r(f) \cdot c_r.$$

For functions in $\mathcal{B}^u$ the mean-values $M(f \cdot c_r)$ do exist, and therefore the RAMANUJAN expansion (1.2) of functions f in $\mathcal{B}^u$ exists. Of course, general theorems on the pointwise convergence of arithmetical functions having a RAMANUJAN expansion (1.2) are highly desirable. We shall return to this subject later (Chapter VIII). For bounded arithmetical functions there is *some* convergent series $f(n) = \sum_{1 \le r < \infty} b_r \cdot c_r(n)$, as was proved by J. SPILKER [1980]. A simpler proof, valid for every arithmetical function, was given by A. HILDEBRAND [1984].

Theorem 1.1. *If* f *is an arithmetical function, then there are complex coefficients* b_r, *so that, for any* n,

(1.3) $$f(n) = \sum_{1 \le r < \infty} b_r \cdot c_r(n).$$

We reproduce HILDEBRAND's **proof**. Put $r^* = 1$ if $r = 1$, and $r^* = r \cdot \prod_{p|r} p$ if $r > 1$; so r^* contains every prime factor to at least the second power if $r > 1$. The formulae for the values of the RAMANUJAN sum c_r, where the index r is a prime power (see I, (3.4)) imply that $c_{r^*}(n) \neq 0$ is possible at most for indices r dividing n. So the sum $\sum b_{r^*} \cdot c_{r^*}(n)$ is a finite sum $\sum_{r \le n} b_{r^*} \cdot c_{r^*}(n)$. We try to choose the coefficients b_{r^*} in such a manner that

(1.4) $$f(n) = \sum_{r|n, r\le n} b_{r^*} \cdot c_{r^*}(n)$$

for every positive integer. This is possible since the system (1.4) of linear equations can be solved recursively by

$$b_{1^*} = f(1),$$
$$b_{n^*} = \{c_{n^*}(n)\}^{-1} \cdot \{f(n) - \textstyle\sum_{r|n, r<n} b_{r^*} \cdot c_{r^*}(n)\}, \quad n > 1.$$

Here,

$$|c_{n^*}(n)| = n \neq 0,$$

and thus the system (1.4) is solvable. □

Theorem 1.1 is not very interesting, because the coefficients b_r are not the "natural ones". Convergence of the RAMANUJAN expansion (1.2) for a large class of functions was proved by A. HILDEBRAND [1984].

Theorem 1.2. *If* f *is an arithmetical function in* $\mathcal{B}^u$, *then the* RAMANUJAN *expansion*

(1.5) $$\sum_{1\le r<\infty} a_r \cdot c_r(k) = f(k)$$

is pointwise convergent for k = 1, 2, ...

Closely related to this result are the following three theorems.

Theorem 1.3. *If* f *is an arbitrary arithmetical function, for which* $\|f\|_u$ *is bounded and for which all the coefficients* $a_r(f)$ *exist, then for every* k

(1.6) $$\sup_{Q\ge 1} \Big| \sum_{r\le Q} a_r(f) \cdot c_r(k) \Big| \le c(k) \cdot \|f\|_u.$$

Theorem 1.4. *For any finite sequence* b_r, $r \le Q_0$, *of complex numbers the estimate*

(1.7) $$\max_{Q\le Q_0} \Big| \sum_{r\le Q} b_r \cdot c_r(k) \Big| \le c(k) \cdot \max_n \Big| \sum_{r\le Q_0} b_r \cdot c_r(n) \Big|$$

is true.

We define the *kernel function* $S_{Q,k}(n)$ by

(1.8) $$S_{Q,k}(n) = \textstyle\sum_{r\le Q} \{\varphi(r)\}^{-1} \cdot c_r(k) \cdot c_r(n).$$

Then the partial sums of the RAMANUJAN expansion may be expressed as

(1.9) $$\sum_{r\le Q} a_r(f)\cdot c_r(k) = M\,(\,f\cdot S_{Q,k}\,).$$

Theorem 1.5. *There is some constant* c(k), *depending on* k, *such that the estimate*

(1.10) $$\|\,S_{Q,k}\,\|_1 \le c(k)$$

holds for every $Q \ge 1$, *where* $\|f\|_1 = \limsup_{x\to\infty} x^{-1}\cdot\sum_{n\le x} |f(n)|$.

V.2. EQUIVALENCE OF THEOREMS 1.2, 1.3, 1.4, 1.5

It is not difficult to see that the theorems given in section V.1 (with the exception of Theorem 1.1) are equivalent, and it is not too difficult to see this. We prove this equivalence as follows:

$$\boxed{5}\Rightarrow\boxed{3}\Rightarrow\boxed{4}\Rightarrow\boxed{5}\,,\ \boxed{3}\Rightarrow\boxed{2}\Rightarrow\boxed{4}\,.$$

The implication $\boxed{5}\Rightarrow\boxed{3}$ is obvious, using the estimate $\left|\sum_{n\le x} f(n)\cdot g(n)\right| \le \|f\|_u\cdot\sum_{n\le x}|g(n)|$:

$$\left|\sum_{r\le Q} a_r(f)\cdot c_r(k)\right| = \left|\sum_{r\le Q} M(f\cdot c_r)\cdot\{\varphi(r)\}^{-1}\cdot c_r(k)\right|$$

$$= \left|M(f\cdot S_{Q,k})\right| \le \|\,S_{Q,k}\,\|_1\cdot\|f\|_u. \qquad \square$$

$\boxed{3}\Rightarrow\boxed{4}$. Put $g(k) = \sum_{r\le Q_0} b_r\cdot c_r(k)$; then the RAMANUJAN coefficients of g are $a_r(g) = b_r$, and so we obtain for $Q \le Q_0$

$$\left|\sum_{r\le Q} b_r\cdot c_r(k)\right| = \left|\sum_{r\le Q} a_r(g)\cdot c_r(k)\right| \le c(k)\cdot\|g\|_u,$$

according to $\boxed{3}$, and

$$\|g\|_u = \max_n \left|\sum_{r\le Q_0} b_r\cdot c_r(n)\right|. \qquad \square$$

[4] ⇒ [5] . The sign-function[1] $f_{Q,k}(n) = \mathrm{sign}\big(S_{Q,k}(n)\big)$ is even modulo (Q!); so, putting R = Q!, it has the expansion

$$f_{Q,k} = \sum_{r\le R} d_r \cdot c_r, \text{ where } d_r = a_r(\, f_{Q,k}\,).$$

With the definition (1.8) of the kernel function $S_{Q,k}$ we obtain

$$\| S_{Q,k} \|_1 = M(|S_{Q,k}|) = M(\, f_{Q,k} \cdot S_{Q,k}\,)$$

$$= \sum_{r\le Q} \{\varphi(r)\}^{-1} \cdot c_r(k) \cdot M(\, f_{Q,k} \cdot c_r\,)$$

$$= \sum_{r\le Q} a_r(\, f_{Q,k}) \cdot c_r(k),$$

and, using [4], this is

$$\le c(k) \cdot \max_n |\, f_{Q,k}(n)\,| \le c(k). \qquad \square$$

[3] ⇒ [2] . Given a function f in $\mathcal{B}^u$, choose approximating functions f_n from $\mathcal{B}$ with the property $\|f-f_n\|_u \to 0$. From [3] we obtain for every k

$$\sup_{Q\ge 1} \Big|\sum_{r\le Q} a_r(f-f_n)\cdot c_r(k)\Big| \le c(k) \cdot \|\, f-f_n\,\|_u.$$

If Q is sufficiently large, then

$$\sum_{r\le Q} a_r(f)\cdot c_r(k) = \sum_{r\le Q} a_r(f-f_n)\cdot c_r(k) + \sum_{r\le Q} a_r(f_n)\cdot c_r(k)$$

$$= \sum_{r\le Q} a_r(f-f_n)\cdot c_r(k) + f_n(k),$$

and thus

$$\Big|f(k) - \sum_{r\le Q} a_r(f)\cdot c_r(k)\Big| \le |\, f(k) - f_n(k)\,| + \Big|\sum_{r\le Q} a_r(f-f_n)\cdot c_r(k)\Big|$$

$$\le (1 + c(k))\cdot \|\, f-f_n \|_u. \qquad \square$$

[2] ⇒ [4] . This implication will be proved by contradiction. Assume that [4] is false, so that

$$\exists\, k\ \forall\, c(k)\ \exists\, Q_0\ \exists\, \{b_r\} : \max_{Q\le Q_0} \Big|\sum_{r\le Q} b_r\cdot c_r(k)\Big| > c(k)\cdot \max_n \Big|\sum_{r\le Q_0} b_r\cdot c_r(n)\Big|.$$

---

[1] $\mathrm{sign}(x) = 1$ if $x > 0$, $\mathrm{sign}(x) = -1$ if $x < 0$, $\mathrm{sign}(0) = 0$.

Therefore, we obtain the existence of

- integers $Q_n \le Q_{n+1} \le Q_{n+1}$, $Q_n \nearrow \infty$,
- integers $M_n \nearrow \infty$, so that $\sum_{n>N} \{M_n\}^{-\frac{1}{2}} \le Q_N^{-2}$,
- even functions $f_n = \sum_{r \le Q_n'} a_r(f_n) \cdot c_r$, satisfying $\|f_n\|_u = 1$, with "large" partial sums at the point k,

$$\left| \sum_{r \le Q_n'} a_r(f_n) \cdot c_r(k) \right| \ge M_n .$$

Define a sequence of even functions,

$$F_N = \sum_{n \le N} \{M_n\}^{-\frac{1}{2}} \cdot f_n .$$

For $\|f_n\|_u = 1$ and $\sum \{M_n\}^{-\frac{1}{2}} < \infty$, this sequence is a $\|.\|_u$-CAUCHY-sequence with limit F in $\mathcal{B}^u$. Then $\|F - F_N\|_u \le \sum_{n>N} \{M_n\}^{-\frac{1}{2}} \le Q_N^{-2}$. Our goal is to show that the RAMANUJAN expansion of F is divergent at the point k. The RAMANUJAN coefficients of $F = \lim_{N \to \infty} F_N$ are

$$a_r(F) = a_r(F-F_N) + a_r(F_N) = a_r(F-F_N) + \sum_{n \le N} \{M_n\}^{-\frac{1}{2}} \cdot a_r(f_n),$$

and so, isolating the single summand with n = N, we obtain

$$\left| \sum_{r \le Q_N'} a_r(F) \cdot c_r(k) \right| \ge \{M_N\}^{-\frac{1}{2}} \cdot \left| \sum_{r \le Q_N'} a_r(f_N) \cdot c_r(k) \right|$$

$$- \sum_{n<N} \{M_n\}^{-\frac{1}{2}} \cdot \left| \sum_{r \le Q_N'} a_r(f_n) \cdot c_r(k) \right| - \left| \sum_{r \le Q_N'} a_r(F-F_N) \cdot c_r(k) \right|$$

$$\ge \{M_N\}^{-\frac{1}{2}} \cdot \left| \sum_{r \le Q_N'} a_r(f_N) \cdot c_r(k) \right| - \sum_{n \le N} \{M_n\}^{-\frac{1}{2}} \cdot |f_n(k)| - \|F-F_N\|_u \cdot \sum_{r \le Q_N'} |c_r(k)| .$$

Using the fact that partial sums $\sum_{r \le Q_N'} a_r(f_N) \cdot c_r(k)$ are large and that $|c_r(k)| \le \varphi(r) \le r$, this is

$$\ge \{M_N\}^{\frac{1}{2}} - \sum_{n \le N} \{M_n\}^{-\frac{1}{2}} \cdot 1 - \|F-F_N\|_u \cdot (Q_N')^2$$

$$\ge \{M_N\}^{\frac{1}{2}} - \sum_{n \le N} \{M_n\}^{-\frac{1}{2}} - (Q_N')^2 \cdot \sum_{n>N} \{M_n\}^{-\frac{1}{2}}$$

$$\ge \{M_N\}^{\frac{1}{2}} - \mathcal{O}(1) \quad \to \infty,$$

as $N \to \infty$, by choice of the integers M_N. □

V.3. SOME LEMMATA

In order to prove Theorem 1.5, claiming that the $\|.\|_1$-norm of the kernel-function $S_{Q,k}$ is bounded [by a constant depending at most on k], some lemmas are necessary.

First of all, it is clear [see II, Theorem 3.1] that for a non-negative multiplicative function f, satisfying

$$0 \le f(p^k) \le \gamma_1 \cdot \gamma_2{}^k, \text{ where } 0 < \gamma_2 < 2,$$

the sum $\sum_{n\le x} f(n)$ can be estimated by

$$\sum_{n\le x} f(n) \le c_1(\gamma_1,\gamma_2) \cdot x\cdot(\log x)^{-1} \cdot \exp\Big(\sum_{p\le x} p^{-1}\cdot f(p) \Big). \tag{3.1}$$

Lemma 3.1. *Uniformly in* $x \ge 1$ *and* $k \in \mathbb{N}$ *the asymptotic formula*

$$\sum_{n\le x, \gcd(n,k)=1} \frac{\mu^2(n)}{\varphi(n)} = \frac{\varphi(k)}{k} \cdot \Big\{ \log x + C + h(k) \Big\} + O\Big(\frac{\psi(k)}{\sqrt{x}}\Big)$$

holds, where $\mathscr{C}$ *is* EULER*'s constant,*

$$C = \mathscr{C} + \sum_p \{p(p-1)\}^{-1} \cdot \log(p),$$

$$h(k) = \sum_{p|k} p^{-1}\cdot\log(p) \text{ is strongly additive,}$$

$$\psi(k) = \sum_{d|k} d^{-\frac{1}{2}} \cdot \mu^2(d) \text{ is multiplicative.}$$

We remark that in H.-E. RICHERT & H. HALBERSTAM [1974] the lower estimate $S_k(x) \ge k^{-1}\cdot\varphi(k) \cdot \log x$ is given; this estimate is rather easily accessible.

Proof of Lemma 3.1. Put $f_k(n) = n \cdot \dfrac{\mu^2(n)}{\varphi(n)}$ if $\gcd(n,k) = 1$, and $f_k(n) = 0$ otherwise. Write $f_k = 1 * g_k$. From the Relationship Theorem (see Chapter III) or, simpler, directly from the values

$$g_k(p^m) = \begin{cases} (p-1)^{-1}, & \text{if } m = 1,\ p \nmid k, \\ -(p-1)\cdot p^{-1}, & \text{if } m = 2,\ p \nmid k, \\ -1, & \text{if } m = 1,\ p|k, \\ 0, & \text{if } m > 2 \text{ or } (m = 2 \text{ and } p|k), \end{cases}$$

we obtain $\sum_{n=1}^{\infty} n^{-1}\cdot|g_k(n)| < \infty$, and $\sum_{n=1}^{\infty} n^{-1}\cdot g_k(n) = k^{-1}\cdot\varphi(k)$. Therefore,

$$S_k(x) = \sum_{n\le x,\gcd(n,k)=1} \frac{\mu^2(n)}{\varphi(n)} = \sum_{n\le x} n^{-1}\cdot f_k(n)$$

$$= \sum_{n\le x} n^{-1}\cdot g_k(n)\cdot \sum_{m\le x/n} m^{-1}$$

$$= \sum_{n\le x} n^{-1}\cdot g_k(n)\cdot \left\{\log(x/n) + \mathscr{C} + \mathcal{O}(n/x)\right\}$$

$$= \Sigma_1 + \Sigma_2 + \Sigma_3, \text{ say.}$$

Turning now to the estimate

$$\sum_{n\le u} \{\varphi(n)\}^{-1}\cdot n = \mathcal{O}(u),$$

which follows from (3.1) (or directly by elementary considerations) and to the fact that $g_k(n) \neq 0$ is possible only if $n = n_1\cdot n_2\cdot n_3^2$, where n_j is squarefree, $n_1|k$, $\gcd(n_2n_3,k) = 1$, in which case the formula

$$g_k(n_1n_2n_3^2) = \mu(n_1)\cdot\mu(n_3)\cdot n_3\cdot\{\varphi(n_2)\cdot\varphi(n_3)\}^{-1},$$

holds, we obtain the estimate

$$\sum_{n\le t} |g_k(n)| \le \sum_{n_1|k} \mu^2(n_1)\cdot \sum_{n_2\le t/n_1} \{\varphi(n_2)\}^{-1}\cdot \sum_{n_3\le\sqrt{(t/n_1n_2)}} \{\varphi(n_3)\}^{-1}\cdot n_3$$

$$\ll \sum_{n_1|k} \mu^2(n_1)\cdot \sum_{n_2\le t/n_1} \{\varphi(n_2)\}^{-1}\cdot(t/n_1n_2)^{\frac12}$$

$$\ll t^{\frac12}\cdot\sum_{n_1|k} \mu^2(n_1)\cdot n_1^{-\frac12} = t^{\frac12}\cdot\psi(k).$$

Partial summation (see I.1) gives

$$\sum_{n>u} n^{-1}\,|g_k(n)| \ll \psi(k)\cdot u^{-\frac12},$$

and thus

$$\Sigma_3 \ll x^{-1}\cdot\sum_{n\le x} |g_k(n)| \ll x^{-\frac12}\cdot\psi(k),$$

and

$$\sum_{n\le x} \frac{g_k(n)}{n} = k^{-1}\cdot\varphi(k) + \mathcal{O}(x^{-\frac12}\cdot\psi(k)).$$

Finally, using partial summation,

$$\sum_{n\le x} \frac{g_k(n)}{n} \cdot \log(x/n) = \int_1^x u^{-1} \cdot \sum_{n\le u} \frac{g_k(n)}{n}\, du$$

$$= \sum_{n=1}^{\infty} n^{-1} \cdot g_k(n) \cdot \log x - \int_1^x u^{-1} \cdot \sum_{n>u} n^{-1} \cdot g_k(n)\, du$$

$$= k^{-1} \cdot \varphi(k) \cdot \log x - \int_1^{\infty} u^{-1} \cdot \sum_{n>u} n^{-1} \cdot g_k(n)\, du + \mathcal{O}\big(x^{-\frac{1}{2}} \cdot \psi(k)\big).$$

The integral from 1 to infinity is equal to $\sum_{n=1}^{\infty} n^{-1} \cdot g_k(n) \cdot \log(n)$, which can be evaluated in the usual manner, replacing $\log n$ by $\sum_{p^k \| n} \log(p^k)$ and inverting the order of summation. This calculation is a little laborious and is left as Exercise 3. The result is

$$\int_1^{\infty} u^{-1} \cdot \sum_{n>u} n^{-1} \cdot g_k(n) du = - \frac{\varphi(k)}{k} \cdot \Big\{ \sum_p \{p(p-1)\}^{-1} \cdot \log(p) + \sum_{p|k} p^{-1} \cdot \log(p) \Big\}.$$

This formula concludes the proof of Lemma 3.1. □

The proof of Theorem 1.5 rests on estimates of the following incomplete sums over the MÖBIUS function:

(3.2) $$M(n,z) = \sum_{d|n, d\le z} \mu(d)$$

and

(3.3) $$M_1(n,z) = \sum_{d|n, d\le z} \mu(d) \cdot \log(z/d) = \int_1^z u^{-1} \cdot M(n,u) du.$$

Lemma 3.2. *Uniformly in* $z \ge 1$

$$\| n \mapsto M_1(n,z) \|_1 \ll 1.$$

Lemma 3.3. *Uniformly in* $z \ge x \ge 1$

$$\lim_{N \to \infty} N^{-1} \cdot \sum_{n \le N, p_{\min}(n) > x} | M(n,z) | \ll \frac{1}{\log(2x)} \cdot \Big(\frac{\log(2x)}{\log(2z)} \Big)^{\frac{1}{2}\delta},$$

where $p_{\min}(n)$ *is the least prime factor of* n, $p_{\min}(1) = \infty$, *and where*

(3.4) $$\delta = 1 - \log(e \cdot \log 2)/\log(2) = 0.0860713\ldots .$$

The more difficult result is the second one; Lemma 3.2 can be deduced from Lemma 3.3 in the following way. First, for $p^k \| n$ and $u \ge 1$, there is an identity

$$M(n,u) = M(p^{-k}\cdot n,u) - M(p^{-k}\cdot n,p^{-1}\cdot u), \tag{3.5}$$

and so

$$M_1(n,z) = \int_{z/p}^{z} u^{-1} \cdot M(p^{-k}\cdot n,u)\, du. \tag{3.6}$$

For the proof of Lemma 3.2 we have to estimate the sum $\sum_{n\le N} |M_1(n,z)|$. We split this sum $\sum_{n\le N} |M_1(n,z)|$ according to the condition $p_{min}(n) > z$ [resp. $\le z$] and use (3.6) in the second sum $\left(\text{with } p = p_{min}(n), (p_{min}(n))^k \| n,\ n' = n/(p_{min}(n))^k\right)$:

$$N^{-1} \cdot \sum_{n\le N} |M_1(n,z)| \le N^{-1} \cdot \sum_{n\le N, p_{min}(n)>z} |M_1(n,z)| + N^{-1} \cdot \sum_{n\le N, p_{min}(n)\le z} \int_{z/p}^{z} u^{-1} \cdot |M(n',u)|\, du.$$

Ordering according to $p = p_{min}(n) \le z$, we obtain, after replacing n' by n,

$$\left\{\begin{aligned} N^{-1} \cdot \sum_{n\le N} |M_1(n,z)| &\le N^{-1} \cdot \sum_{n\le N, p_{min}(n)>z} |M_1(n,z)| \\ &+ \sum_{p\le z} \sum_{k\ge 1} p^{-k} \cdot \int_{z/p}^{z} u^{-1} \cdot p^k \cdot N^{-1} \cdot \sum_{n\le N/p^k, p_{min}(n)>p} |M(n,u)|\, du. \end{aligned}\right. \tag{3.7}$$

In the first sum, according to the condition $p_{min}(n) > z$ and the definition of $M_1(n,z)$ there is only one divisor d of n with $d \le z$, namely $d = 1$, so in this sum $M_1(n,z) = \log z$. The sum $\sum_{n\le N, p_{min}(n)>z} 1$ equals $\sum_{n\le N, \gcd(n,k)=1} 1$, where $k = \prod_{p\le z} p$, and this sum is

$$\sum_{n\le N, \gcd(n,k)=1} 1 = \sum_{d|k} \mu(d) \cdot \left(\frac{N}{d} + \Theta(d)\right) = N \cdot \frac{\varphi(k)}{k} + R, \tag{3.8}$$

where $|\Theta(d)| \le 1$, and $|R| \le \tau(k)$. So, for $N \to \infty$, the first sum on the right- hand side of (3.7) approaches

$$\lim_{N\to\infty} N^{-1} \cdot \sum_{n\le N, p_{min}(n)>z} \log z = \prod_{p\le z} (1-p^{-1}) \cdot \log z \ll 1.$$

Using Lemma 3.3 in the second sum, we obtain Lemma 3.2 after a short calculation. □

The sum appearing in Lemma 3.3 is estimated by the CAUCHY-SCHWARZ inequality:

$$\left[\sum_{\substack{n \le N \\ p_{min}(n) > x}} |M(n,z)| \right]^2 \le \sum_{\substack{n \le N \\ p_{min}(n) > x}} |M(n,z)|^2 \cdot \sum_{\substack{n \le N,\ M(n,z) \ne 0 \\ p_{min}(n) > x}} 1.$$

So, for the proof of Lemma 3.3 it is sufficient to deduce the following two lemmas:

Lemma 3.4. *Uniformly in* $z \ge x \ge 1$

$$\lim_{N \to \infty} N^{-1} \cdot \sum_{\substack{n \le N \\ p_{min}(n) > x}} |M(n,z)|^2 \ll \{ \log 2x \}^{-1}.$$

Lemma 3.5. *Uniformly in* $z \ge x \ge 1$

$$\lim_{N \to \infty} N^{-1} \cdot \sum_{\substack{n \le N,\ M(n,z) \ne 0 \\ p_{min}(n) > x}} 1 \ll \{ \log 2x \}^{-1} \cdot \left(\frac{\log 2x}{\log 2z} \right)^{\delta},$$

where δ *was defined in Lemma 3.3.*

The proof of these two lemmas is given in section 5.

V. 4. PROOF OF THEOREM 1.5

In order to prove Theorem 1.5 (and thus the other theorems of this chapter) first we have to transform the sum defining the kernel $S_{Q,k}(n)$, using $c_r(k) = \sum_{k' | \gcd(r,k)} k' \cdot \mu(r/_{k'})$. We obtain

$$S_{Q,k}(n) = \sum_{k'|k} k' \cdot \sum_{r \le Q, k'|r} \{\varphi(r)\}^{-1} \cdot c_r(n) \cdot \mu(r/k')$$

$$= \sum_{k'|k} k' \cdot \sum_{r \le Q/k'} \{\varphi(rk')\}^{-1} \cdot c_{rk'}(n) \cdot \mu(r).$$

If r is squarefree, then factorize $r = r' \cdot r''$, where $r' | k'$ and $\gcd(r'', k') = 1$, in order to obtain

$$S_{Q,k}(n) = \sum_{k'|k} k' \sum_{\substack{r' \le Q/k' \\ r'|k'}} \frac{c_{r'k'}(n)\cdot\mu(r')}{\varphi(r'k')} \sum_{\substack{r'' \le Q/k'r' \\ \gcd(r'',k')=1}} \frac{\mu(r'')\cdot c_{r''}(n)}{\varphi(r'')}.$$

Using the inequality $|c_r(n)| \le \varphi(r)$ and the abbreviation

(4.1) $$T_{z,k'}(n) = \sum_{r\le z,\gcd(r,k')=1} \{\varphi(r)\}^{-1} \cdot \mu(r)\cdot c_r(n),$$

where $z > 0$, the estimate

$$|S_{Q,k}(n)| \le \sum_{k'|k} k'\cdot \sum_{r'|k'} |T_{Q/k'r',k'}(n)|$$

follows, and thus

(4.2) $$\|S_{Q,k}\|_1 \le \sum_{k'|k} k'\cdot \sum_{r'|k'} \|T_{Q/k'r',k'}\|_1.$$

For (4.2) it suffices to show the estimate

(4.3) $$\sup_{z\ge1} \|T_{z,k}\|_1 < \infty .$$

Replacing $c_r(n)$ by the usual sum over divisors of gcd(r,n) and inverting the order of summation, we begin with

$$T_{z,k}(n) = \sum_{d|n} d \cdot \sum_{r\le z,d|r,\gcd(r,k)=1} \{\varphi(r)\}^{-1}\cdot\mu(r)\cdot\mu(r/d)$$

$$= \sum_{\substack{d|n,d\le z \\ \gcd(d,k)=1}} \{\varphi(d)\}^{-1}\cdot\mu(d)\cdot d \cdot \sum_{\substack{r'\le z/d \\ \gcd(r',dk)=1}} \{\varphi(r')\}^{-1}\cdot\mu^2(r').$$

The inner sum is known from Lemma 3.1. Inserting the result, we arrive at

$$T_{z,k}(n) = \frac{\varphi(k)}{k}\cdot F^{(1)}_{z,k}(n) + \frac{\varphi(k)}{k}\cdot\big(C+h(k)\big)\cdot F^{(2)}_{z,k}(n) + \frac{\varphi(k)}{k}\cdot F^{(3)}_{z,k}(n) + O\left(\psi(k)\cdot F^{(4)}_{z,k}(n)\right),$$

with the abbreviations

$$F^{(1)}_{z,k}(n) = \sum_{d|n,d\le z,(d,k)=1} \mu(d)\cdot\log(z/d),$$

$$F^{(2)}_{z,k}(n) = \sum_{d|n,d\le z,(d,k)=1} \mu(d),$$

$$F^{(3)}_{z,k}(n) = \sum_{d|n,d\le z,(d,k)=1} \mu(d)\cdot h(d),$$

and

$$F^{(4)}_{z,k}(n) = z^{-\frac{1}{2}} \cdot \sum_{d|n, d\le z, (d,k)=1} \{\varphi(d)\}^{-1} \cdot d^{3/2}\psi(d).$$

Thus, in order to prove Theorem 1.5, it is sufficient to prove

$$\sup_{z\ge 1} \| F^{(i)}_{z,k} \|_1 < \infty \ , \ i = 1,\ 2,\ 3,\ 4.$$

The treatment of the [semi-] norm $\| F^{(4)}_{z,k}\|_1$ is easy: we estimate

$$N^{-1} \cdot \sum_{n\le N} |F^{(4)}_{z,k}(n)| = N^{-1} \cdot z^{-\frac{1}{2}} \cdot \sum_{\substack{d\le z \\ (d,k)=1}} \{\varphi(d)\}^{-1} \cdot d^{3/2}\, \psi(d) \cdot \sum_{n\le N/d} 1$$

$$\le \ z^{-\frac{1}{2}} \cdot \sum_{d\le z} \{\varphi(d)\}^{-1} \cdot d^{\frac{1}{2}}\, \psi(d).$$

Partial summation (beginning with $\sum_{n\le N} \{\varphi(n)\}^{-1} \cdot n \cdot \psi(n) \ll N$) gives the estimate $\mathcal{O}(1)$ for the last expression, uniformly in z.

Next we show that, *without loss of generality, one may assume that* $k = 1$.

For the remainder of section 4, we write $\mu_k = \mu \cdot \chi_k$ **for short**, where χ_k is the characteristic function of the set of integers which are coprime with k. μ_k is multiplicative, and has a representation as a convolution $\mu_k = \mu * h_k$, where h_k is multiplicative, and $h_k(p^m) = 1$ if $p|k$, and zero otherwise. The series $\sum_{n=1}^{\infty} n^{-1} \cdot h_k(n)$ equals $\{\varphi(k)\}^{-1} \cdot k$. Using this notation,

$$F^{(2)}_{z,k}(n) = \sum_{d|n, d\le z} \mu_k(d) = \sum_{d|n, d\le z} \sum_{d'\cdot d''=d} \mu(d') \cdot h_k(d'')$$

$$= \sum_{d''|n, d''\le z} h_k(d'') \cdot \sum_{d'|(n/d''), d'\le z/d''} \mu(d')$$

$$= \sum_{d''|n, d''\le z} h_k(d'') \cdot F^{(2)}_{z/d'',1}(n/d''),$$

whence

$$\|F^{(2)}_{z,k}\|_1 \le \sum_{d\le z} d^{-1} \cdot h_k(d) \cdot \| F^{(2)}_{z/d,1}\|_1 \le \{\varphi(k)\}^{-1} \cdot k \cdot \sup_{w\ge 1} \| F^{(2)}_{w,1}\|_1.$$

Similarly, a corresponding result is true when the upper index (2) is replaced by (1). For the upper index (3), a careful calculation gives

$$F^{(3)}_{z,k}(n) = \sum_{d|n, d\le z, (d,k)=1} \mu(d) \cdot \sum_{p|d} p^{-1} \cdot \log(p)$$

$$= \sum_{p|n, p\le z, p\nmid k} p^{-1}\cdot\log(p) \cdot \sum_{d|n, d\le z, d\equiv 0\,(p), (d,k)=1} \mu(d)$$

$$= - \sum_{p|n, p\le z, p\nmid k} p^{-1}\cdot\log(p) \cdot \sum_{d|n/p, d\le z/p, (d,kp)=1} \mu(d)$$

$$= - \sum_{p|n, p\le z, p\nmid k} p^{-1}\cdot\log(p) \cdot F^{(2)}_{z/p,kp}(n/p),$$

and a short calculation gives

$$\| F^{(3)}_{z,k} \|_1 \le \sum_{p\le z, p\nmid k} p^{-2}\cdot\log(p) \cdot \| F^{(2)}_{z/p,kp} \|_1$$

$$\le \frac{k}{\varphi(k)} \cdot \sum_p \frac{\log(p)}{p\cdot(p-1)} \cdot \sup_{w\ge 1} \|F^{(2)}_{w,1}\|_1.$$

So, finally, the assertion of Theorem 1.5 is reduced to the problem of a uniform estimation for the following "incomplete" sums over the MÖBIUS function:

$$M(n,z) = \sum_{d|n, d\le z} \mu(d) \left[= F^{(2)}_{z,1}(n) \right],$$

and

$$M_1(n,z) = \sum_{d|n, d\le z} \mu(d) \cdot \log(z/d) = \int_1^z u^{-1}\cdot M(n,u)du \left[= F^{(1)}_{z,1}(n) \right].$$

Thus Theorem 5 follows from Lemma 3.2 and Lemma 3.3 (see section 3). □

V.5. PROOF OF LEMMAS 3.4 AND 3.5

The proof of Theorem 1.5 will be finished as soon as we have proved Lemmas 3.4 and 3.5. For this purpose we need the following result on the MÖBIUS function:

Uniformly in $x \ge 1$, $t \ge 1$, *and* $d \in \mathbb{N}$ *the estimate*

$$\text{(5.1)} \qquad \sum^{\#}_{n\le t, (n,d)=1} n^{-1}\cdot\mu(n) \ll \psi(d) \cdot \min\left(1, \frac{\log x}{\log 2t}\right)$$

holds true, where $^{\#}$ *stands for the condition* $p_{min}(n) > x$, *and where*

(5.2) $$\psi(d) = \sum_{d|k} d^{-\frac{1}{2}} \cdot \mu^2(d) = \prod_{p|k} (1 + p^{-\frac{1}{2}}).$$

Using the notation $P(x) = \prod_{p \le x} p$, the absolute value of the sum

$$\sum^{\#}_{n \le t, (n,d)=1} n^{-1} \cdot \mu(n)$$

is equal to

(5.3) $$\begin{cases} \left|\sum_{n \le t, (n,d)=1} n^{-1} \cdot \mu(n) \cdot \sum_{m | \gcd(n, P(x))} \mu(m)\right| \\ \qquad \le \sum_{m | P(x), m \le t} m^{-1} \cdot \mu^2(m) \cdot \left|\sum_{n \le t/m, (n,dm)=1} n^{-1} \cdot \mu(n)\right|. \end{cases}$$

Using the estimate $\sum_{n \le u} n^{-1} \cdot \mu(n) \ll \log^{-2}(2u)$ in $u \ge 1$, which is a little stronger than the prime number theorem, the inner sum in (5.3), with slightly changed notation ($u = t/m$, $dm = k$), is equal to

$$\left|\sum_{n \le u, (n,k)=1} n^{-1} \cdot \mu(n)\right| = \left|\sum_{d \le u, d | k^\infty} d^{-1} \cdot \sum_{n \le u/d} n^{-1} \cdot \mu(n)\right|$$
$$\ll \sum_{d \le u, d | k^\infty} d^{-1} \cdot \log^{-2}(2u/d);$$

the notation $d | k^\infty$ means: any prime divisor of d is a prime divisor of k. Splitting the sum $\sum_{d \le u, d | k^\infty}$ into $\sum_{d \le \sqrt{u}}$ and $\sum_{d > \sqrt{u}}$, the last expression is

$$\ll \log^{-2}(2u) \cdot \sum_{d | k^\infty} d^{-1} + \sum_{d | k^\infty} d^{-1} \cdot \left(d/\sqrt{u} \right)^{\frac{1}{3}} \ll \psi(k) \cdot \log^{-2}(2u),$$

since

$$\sum_{d | k^\infty} d^{-1} = \prod_{p | k} (1 + (p-1)^{-1}) \le 2 \cdot \prod_{p | k} (1 + p^{-\frac{1}{2}}).$$

and

$$\sum_{d | k^\infty} d^{-\frac{2}{3}} = \prod_{p | k} (1 + (p^{\frac{2}{3}}-1)^{-1}) \le 2 \cdot \prod_{p | k} (1 + p^{-\frac{1}{2}}).$$

Therefore,

$$\left|\sum_{n \le u, (n,k)=1} n^{-1} \cdot \mu(n)\right| \ll \psi(k) \cdot \log^{-2}(2u).$$

Inserting this result into (5.3) we obtain

$$\left|\sum^{\#}_{n \le t, (n,d)=1} n^{-1} \cdot \mu(n)\right| \ll \sum_{m | P(x), m \le t} m^{-1} \cdot \mu^2(m) \cdot \psi(dm) \big/ \log^2(2t/m)$$

$$\leq \psi(d) \cdot \sum_{m|P(x),m\leq t} m^{-1}\cdot\psi(m)/\log^2(2t/m).$$

By (3.1)

$$\sum_{m\leq u,m|P(x)} \psi(m) \ll u\cdot\min\big(1, \log(x)/\log(2u)\big),$$

and partial summation immediately leads to assertion (5.1).

Now we come to the **proof of Lemma 3.4.** Uniformly in $z \geq x \geq 1$ we have to estimate

$$\lim_{N\to\infty} N^{-1} \cdot \sum_{n\leq N,p_{min}(n)>x} \{M(n,z)\}^2.$$

The sum $\sum_{n\leq N,p_{min}(n)>x} \{M(n,z)\}^2$ is equal to (remember that $^{\#}$ means the minimal prime divisor of the variable[s] of summation is $> x$)

$$(5.4)\quad \begin{cases} \sum_{d_1,d_2\leq z} \mu(d_1)\cdot\mu(d_2) \cdot \lim_{N\to\infty} N^{-1}\cdot\sum^{\#}_{n\leq N,n\equiv 0 \bmod \mathrm{lcm}[d_1,d_2]} 1 \\ \qquad = V(x) \cdot \sum^{\#\#}_{d_1,d_2\leq z} \mu(d_1)\cdot\mu(d_2)\cdot\{\mathrm{lcm}[d_1,d_2]\}^{-1}, \end{cases}$$

where

$$V(x) = \prod_{p\leq x} (1 - p^{-1}) \ll (\log 2x)^{-1}.$$

The argument needs an asymptotic evaluation of $\sum_{m\leq M,p_{min}(m)>x} 1$, which was given in (3.8). Using

$$\{\mathrm{lcm}[d_1,d_2]\}^{-1} = \{d_1d_2\}^{-1}\cdot\sum_{d|d_1,d|d_2} \varphi(d),$$

we obtain

$$\sum^{\#\#}_{d_1,d_2\leq z} \mu(d_1)\cdot\mu(d_2)\cdot\{\mathrm{lcm}[d_1,d_2]\}^{-1} = \sum_{d\leq z} \varphi(d)\cdot\Big(\sum^{\#}_{n\leq z,d|n} n^{-1}\cdot\mu(n)\Big)^2$$

$$= \sum^{\#}_{d\leq z} d^{-2}\varphi(d)\mu^2(d)\cdot\Big(\sum^{\#}_{n\leq z/d,(n,d)=1} n^{-1}\mu(n)\Big)^2.$$

The estimate of the inner sum was given at the beginning of this section. Inserting the result, we obtain

$$\sum^{\#\#}_{d_1,d_2\leq z} \mu(d_1)\cdot\mu(d_2)\cdot\{\mathrm{lcm}[d_1,d_2]\}^{-1} \ll \sum^{\#}_{d\leq z} \frac{\varphi(d)}{d^2}\cdot\psi^2(d)\cdot\min\Big(1, \frac{\log x}{\log^2(2z/d)}\Big).$$

If $z \leq x$, then replace min(...) by 1. If $z > x$, then split the interval $d\leq z$ into $d \leq z/x$ and $z/x < d \leq z$. Replacing min(...) by 1 in the second sum and by $\log^2 x/\log^2(2z/d)$ in the first sum, we obtain

$$\sum^{\#}_{d_1,d_2\le z} \mu(d_1)\cdot\mu(d_2)\cdot\{\ \mathrm{lcm}[d_1,d_2]\}^{-1} \ll \log^2 x\cdot \Sigma_1 + \Sigma_2,$$

where

$$\Sigma_1 = \sum^{\#}_{d\le z/x} d^{-2}\cdot\frac{\varphi(d)\cdot\psi^2(d)}{\log^2(2z/d)},\qquad \Sigma_2 = \sum^{\#}_{z/x<d\le z} d^{-2}\varphi(d)\cdot\psi^2(d).$$

From (3.1), $\sum^{\#}_{n\le u} n^{-1}\cdot\varphi(n)\cdot\psi^2(n) \ll 1 + u/\log(2x)$ if $u \ge 1$, and so, by partial summation,

$$\Sigma_2 = \sum^{\#}_{z/x<d\le z} d^{-2}\varphi(d)\cdot\psi^2(d) = z^{-1}\cdot S(z) - (x/z)\cdot S(z/x) + \int_{z/x}^{z} u^{-2}\cdot S(u)du \ll 1.$$

Σ_1 contains an additional factor $\log^2(2z/d)$ in the denominator; partial summation leads to

$$\Sigma_1 \ll \{\log^2 2x\}^{-1}.$$

Thus Lemma 3.4 is proved. □

A modification of a method due to ERDÖS and HALL is used to prove Lemma 3.5. Recall the notation $M(n,z) = \sum_{d|n,d\le z}\mu(d)$. Then, in $\frac{1}{2} < y < 2$, uniformly in $1 \le x \le z$, with an $\mathcal{O}$-constant depending at most on y, the relation

$$\text{(5.5)}\qquad \lim_{N\to\infty} N^{-1}\cdot \sum_{n\le N,\,x<p_{min}(n)\le z} |\,M(n,z)|\cdot y^{\Omega(n;x,y)} \ll_y \frac{1}{\log 2x}\left(\frac{\log 2z}{\log 2x}\right)^{2(y-1)}$$

is true, where

$$\Omega(n;x,z) = \sum_{x<p\le z,m,p^m|n} 1$$

is a completely additive function of n.

The proof of (5.5) begins with

$$|M(n,z)| = |\sum_{d\le z,d|n}\mu(d)| \le \sum_{d|n',z/p(n)<d\le z} 1\ ,$$

where $n' = n/\{p(n)\}^m$, where $p(n)^m \| n$ and $p(n) = p_{min}(n)$. This follows from the identity

$$M(n,u) = M(n/p^m,u) - M(n/p^m,u/p).$$

We now obtain

$$(5.6)\quad \begin{cases} \lim_{N\to\infty} N^{-1}\cdot\sum_{n\le N, x<p_{min}(n)\le z} |M(n,z)| \cdot y^{\Omega(n;x,y)} \\ \le \lim_{N\to\infty} N^{-1}\cdot\sum_{n\le N, x<p_{min}(n)\le z} \left(\sum_{d|n', z/p(n)<d\le z} 1\right) \cdot y^{\Omega(n;x,y)} \\ = \sum_{x<p\le z} \sum_{z/p<d\le z} \lim_{N\to\infty} N^{-1}\cdot \sum_{n\le N, p(n)=p, d|n'} y^{\Omega(n;x,y)}. \end{cases}$$

The limit in the last line is $\neq 0$ at most if $p_{min}(d) > p$. Using (3.1), in this case we obtain for this limit

$$\sum_{m\ge1} p^{-m} d^{-1}\cdot y^{\Omega(p^m d;x,z)} \cdot \lim_{N\to\infty} N^{-1}\cdot \sum_{n\le N,\, p(n)>p} y^{\Omega(n;x,y)}$$

$$\ll_y \left\{ y^{\Omega(d)}/\left(pd\cdot(\log p)^y\right) \right\} \cdot (\log 2z)^{y-1}.$$

Therefore, the sum to be estimated in (5.6) is

$$\ll_y (\log 2z)^{y-1} \cdot \sum_{x<p\le z} (p\cdot\log p)^{-y} \cdot \sum_{z/p<d\le z, p_{min}(d)>y} d^{-1}\cdot y^{\Omega(d)}.$$

Using, for $u \ge 1$, the estimate

$$S_p(u) = \sum_{d\le u, p_{min}(d)>p} y^{\Omega(d)} \ll_y \frac{u}{\log(2u)} \cdot \left(1 + \frac{\log u}{\log p}\right)^y,$$

which follows from (3.1), we obtain for the inner sum, using partial summation,

$$\sum_{z/p<d\le z, p_{min}(d)>y} d^{-1}\cdot y^{\Omega(d)} \ll (\log(2z))^{y-1} \cdot (\log p)^{1-y}.$$

Thus the left-hand side of (5.5) is

$$\ll_p (\log(2z))^{2(y-1)} \cdot \sum_{x<p\le z} \left(p\cdot(\log p)^{2y-1}\right)^{-1}.$$

Finally, partial summation shows that the last sum is

$$\ll_y \left((\log(2x))^{2y-1}\right)^{-1}.$$

This is (5.5). □

Proof of Lemma 3.5. The density to be estimated is

$$\lim_{N\to\infty} N^{-1} \cdot \sum_{n\le N, p_{min}(n)>x, M(n,z)\neq 0} 1$$

$$= \lim_{N\to\infty} N^{-1}\cdot\left\{\Sigma_1\, 1 + \Sigma_2\, 1 + \Sigma_3 1\right\},$$

where, using parameters $\varkappa > 1$, $1 < y_1 < 2$, $\frac{1}{2} < y_2 < 1$, the conditions of summation are

(1) $n \le N$, $p_{min}(n) > z$ in Σ_1,

(2) $n \le N$, $x < p_{min}(n) \le z$, $\Omega(n;x,z) > \varkappa\cdot\log(\log(2z)/\log(2x))$ in Σ_2,

(3) $M(n,z) \ne 0$, $x < p_{min}(n) \le z$, $\Omega(n;x,z) \le \varkappa\cdot\log(\log(2z)/\log(2x))$ in Σ_3.

Obviously $\lim\limits_{N\to\infty} N^{-1}\cdot\Sigma_1 = \prod\limits_{p\le z}(1-p^{-1})$. This is

$$\ll (\log 2z)^{-1} \le (\log 2x)^{-1}\cdot(\log(2x)/\log(2z))^{\delta}.$$

Enlarge Σ_2 by replacing 1 by $y_1^{\Omega(n;x,z)-\varkappa\cdot\log(\log(2z)/\log(2x))}$ and deleting the condition of summation for $\Omega(n;x,z)$. Then (3.1) leads to

$$\Sigma_2 \ll_{y_1} (\log 2z)^{-1}\cdot\exp\Big\{\sum_{x<p\le z} p^{-1}\cdot y_1\cdot(\log(2z)/\log(2x))^{-\varkappa\cdot\log y_1}\Big\}$$

$$\ll \left(\log 2x\right)^{-1}\cdot\left(\log(2z)/\log(2x)\right)^{y_1-1-\varkappa\cdot\log y_1}.$$

We replace the constant 1 in Σ_3 by

$$1 \le |M(n,z)|\cdot y_2^{\Omega(n;x,z)-\varkappa\cdot\log(\log(2z)/\log(2x))}$$

and delete the condition of summation for $\Omega(n;x,z)$. Then (5.5) gives

$$\Sigma_3 \ll_{y_2} (\log 2x)^{-1}\cdot(\log(2z)/\log(2x))^{2(y_2-1)-\varkappa\cdot\log y_2}.$$

Now we choose y_1, y_2 in such a way that the exponents of $(\log(2z)/\log(2x))$ become minimal, and then we choose $\varkappa$ in an optimal manner:

$$y_1 = \varkappa,\ y_2 = \tfrac{1}{2}\varkappa,\ \varkappa = (\log 2)^{-1}$$

leads to an exponent $(\log 2)^{-1}\cdot\log(e\cdot\log 2) - 1$, and this is the assertion of Lemma 3.5. Thus the proof of HILDEBRAND's Theorem is finished.[1] □

---

[1] Some of the techniques used here appear in more refined form in HALL–TENENBAUM's proof of the ERDÖS conjecture. All integers except a set of density zero have two divisors d, d′ satisfying $d < d' \le 2d$. See the monograph *"Divisors"* by HALL–TENENBAUM [1988].

V.6. EXERCISES

1) According to Theorem 1.3, the partial sums of the RAMANUJAN expansion

(∗) $$\sup_{Q \in \mathbb{N}} \left| \sum_{r \le Q} a_r(f) \cdot c_r(k) \right| < \infty$$

are bounded for every function f in $\mathcal{B}^u$ and every integer k. Prove Theorem 1.5 by using (∗) and the uniform boundedness principle (see, for example, HEWITT & STROMBERG [1965] (14.23)).

2) The analogue of Theorem 1.5 for uniformly-limit-periodic functions is wrong. More exactly: for every $f \in \mathcal{D}^u$ and every $k \in \mathbb{N}$

$$\sum_{r \le Q} \sum_{1 \le a \le r, \gcd(a,r)=1} \hat{f}\,(a/r) \cdot e_{a/r}(k) = M(f \cdot \tilde{S}_{Q,k}),$$

where

$$\tilde{S}_{Q,k}(n) = \sum_{r \le Q} \sum_{1 \le a \le r, \gcd(a,r)=1} e_{-a/r}(n-k).$$

Prove that the sequence $Q \mapsto \| \tilde{S}_{Q,k}(n) \|_1$ is **not** bounded.

3) The function g_k was defined at the beginning of the proof of Lemma 3.1. Evaluate the integral

$$\int_1^\infty u^{-1} \cdot \sum_{n>u} n^{-1} \cdot g_k(n)\, du,$$

which occurred at the end of the proof of Lemma 3.1.

4) The estimates in Theorems 1.2, 1.3, 1.4 and 1.5 are **not** uniform in k. Prove this (for Theorem 1.4) by the following example: if $N \in \mathbb{N}$, $P(N) = \prod_{N<p\le 2N} p$, then use

$$b_r^{(N)} = \mu(r) \big/ \varphi(r), \text{ if } r \mid P(N), \text{ and } b_r^{(N)} = 0, \text{ if } r \nmid P(N).$$

5) Construct an arithmetical function belonging to every space $\mathcal{B}^q$, where $q \ge 1$, but whose RAMANUJAN expansion diverges for every integer n.

Chapter VI

Almost-Periodic and Almost-Even Arithmetical Functions

Abstract. *In this chapter, starting again with the spaces* $\mathcal{B}$, $\mathcal{D}$, *and* $\mathcal{A}$, *a completion with respect to the semi-norm* $\|f\|_q = \limsup_{x \to \infty} x^{-1} \cdot \sum_{n \le x} |f(n)|^q$, $q \ge 1$, *gives the spaces* $\mathcal{B}^q$, $\mathcal{D}^q$, *and* $\mathcal{A}^q$ *of q-even, q-limit-periodic and q-almost-periodic arithmetical functions. Following J. KNOPFMACHER, it is shown that these spaces are BANACH spaces. Next, the properties of these spaces are derived: functions in* $\mathcal{A}^1$ *have mean-values, Ramanujan coefficients, Fourier-coefficients,* $f \in \mathcal{A}^1$ *implies* $|f| \in \mathcal{A}^1$. *If* f, g *are in* $\mathcal{A}^1$ *and real-valued, then* max(f,g), min(f,g) *are also in* $\mathcal{A}^1$, *etc. The PARSEVAL equation is given with two different proofs, and a result due to A. HILDEBRAND on the approximatibility of functions in* $\mathcal{B}^1$ *by partial sums of the RAMANUJAN expansion is given. Furthermore, a theory of integration is sketched, and many arithmetical applications (mean-values, and the behaviour of power series with multiplicative coefficients) are given.*

VI.1. BESICOVICH NORM, SPACES OF ALMOST-PERIODIC FUNCTIONS

Chapter IV dealt with uniformly almost-periodic arithmetical functions; the BANACH-algebras considered there arose from the algebra

$$\mathcal{B} = \mathrm{Lin}_{\mathbb{C}} \, [c_r, \; r = 1, 2, \ldots]$$

of linear combinations of RAMANUJAN sums, respectively from the algebra $\mathcal{D}$ of periodic arithmetical functions,

$$\mathcal{D} = \mathrm{Lin}_{\mathbb{C}} \, [e_{a/r}, \; r = 1, 2, \ldots, \; 1 \le a \le r, \; \gcd(a,r) = 1],$$

where e_β stands for the function e_β: $n \mapsto \exp(2\pi i \cdot \beta \cdot n)$, and from the algebra of linear combinations of the functions e_β, $\beta \in \mathbb{R} \bmod \mathbb{Z}$,

$$\mathcal{A} = \mathrm{Lin}_{\mathbb{C}} \, [e_\beta, \; \beta \in \mathbb{R}/\mathbb{Z}].$$

These spaces were enlarged by the use of the supremum-norm

$$\|f\|_u = \sup_{n \in \mathbb{N}} |f(n)|, \tag{1.1}$$

and properties of the resulting spaces [in fact, these spaces turned out to be BANACH-algebras]

$$\mathcal{B}^u = \|\,.\,\|_u\text{-closure}(\mathcal{B}), \quad \mathcal{D}^u = \|\,.\,\|_u\text{-closure}(\mathcal{D}), \quad \text{and } \mathcal{A}^u = \|\,.\,\|_u\text{-closure}(\mathcal{A}) \tag{1.2}$$

were given in Chapter IV. However, these algebras were rather small, for example the "smooth" arithmetical function $n \mapsto n^{-1}\varphi(n)$ is not in $\mathcal{B}^u$ (IV,Theorem 4.10). The "graph" of this function is given in Figure VI.1.

Of course, there are other norms on subspaces of the vector-space of arithmetical functions, and an enlargement similar to the one used above leads to other spaces which seem to be of greater importance in number theory. We use, for $q \ge 1$, the BESICOVICH [semi-] norms

$$\| f \|_q = \{ \limsup_{x \to \infty} \; x^{-1} \cdot \sum_{n \le x} | f(n) |^q \}^{1/q}. \tag{1.3}$$

With the exception of the condition $\{ \| f \|_q = 0 \Rightarrow f = 0 \}$, which is **not** true [for example, $\varepsilon = 1 * \mu \neq 0$, but $\|\varepsilon\|_q = 0$], the properties of a

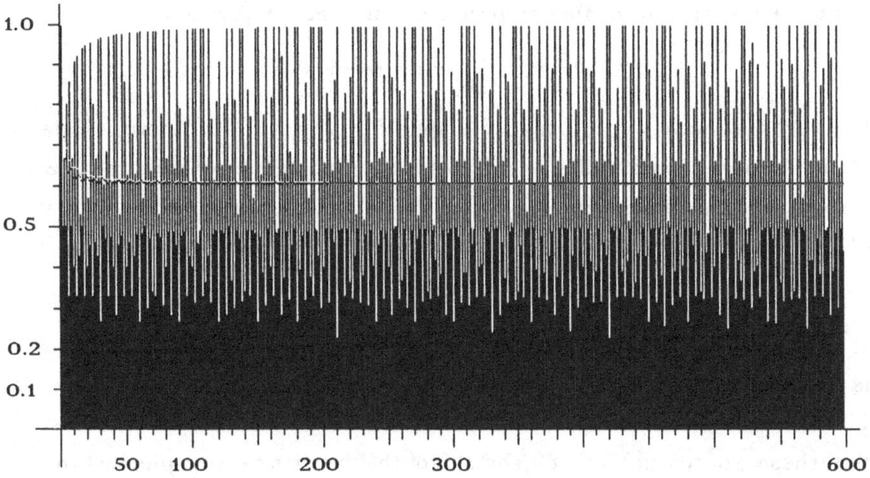

F i g u r e VI.1.

Values of the function $n \mapsto n^{-1} \cdot \varphi(n)$ in $1 \le n \le 598$. The mean-value-function $N \mapsto N^{-1} \cdot \sum_{n \le N} n^{-1} \cdot \varphi(n)$ is plotted in the same range. Its limit is $6 \cdot \pi^{-2}$.

norm [listed in Chapter IV, section 1] hold true for $\|.\|_q$, as is easily proved; for example, the proof of the triangle inequality [or MINKOWSKI's inequality] uses

$$(1.4) \quad \{ \sum_{n \le x} |f(n) + g(n)|^q \}^{1/q} \le \{ \sum_{n \le x} |f(n)|^q \}^{1/q} + \{ \sum_{n \le x} |g(n)|^q \}^{1/q}.$$

This follows from HÖLDER's inequality

$$\left| \sum_{n \in \mathcal{F}} a_n \cdot b_n \right| \le \left\{ \sum_{n \in \mathcal{F}} |a_n|^q \right\}^{1/q} \cdot \left\{ \sum_{n \in \mathcal{F}} |b_n|^{q'} \right\}^{1/q'}$$

for sums. The summation runs over a [finite] set $\mathcal{F}$; q is restricted to $1 < q < \infty$, and $q^{-1} + q'^{-1} = 1$. An enlargement of the spaces $\mathcal{B}$, $\mathcal{D}$, $\mathcal{A}$ by forming closures with respect to $\|f\|_q$, as mentioned above, gives

- the space of **q-almost-even arithmetical functions**,

$$\mathcal{B}^q = \|.\|_q \text{ - closure of } \mathcal{B},$$

- the space of **q-limit-periodic arithmetical functions**,

$$\mathcal{D}^q = \|.\|_q \text{ - closure of } \mathcal{D},$$

- and the space of **q-almost-periodic arithmetical functions,**

$$\mathcal{A}^q = \|.\|_q\text{-closure of } \mathcal{A}.$$

So, for a function f in, say, $\mathcal{B}^q$, and for any $\varepsilon > 0$, there exists a [finite] linear combination t of RAMANUJAN sums ε-close to f with respect to the semi-norm $\|.\|_q$, thus $\|f-t\|_q < \varepsilon$. In the sequel, we will often speak, inaccurately, of "the norm" $\|.\|_q$ instead of using the correct term "semi-norm" $\|.\|_q$.

It is clear that $\mathcal{B}^q$, $\mathcal{D}^q$, and $\mathcal{A}^q$ are $\mathbb{C}$-vector-spaces; heuristically, one is inclined to expect that properties of $\mathcal{B}$, $\mathcal{D}$ and $\mathcal{A}$ are also valid in $\mathcal{B}^q$, $\mathcal{D}^q$ and $\mathcal{A}^q$, and this principle often turns out to be successful. But these spaces are **not** algebras [so the heuristic principle just mentioned is sometimes not applicable], neither with convolution nor with the pointwise product. For example, $1 \in \mathcal{B}$, but $\tau = 1 * 1$ is not in $\mathcal{A}^1$ because functions in $\mathcal{A}^1$ have a mean-value (as will shortly be established in Theorem 1.2); τ does not have a mean-value. The function $f(n) = \{\log p\}^{2/3}$ if $n = p$ is a prime, and zero otherwise, is in $\mathcal{A}^1$. Due to the scarcity of the primes (because of weak versions of the prime number theorem the result $\lim_{x \to \infty} x^{-1} \cdot \pi(x) \cdot (\log x)^{2/3} = 0$ is true) this function is arbitrarily near to zero with respect to $\|.\|_1$, but the [pointwise] square $f \cdot f$ does not possess a mean-value and so f^2 is not in $\mathcal{A}^1$.

The [obvious] inclusion relations $\mathcal{B} \subset \mathcal{D} \subset \mathcal{A}$ imply

$$\mathcal{B}^q \subset \mathcal{D}^q \subset \mathcal{A}^q \text{ , where } q \ge 1.$$

For $r < q$, HÖLDER's inequality gives

$$\sum_{n \le N} |f(n)|^r \le \Big\{ \sum_{n \le N} |f(n)|^q \Big\}^{r/q} \cdot \Big\{ \sum_{n \le N} 1 \Big\}^{q/(q-r)},$$

therefore,

$$\|f\|_r \le \|f\|_q \quad \text{if } r \le q, \tag{1.5}$$

and so

$$\mathcal{B}^q \subset \mathcal{B}^r,\ \mathcal{D}^q \subset \mathcal{D}^r, \text{ and } \mathcal{A}^q \subset \mathcal{A}^r \text{ , if } r \le q.$$

Thus we obtain Figure VI.2, showing the inclusion relations between the various spaces defined up to now.

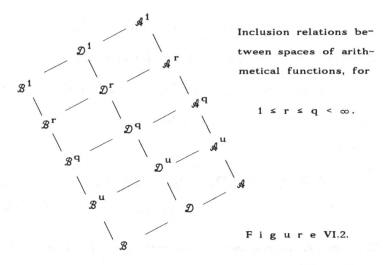

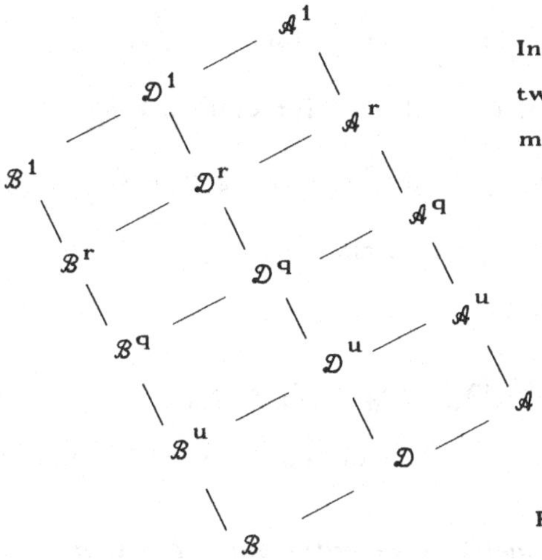

Inclusion relations between spaces of arithmetical functions, for

$1 \le r \le q < \infty$.

Figure VI.2.

A general, simple, but none the less useful principle is given as the following lemma.

Lemma 1.1. *Suppose* $\{\Lambda_n\}$ *is a sequence of linear functionals from a complex vector-space* X (*with semi-norm* $\|.\|$) *into* $\mathbb{C}$. *If for every* $x \in X$

(1.6.1) $$\limsup_{n\to\infty} |\Lambda_n(x)| \le c \cdot \|x\|,$$

and

(1.6.2) *there is a* <u>*dense*</u> *subset* E *of* X *such that the sequence* $\{\Lambda_n(x)\}_{n=1,2,\ldots}$ *is convergent for every* x *in* E,

then

(1.7.1) *the sequence* $\{\Lambda_n(x)\}_{n=1,2,\ldots}$ *converges for every* x *in* X,

(1.7.2) *the map* $\Lambda\colon x \mapsto \lim_{n\to\infty} \Lambda_n(x)$ *is a continuous linear functional on* X, *and* $\|\Lambda\| \le c$.

Remark. This lemma and its proof are modelled after Exercise 18 in W. RUDIN [1966], p. 116.

Proof. (1) Let $\varepsilon > 0$, $x \in X$, and let $\{x_m\}_{m=1,2,\dots}$ be a sequence of elements of E converging to x. Put

$$c_m = \lim_{n \to \infty} \Lambda_n(x_m), \text{ for } m = 1, 2, \dots .$$

There exist integers m, n_0 such that for every $n \ge n_0$

$$c \cdot \|x-x_m\| < \frac{1}{8} \cdot \varepsilon,\ |\Lambda_n(x-x_m)| \le c \cdot \|x-x_m\| + \frac{1}{8} \cdot \varepsilon, \text{ and } |\Lambda_n(x_m)-c_m| < \frac{1}{4}\varepsilon.$$

Then, for every k, $n \ge n_0$ we obtain

$$|\Lambda_n(x) - \Lambda_k(x)|$$
$$\le |\Lambda_n(x-x_m)| + |\Lambda_n(x_m)-c_m| + |\Lambda_k(x-x_m)| + |\Lambda_k(x_m)-c_m| < \varepsilon.$$

This proves (1.7.1). Then (1.7.2) is obvious from (1.7.1) and (1.6.1). □

Theorem 1.2. *Assume that* f *is an arithmetical function in* $\mathcal{A}^1$. *Then the mean-value*

(1.8.1) $$M(f) = \lim_{x \to \infty} x^{-1} \cdot \sum_{n \le x} f(n),$$

the FOURIER-*coefficients*

(1.8.2) $$\hat{f}(\alpha) = M(f \cdot e_{-\alpha}),\ \alpha \textit{ real},$$

and the RAMANUJAN *(or* RAMANUJAN-FOURIER*) coefficients*

(1.8.3) $$a_r(f) = \{\varphi(r)\}^{-1} \cdot M(f \cdot c_r),$$

$r = 1, 2, \dots,$ *exist.*

Proof. For the mean-value, consider the maps

$$\Lambda_N : \mathcal{A}^1 \to \mathbb{C},\ \Lambda_N(f) = N^{-1} \cdot \sum_{n \le N} f(n).$$

These maps are linear and $\limsup_{N \to \infty} |\Lambda_N(f)| \le \|f\|_1$. The mean-value $M(f) = \lim_{N \to \infty} \Lambda_N(f)$ exists for every function in $\mathcal{A}$. This vector-space is $\|.\|_1$-dense in $\mathcal{A}^1$; therefore Lemma 1.1 gives the existence of M(f) for every function f in $\mathcal{A}^1$. The assertions for FOURIER and RAMANUJAN coefficients follow from the fact that $f \cdot e_{-\alpha}$ and $f \cdot c_r$ are in $\mathcal{A}^1$. □

Another application of Lemma 1.1 is provided by the following theorem.

Theorem 1.3. *Let* f *be an arithmetical function in* $\mathcal{D}^q$, *where* $q \ge 1$, *and* g *is another arithmetical function. If* g *is bounded, in the case where* $q = 1$ *[respectively if* $\|g\|_{q'} < \infty$ *in the case where* $q > 1$*], where* $q'^{-1} + q^{-1} = 1$ *(as usual), and if*

$$(1.9) \qquad \lim_{x \to \infty} x^{-1} \cdot \sum_{n \le x,\ n \equiv a \bmod r} g(n)$$

exists for every pair a, r *of integers, then the mean-value*

$$(1.10) \qquad M(f \cdot g) = \lim_{x \to \infty} x^{-1} \cdot \sum_{n \le x} f(n) \cdot g(n)$$

exists.

Proof. We apply Lemma 1.1 using the maps Λ_N defined on $\mathcal{D}^q$ by

$$\Lambda_N(f) = N^{-1} \cdot \sum_{n \le N} f(n) \cdot g(n).$$

HÖLDER's inequality gives

$$\limsup_{N \to \infty} |\Lambda_N(f)| \le c \cdot \|f\|_q, \text{ where } c = \begin{cases} \|g\|_u, & \text{if } q = 1, \\ \|g\|_{q'}, & \text{if } q > 1. \end{cases}$$

The value of Λ_N at $e_{b/r}$ is

$$\Lambda_N(e_{b/r}) = \sum_{0 \le a < r} \Big(e_{b/r}(a) \cdot N^{-1} \cdot \sum_{n \le N,\ n \equiv a \bmod r} g(n) \Big).$$

Using (1.9), we see that the sequence $\{\Lambda_N(f)\}_{N=1,2,\dots}$ is convergent for every function f in $\mathcal{D}$. Hence, the existence of $M(f \cdot g)$ follows from Lemma 1.1. □

Remark. This proof does not seem to work for functions $f \in \mathcal{A}^1$.

Examples.

(1) For an integer $r > 0$ let g be the indicator-function of the set $r \cdot \mathbb{N}$. Then, for every function $f \in \mathcal{D}^1$, the "mean-value with divisor-condition"

$$M_r(f) = \lim_{x \to \infty} x^{-1} \cdot \sum_{n \le x, r|n} f(n)$$

exists. The same argument shows that

$$\lim_{x \to \infty} x^{-1} \cdot \sum_{n \le x, n \equiv s \bmod r} f(n)$$

exists for every residue-class s mod r.

(2) For the function $g = \mu^2$, the square of the MÖBIUS function, assumption (1.9) is easily checked by [well-known] elementary calculations, relying on the convolution-representation $\mu^2(n) = \sum_{d^2|n} \mu(d)$ (see II § 2).[1)] Therefore, according to Theorem 1.3, the "mean-value on squarefree numbers"

$$\lim_{x \to \infty} x^{-1} \cdot \sum_{n \le x,\ n \text{ squarefree}} f(n) = \lim_{x \to \infty} x^{-1} \cdot \sum_{n \le x} \mu^2(n) \cdot f(n)$$

exists for functions $f \in \mathcal{D}^1$.

(3) For $g = \mu$ assumption (1.9), in the case where $\gcd(a,r) = 1$, is guaranteed by properties of DIRICHLET's L-functions. The case where $\gcd(a,r) = d > 1$ can be reduced to the case mentioned above as follows: if $d = p_1 \cdot p_2 \cdot \ldots \cdot p_k$ with distinct primes, and $a = a' \cdot d$, $r = r' \cdot d$, then

$$\lim_{x \to \infty} x^{-1} \cdot \sum_{n \le x,\, n \equiv a \bmod r} \mu(n) = (-1)^k \cdot d^{-1} \lim_{y \to \infty} y^{-1} \cdot \sum_{m \le y,\, m \equiv a' \bmod r'} \mu(m)$$

exists. So (1.9) is valid for $g = \mu$, and, according to Theorem 1.3, the mean-value

$$\lim_{x \to \infty} x^{-1} \cdot \sum_{n \le x} \mu(n) \cdot f(n)$$

exists for every function $f \in \mathcal{D}^1$.

The spaces $\mathcal{B}^q$, $\mathcal{D}^q$, and $\mathcal{A}^q$ are complete: any $\|.\|_q$-CAUCHY-sequence of functions in, say $\mathcal{B}^q$, has a limit, which is again in $\mathcal{B}^q$. Thus, following J. KNOPFMACHER [1976], we prove the completeness of these spaces.

Theorem 1.4 [J. KNOPFMACHER]. *For* $q \ge 1$, *the normed spaces* $\mathcal{B}^q$, $\mathcal{D}^q$, *and* $\mathcal{A}^q$ *are complete.*

Proof. The spaces $\mathcal{B}^q$, $\mathcal{D}^q$, and $\mathcal{A}^q$ are closed subsets of the vector-space

$$V^q = \{\, f\colon \mathbb{N} \to \mathbb{C};\ \|f\|_q < \infty \,\}.$$

So it suffices to prove the completeness of V^q, where $q \ge 1$.

---

1) Another possibility for proving this is: the function $g = \mu^2$ is in $\mathcal{D}^1$, and so assumption (1.9) is valid by example 1.

Let $\{f_k\}_k$ be a CAUCHY sequence in V^q. There exists a sequence $\{\varepsilon_k\}_k$ of real positive numbers, converging to zero, with the property

$$\| f_\ell - f_k \|_q^q < \varepsilon_k \text{ for every } \ell > k.$$

We are going to construct a sequence $\{x_m\}_m$ of non-negative numbers with the properties

(1.11)
(a) $0 = x_0 < x_1 < x_2 < \ldots,\ x_m \to \infty$,
(b) the function f, defined by $f(n) = f_\ell(n)$, if $x_{\ell-1} < n \le x_\ell$, satisfies

$$x^{-1} \cdot \sum_{n\le x} |f(n) - f_k(n)|^q < 2\,\varepsilon_k$$

for every k in $\mathbb{N}$, and for every $x \ge x_k$.

These properties imply $\| f - f_k \|_q \le 2\varepsilon_k$ for every k; therefore $\| f \|_q < \infty$, so $f \in V^q$ and f is the limit of the sequence f_k. So it remains to give the [inductive] **construction** of a sequence $\{x_m\}_{m=1,2,\ldots}$ with properties (1.11).

For any integer k, $0 < k < \ell$, there is a real number $x_{k\ell} > 0$ such that

(1.12) $$x^{-1} \sum_{n\le x} |f_\ell(n) - f_k(n)|^q < \varepsilon_k, \text{ for every } x \ge x_{k\ell}.$$

Put $x_0 = 0$, $x_1 = \max\{x_{12}, 1\}$. Then

(A_{12}) $$x^{-1} \sum_{n\le x} |f_2(n) - f_1(n)|^q < \varepsilon_1, \text{ for every } x \ge x_1,$$

(D_1) $$x_1 \ge x_0 + 1.$$

We assume now that $0 < x_1 < x_2 < \ldots < x_m$ are chosen with the properties

$(A_{k\ell})$ $k < \ell \le m+1$: $$x^{-1} \sum_{n\le x} |f_\ell(n) - f_k(n)|^q < \varepsilon_k, \text{ for every } x \ge x_{\ell-1},$$

$(B_{k\ell})$ $k < \ell \le m$: $$\sum_{x_{\ell-1}<n\le x_\ell} |f_\ell(n) - f_k(n)|^q < \varepsilon_k \cdot (x_\ell - x_{\ell-1}),$$

(C_k) $k \le m$: $$\sum_{\ell<k} \sum_{x_{\ell-1}<n\le x_\ell} |f_\ell(n) - f_k(n)|^q < \varepsilon_k \cdot x_k,$$

(D_k) $k \le m$: $$x_k \ge x_{k-1} + 1.$$

Given any k, $0 < k < m+1$, there exists a positive $\varepsilon_k' < \varepsilon_k$, such that

$$x^{-1} \cdot \sum_{n \le x} |f_{m+1}(n) - f_k(n)|^q < \varepsilon_k' \text{, for all } x \ge x_{k,m+1}.$$

If $x_k' \ge \max\left\{ x_m \cdot \left(1 - \varepsilon_k'/\varepsilon_k\right)^{-1}, x_{k,m+1}\right\}$, then

$$\sum_{x_m < n \le x_k'} |f_{m+1}(n) - f_k(n)|^q \le \sum_{n \le x_k'} |f_{m+1}(n) - f_k(n)|^q$$

$$< \varepsilon_k' \cdot x_k' \le \varepsilon \cdot \left(x_k' - x_m\right).$$

Now take x_{m+1} so large that for $k < m+1$

$$(B_{k,m+1}) \qquad \sum_{x_m < n \le x_{m+1}} |f_{m+1}(n) - f_k(n)|^q < \varepsilon_k \cdot (x_{m+1} - x_m)$$

holds, and for $k < m+2$:

$$(A_{k,m+2}) \qquad x^{-1} \cdot \sum_{n \le x} |f_{m+2}(n) - f_k(n)|^q < \varepsilon_k, \ \forall\, x \ge x_{m+1},$$

$$(C_{m+1}) \qquad \sum_{\ell < m+1} \sum_{x_{\ell-1} < n \le x_\ell} |f_\ell(n) - f_{m+1}(n)|^q < \varepsilon_{m+1}\, x_{m+1},$$

$$(D_{m+1}) \qquad x_{m+1} \ge x_m + 1.$$

Thus, we obtain a sequence $\{x_m\}$ with properties (a), and $(A_{k,\ell})$, $(B_{k,\ell})$, (C_k) for all $k < \ell$. The function f, defined as in (b), has the desired property

$\forall\, k \in \mathbb{N}$, $\forall\, x \le x_k$, and $m \in \mathbb{N}$ defined by $x_m \le x < x_{m+1}$:

$$\sum_{n \le x} \left|f(n) - f_k(n)\right|^q = \sum_{\ell < k} \sum_{x_{\ell-1} < n \le x_\ell} \left|f_\ell(n) - f_k(n)\right|^q$$

$$+ \sum_{k < \ell \le m} \sum_{x_{\ell-1} < n \le x_\ell} \left|f_\ell(n) - f_k(n)\right|^q + \sum_{x_m < n \le x} \left|f_{m+1}(n) - f_k(n)\right|^q$$

$$< \varepsilon_k \cdot x_k + \varepsilon_k \cdot \sum_{k < \ell \le m} (x_\ell - x_{\ell-1}) + \varepsilon_k \cdot x \le 2\, \varepsilon_k\, x.$$

This proves the completeness of V^q. □

The **null-spaces** are defined as follows:

$$\mathcal{N}(\mathcal{A}^q) = \{\, f \in \mathcal{A}^q,\ \| f \|_q = 0 \,\},$$

and similarly $\mathcal{N}(\mathcal{D}^q)$ and $\mathcal{N}(\mathcal{B}^q)$; these are subspaces of $\mathcal{A}^q$, resp. $\mathcal{D}^q$, resp. $\mathcal{B}^q$. The null-spaces are closed; the limit f of a $\|.\|_q$-convergent sequence of functions in, say, $\mathcal{N}(\mathcal{A}^q)$, is in $\mathcal{A}^q$ and has norm $\| f \|_q = 0$.

We denote the quotient spaces $\mathcal{A}^q/\mathcal{N}(\mathcal{A}^q)$, etc., by

$$A^q = \mathcal{A}^q/\mathcal{N}(\mathcal{A}^q),\ D^q = \mathcal{D}^q/\mathcal{N}(\mathcal{D}^q), \text{ and } B^q = \mathcal{B}^q/\mathcal{N}(\mathcal{B}^q).$$

There is a canonical quotient map π,

$$\pi = \pi_{\mathcal{A}}{}^q\colon \mathcal{A}^q \to \mathcal{A}^q/\mathcal{N}(\mathcal{A}^q),\ \pi_{\mathcal{A}}{}^q(f) = f + \mathcal{N}(\mathcal{A}^q).$$

The quotient norm is defined by (see, for example, RUDIN [1966], 18.15)

$$\|\pi(f)\|_q = \inf \{ \| f+g \|_q, g \in \mathcal{N} \} = \| f \|_q.$$

Then A^q, D^q, B^q are BANACH-spaces.

In Chapter IV, Theorem 2.9, a uniqueness theorem was proved for functions in $\mathcal{D}^u$. In $\mathcal{A}^1$, a theorem of this kind is not true. However, arithmetical properties such as additivity or multiplicativity have consequences on the uniqueness of functions in $\mathcal{A}^1$. As examples, we prove the following theorems.

Theorem 1.5. *(Uniqueness theorem for additive functions). Assume that* f *and* g *are additive functions in* $\mathcal{A}^1$. *If* $\| f - g \|_1 = 0$, *then* $f = g$ *identically.*

Theorem 1.6. *(Uniqueness theorem for multiplicative functions). Assume* f *and* g *are multiplicative, and both are in* $\mathcal{D}^1$, $\|f\|_1 \neq 0$, $\sum_{k\geq 1} p^{-k}\cdot|f(p^k)| < \infty$ *for every prime* p, *and* f-g *is in the null-space* $\mathcal{N}(\mathcal{A}^1)$. *Then* $f = g$ *identically.*

Remark 1. The assumption $\|f\|_1 \neq 0$ in Theorem 1.6 is necessary. The functions $f = \varepsilon$, $g(n) = 1$ if $n = 2^k$ for some k, and $g(n) = 0$ otherwise, are both multiplicative and satisfy $\|f - g\|_1 = 0$, but **f ≠ g.**

Remark 2. The finiteness of the norm $\|f\|_q$ for some $q > 1$ implies (see III, Lemma 5.2) $\sum_{k\geq 1} p^{-k}\cdot|f(p^k)| < \infty$ for every prime p. So this condition can be omitted in Theorem 1.6 if $f \in \mathcal{D}^q$ for some $q > 1$ is assumed. We shall see later (VII, Theorem 5.1) that this condition is also superfluous in the case $q = 1$.

Remark 3. If $f \in \mathcal{D}^1$ is multiplicative and non-negative, then the condition $M_p(f) < M(f)$ implies $\sum_{k\geq 1} p^{-k}\cdot|f(p^k)| < \infty$ (see Exercise 4).

Proof of Theorem 1.5. Put $h = f - g$, and let p^k be a fixed prime power. For any integer N, we obtain the lower estimate

$$\sum_{n\leq N} |h(n)| \geq \sum_{n\leq N, p^k \| n} |h(n)| = \sum_{m\leq N/p^k, p\nmid m} |h(p^k) + h(m)|$$

$$\geq |h(p^k)| \cdot \sum_{m\leq N/p^k, p\nmid m} 1 - \sum_{m\leq N/p^k} |h(m)|.$$

Dividing by N, for $N \to \infty$, the inequality

$$\|h\|_1 \geq |h(p^k)| \cdot p^{-k} \cdot (1-p^{-1}) - p^{-k}\cdot \|h\|_1$$

is obtained, and the assumption $\|h\|_1 = 0$ implies $h(p^k) = 0$. □

Proof of Theorem 1.6. Assume that there is an integer n_0 for which $f(n_0) \neq g(n_0)$; then there is a prime-power p^k with $f(p^k) \neq g(p^k)$. Then, for every N,

$$\sum_{n\leq N} |f(n)-g(n)| \geq \sum_{m\leq N/p^k, \gcd(m,p)=1} |f(p^k)\cdot f(m) - g(p^k)\cdot g(m)|$$

$$\geq |f(p^k) - g(p^k)| \cdot \sum_{\substack{m\leq N/p^k \\ p\nmid m}} |f(m)| - |g(p^k)| \cdot \sum_{m\leq N/p^k} |f(m)-g(m)|.$$

Add the term

$$|f(p^k) - g(p^k)| \cdot \sum_{m\leq N/p^k, p|m} |f(m)|,$$

to both sides of this inequality and divide the resulting inequality by $N\ (= p^k\cdot(N/p^k))$. Letting N tend to infinity, we obtain, using the abbreviation $M_p(g) = \lim_{x\to\infty} x^{-1}\cdot \sum_{n\leq x, p|n} g(n)$ [for the existence of this mean-value, see Example 1 following Theorem 1.3],

$$\|f-g\|_1 + |f(p^k) - g(p^k)|\cdot p^{-k} \cdot M_p(|f|)$$

$$\geq |f(p^k) - g(p^k)|\cdot p^{-k} \cdot M(|f|) - |g(p^k)|\cdot p^{-k} \cdot \|f-g\|_1,$$

and therefore $M_p(|f|) \geq M(|f)|$, a strange result certainly, which comes from the assumption $f \neq g$, which is to be refuted. Next

$$\sum_{n\le N, p|n} |f(n)| = \sum_{k\ge 1} \sum_{n\le N, p^k \| n} |f(n)|,$$

$$= \sum_{k\ge 1} |f(p^k)| \left(\sum_{m\le N/p^k} |f(m)| - \sum_{m\le N/p^k, p|m} |f(m)|\right),$$

and so

$$\sum_{k\ge 0} |f(p^k)| \sum_{m\le N/p^k, p|m} |f(m)| = \sum_{k\ge 1} |f(p^k)| \cdot \sum_{m\le N/p^k} |f(m)|.$$

Dividing by N and using the dominated convergence theorem (this is possible for $\gamma = \sum_{k\ge 1} p^{-k} \cdot |f(p^k)| < \infty$), we obtain

$$(1+\gamma) \cdot M_p(|f|) = \gamma \cdot M(|f|),$$

therefore, using the estimate $M_p(|f|) \ge M(|f|)$, proved above,

$$\gamma \cdot (1+\gamma)^{-1} \cdot M(|f|) \ge M(|f|),$$

which contradicts the assumption $\|f\|_1 = M(|f|) > 0$. □

VI.2. SOME PROPERTIES OF SPACES OF q-ALMOST-PERIODIC FUNCTIONS

As mentioned already in section 1, HÖLDER's inequality

$$\| f\cdot g\|_1 \le \|f\|_q \cdot \|g\|_{q'}, \text{ where } q^{-1}+q'^{-1} = 1,$$

implies $\mathcal{A}^q \subset \mathcal{A}^r \subset \mathcal{A}^1$ whenever $1 \le r \le q$ [and there are corresponding results for the other spaces - see Figure VI.2]. Starting with k = 2 (which is HÖLDER's inequality), mathematical induction gives the following

Proposition 2.1. *Assume that*

(2.1) $$1/q_1 + 1/q_2 + \dots + 1/q_k = 1,$$

where $1 < q_\varkappa < \infty$. *Then*

(2.2) $$\| f_1 \cdot \dots \cdot f_k \|_1 \le \|f_1\|_{q_1} \cdot \dots \cdot \|f_k\|_{q_k}.$$

Proposition 2.2. *Assume that all the norms appearing in equation (2.3) below are finite. Then the following assertions are true:*

$$(2.3)\quad \begin{cases} \text{(i) If } r \ge 1,\ q^{-1}+q'^{-1} = 1, \textit{ then } \ \|f\cdot g\|_r \le \|f\|_{rq} \cdot \|g\|_{r\cdot q'}. \\ \text{(ii) If } 1 \le q \le r, \textit{ then } \|f\|_1 \le \|f\|_q \le \|f\|_r \le \|f\|_u. \\ \text{(iii) } \|f\cdot g\|_q \le \|g\|_u \cdot \|f\|_q. \end{cases}$$

Proof. (iii) follows from the definition of $\|.\|_q$; the other inequalities are obtained from HÖLDER's inequality. □

Theorem 2.3. *Assume that* $1 \le q \le r < \infty$, *and* $q^{-1}+q'^{-1} = 1$. *Then*

(1) $\mathcal{B} \subset \mathcal{B}^u \subset \mathcal{B}^r \subset \mathcal{B}^q \subset \mathcal{B}^1$,
$\mathcal{D} \subset \mathcal{D}^u \subset \mathcal{D}^r \subset \mathcal{D}^q \subset \mathcal{D}^1$,
$\mathcal{A} \subset \mathcal{A}^u \subset \mathcal{A}^r \subset \mathcal{A}^q \subset \mathcal{A}^1$.

(2) $\mathcal{B}^q \subset \mathcal{D}^q \subset \mathcal{A}^q \subset \mathcal{A}^1$.

(3) $\mathcal{B}^u\cdot\mathcal{B}^q \subset \mathcal{B}^q$, $\quad \mathcal{D}^u\cdot\mathcal{D}^q \subset \mathcal{D}^q$, $\quad \mathcal{A}^u\cdot\mathcal{A}^q \subset \mathcal{A}^q$.

(4) $\mathcal{B}^q\cdot\mathcal{B}^{q'} \subset \mathcal{B}^1$, $\quad \mathcal{D}^q\cdot\mathcal{D}^{q'} \subset \mathcal{D}^1$, $\quad \mathcal{A}^q\cdot\mathcal{A}^{q'} \subset \mathcal{A}^1$.

(5) *If* $f \in \mathcal{B}^q$, *then* Re(f), Im(f) *and* $|f| \in \mathcal{B}^q$,
if $f \in \mathcal{D}^q$, *then* Re(f), Im(f) *and* $|f| \in \mathcal{D}^q$,
if $f \in \mathcal{A}^q$, *then* Re(f), Im(f) *and* $|f| \in \mathcal{A}^q$.

(6) *If* f, g *are real-valued and both are in* $\mathcal{B}^q$ [*resp.* $\mathcal{D}^q$, *resp.* $\mathcal{A}^q$], *then*

max(f,g) *and* min(f,g) *are in* $\mathcal{B}^q$ [*resp.* $\mathcal{D}^q$, *resp.* $\mathcal{A}^q$].

Proof. Assertions (1) and (2) are clear. For (3), assume that $f \in \mathcal{A}^q$, $g \in \mathcal{A}^u$, $\varepsilon > 0$; choose functions G, F in $\mathcal{A}$ near g, f such that $\|g-G\|_u < \varepsilon/(\|f\|_q+1)$, $\|f-F\|_q < \varepsilon/(\|G\|_u+1)$. Then $F\cdot G$ is in $\mathcal{A}$ and

$$\|g\cdot f-G\cdot F\|_q \le \|(g-G)\cdot f\|_q + \|G\cdot(f-F)\|_q \le \|g-G\|_u\cdot\|f\|_q + \|G\|_u\cdot\|f-F\|_q < 2\varepsilon.$$

(4) If $f \in \mathcal{A}^q$, $g \in \mathcal{A}^{q'}$, $\varepsilon > 0$, choose F, $G \in \mathcal{A}$, $\|f-F\|_q < \varepsilon/(\|g\|_q+1)$, $\|g-G\|_{q'} < \varepsilon/(\|F\|_q+1)$. Then $F\cdot G$ is in $\mathcal{A}$ and (using HÖLDER's inequality)

$$\|f\cdot g-F\cdot G\|_1 \le \|(f-F)\cdot g\|_1 + \|F\cdot(g-G)\|_1 \le \|f-F\|_q\cdot\|g\|_{q'} + \|F\|_q\cdot\|g-G\|_{q'} < 2\varepsilon.$$

(5) The real or imaginary part of a function in $\mathcal{B}$ [resp. $\mathcal{D}$, resp. $\mathcal{A}$] is again in $\mathcal{B}$ [resp. $\mathcal{D}$, resp. $\mathcal{A}$]. If F is in $\mathcal{B}$ or $\mathcal{D}$, then $|F|$ is even or periodic and so, again, is in $\mathcal{B}$ or $\mathcal{D}$. And, using the usual approximation arguments [and the inequality $\| \, |f| - |g| \, \| \le \| f-g \|$], the assertions are proved with the exception that f in $\mathcal{A}^q$ implies $|f|$ in $\mathcal{A}^q$. But in this case the WEIERSTRASS Approximation Theorem gives: if F in $\mathcal{A}$, then $|F|$ in $\mathcal{A}^u$ (by IV. Theorem 2.2), and this is sufficient for a proof of the remaining assertion.

Assertion (6) follows from the formulae

(2.4) $$\max(f,g) = \tfrac{1}{2}(f+g) + \tfrac{1}{2}|f-g|, \quad \min(f,g) = \tfrac{1}{2}(f+g) - \tfrac{1}{2}|f-g|. \qquad \square$$

Theorem 2.4.

(1) *If* f *is in* $\mathcal{A}^1$, *then the mean-value* M(f), *the* FOURIER *coefficients* $\hat{f}(\alpha) = M(f \cdot e_{-\alpha})$ *and the* RAMANUJAN *coefficients*

(2.5) $$a_r(f) = \{\varphi(r)\}^{-1} \cdot M(f \cdot c_r)$$

exist.

(2) *In* $\mathcal{A}^2$ *(and so in the subspaces* $\mathcal{B}^2$, $\mathcal{D}^2$*) there is an inner product*

(2.6) $$\langle f,g \rangle = M(f \cdot \overline{g}),$$

and the CAUCHY-SCHWARZ *inequality*

(2.7) $$|\langle f,g \rangle| \le \|f\|_2 \cdot \|g\|_2$$

holds.

(3) *If f is in* $\mathcal{A}^2$, *then at most denumerably many* FOURIER *coefficients are non-zero, and* BESSEL*'s inequality*

(2.8.1) $$\sum_{\beta \in \mathbb{R}/\mathbb{Z}} |\hat{f}(\beta)|^2 \le \|f\|_2^{\,2}$$

holds. If f *is in* $\mathcal{B}^2$, *then* BESSEL*'s inequality reads*

(2.8.2) $$\sum_{1 \le r < \infty} \varphi(r) \cdot |a_r(f)|^2 \le \|f\|_2^{\,2}.$$

(4) *The maps from* $\mathcal{A}^1$ *to* $\mathbb{C}$, *defined by*

$$f \mapsto M(f), \; f \mapsto \hat{f}(\beta), \; f \mapsto a_r(f),$$

are linear and continuous. Moreover, the first is non-negative.

(5) $$|\hat{f}(\alpha)| \le M(|f|), \ |a_r(f)| \le M(|f|).$$

Remark. Some of the assertions of this theorem have already been shown to be true in section 1 by applying a general principle from functional analysis. In spite of this, we give an ad-hoc proof again here.

Proof. (4) and (5) are obvious. For (1), the existence of the mean-value has already been proved (in section 1); a simple, direct proof for (1) is to be given in Exercise 1. The functions $f \cdot e_\alpha$ and $f \cdot c_r$ are in $\mathcal{A}^1$ again, and so the FOURIER coefficients, which are mean-values, do exist.

(2) The function $f \cdot \overline{g}$ is in $\mathcal{A}^1$, and so its mean-value, which is the inner product, exists. The usual properties of an inner product are easily verified (note: $\langle f, f \rangle = 0$ implies $\|f\|_2 = 0$, but not necessarily $f = 0$). The method of proving the CAUCHY-SCHWARZ inequality is standard in linear algebra. The same is true for BESSEL's inequality.

(3) The functions e_α, $\alpha \in \mathbb{R}/\mathbb{Z}$, are an orthonormal system. Using only finitely many FOURIER coefficients $\hat{f}(\alpha)$, we obtain

$$\begin{aligned} 0 &\le \langle f - \sum \hat{f}(\alpha) \cdot e_\alpha, \ f - \sum \hat{f}(\beta) \cdot e_\beta \rangle \\ &= \langle f,f \rangle - \sum \overline{\hat{f}(\beta)} \cdot \langle f, e_\beta \rangle - \sum \hat{f}(\alpha) \cdot \langle e_\alpha, f \rangle + \sum\sum \hat{f}(\alpha) \overline{\hat{f}(\beta)} \cdot \langle e_\alpha, e_\beta \rangle \\ &= \langle f,f \rangle - \sum |\hat{f}(\alpha)|^2. \end{aligned}$$ □

Corollary 2.5. *Assume that* $q \ge 1$, *and* $k \ge 1$ *is an integer.*

(1) *If* f *is in* $\mathcal{A}^q$ *[resp.* $\mathcal{D}^q$*], then the "shifted" function* $f_{(a)}: n \mapsto f(a+n)$ *is in* $\mathcal{A}^q$ *[resp.* $\mathcal{D}^q$*], where* a *is in* $\mathbb{Z}$.[1)]

(2) *If* f *is in* $\mathcal{A}^q$ *[resp.* $\mathcal{D}^q$*] then, for* $b \in \mathbb{N}$, *the multiplicatively shifted function* $f_{(b\cdot)}: n \mapsto f(b \cdot n)$ *is in* $\mathcal{A}^q$ *[resp.* $\mathcal{D}^q$*].*

(3) *If* $f_1, \ldots, f_k$ *are in* $\mathcal{A}^k$, *then the function* F *is in* $\mathcal{A}^1$, *where*

(2.9) $$F(n) = \prod_{\varkappa=1}^{k} f_\varkappa(b_\varkappa \cdot n + a_\varkappa), \ b_\varkappa \in \mathbb{N}, \ a_\varkappa \in \mathbb{Z}.$$

Proof. (1) Given $\varepsilon > 0$ there is a function F in $\mathcal{A}$, $F = \sum a_\alpha \cdot e_\alpha$, near f, $\|f-F\|_q < \varepsilon$. Then $F_{(a)}(n) = \sum (a_\alpha \cdot e_\alpha(a)) \cdot e_\alpha(n)$, and so $F_{(a)}$ is in $\mathcal{A}$ and

---

1) *We assume that* $f(a+n) = 1$ *as long as* $a+n \le 0$.

$$\|f_{(a)}-F_{(a)}\|_q = \|f-F\|_q < \varepsilon.$$

(2) Choose F as above, put G: $n \mapsto F(b\cdot n)$; then G is in $\mathcal{A}$ [resp. $\mathcal{D}$], and

$$x^{-1}\cdot\sum_{n\le x} |f(b\cdot n) - G(n)|^q \le b\cdot(bx)^{-1}\cdot\sum_{m\le bx} |f(m)-F(m)|^q,$$

and we obtain $\|f_{(b\cdot)}-G\|_q < b^{1/q}\cdot\varepsilon$.

(3) The case where k = 2 is obvious from (1), (2) and Theorem 2.3. The case where k > 2 is left as an exercise. □

Theorem 2.6.

(i) *If* g *is in* $\mathcal{A}^1$, *real-valued and bounded, then, for any* $\varepsilon > 0$ *there exists a function* t *in* $\mathcal{A}^u$ *near* g, $\| g - t \|_1 < \varepsilon$, *with the additional property* $\|t\|_u \le \|g\|_u$.

(ii) *If* $g \in \mathcal{A}^1$ *is bounded, then* $g \in \mathcal{A}^q$ *for every* $q \ge 1$.

(iii) *Assume that* $g \in \mathcal{A}^1$ *has a bounded representative and* f *is in* $\mathcal{A}^q$. *Then the pointwise product* $g\cdot f$ *is in* $\mathcal{A}^q$.

Remark. The same results are true for the other spaces $\mathcal{B}^q$ and $\mathcal{D}^q$. Of course, in (i), in these cases t may be taken to be in $\mathcal{B}$ [resp. in $\mathcal{D}$]. If g is complex-valued, in (i) it is possible to find a t satisfying $\|t\|_u \le \sqrt{\|\mathrm{Re}\,g\|_u^2 + \|\mathrm{Im}\,g\|_u^2} \le 2\,\|g\|_u$.

Corollary 2.7. *If* f *is in* $\mathcal{D}^q$, *then the functions* $\chi\cdot f$, *where* χ *is a* DIRICHLET *character,* $1_k\cdot f$, *where* 1_k *is the characteristic function of the set of integers relatively prime to* k *or the characteristic function of the set of integers congruent to* a mod k, *where* gcd(a,k) = 1, *and the pointwise product* $\mu^2\cdot f$ *are in* $\mathcal{D}^q$ *again.*

Proof of Corollary 2.7. The functions χ and 1_k are periodic and bounded. The function μ^2 is bounded and is in $\mathcal{B}^1$ (this is a consequence of the Relationship Theorem from Chapter III; μ^2 is bounded, therefore in $\mathcal{G}$, and it is related to the constant function $1 \in \mathcal{G}^*$). □

Proof of Theorem 2.6. (i) Given $\varepsilon > 0$, choose a real-valued trigonometric polynomial t^* in $\mathcal{A}$ [resp. $\mathcal{D}$ or $\mathcal{B}$] near f, $\| g-t^*\|_1 < \varepsilon$. Put $t = \max\{ \min(t^*, \|g\|_u), -\|g\|_u\}$. Then $\|g-t\|_1 \le \|g-t^*\|_1 < \varepsilon$, and t is in $\mathcal{A}^u$ [resp. $\mathcal{D}$, resp. $\mathcal{B}$ for the other spaces], and $\|t\|_u \le \|g\|_u$.

(ii) is a special case of (iii).

(iii) Let $\varepsilon > 0$, choose t_1, t_2 in $\mathcal{A}^u$, such that $\|f-t_1\|_q < \varepsilon$, $\|g-t_2\|_1 < \varepsilon^q/(1+\|g\|_u^2+\|t_1\|_u^2)$, and $\|t_2\|_u \le 2\cdot\|g\|_u$. Then an easy computation shows

$$\| g-t_2 \|_q \le \left\{ 4^{q-1} \|g\|_u^{q-1} \cdot \|g-t_2\|_1 \right\}^{1/q}.$$

Therefore,

$$\|fg-t_1t_2\|_q \le \|(f-t_1)\cdot g\|_q + \|t_1\cdot(g-t_2)\|_q \le \|g\|_u\cdot \|(f-t_1)\|_q + \|t_1\|_u\cdot\|g-t_2\|_q$$

$$\le \|g\|_u\cdot\varepsilon + \|t_1\|_u\cdot \text{const}(q)\cdot \|g\|_u^{1-1/q}\cdot\|g-t_2\|_1^{1/q}$$

$$\le \text{const}^*(q, \|g\|_u)\cdot\varepsilon.$$

Since t_1t_2 is in $\mathcal{A}^u$, Theorem 2.6 is proved. □

Theorem 2.8. *If* f *is in* $\mathcal{B}^1$ *[resp.* $\mathcal{D}^1$, *resp.* $\mathcal{A}^1$*] and* $\|f\|_q < \infty$, $q > 1$, *then* f *is in* $\mathcal{B}^r$ *[resp.* $\mathcal{D}^r$, *resp.* $\mathcal{A}^r$ *] for any* r *in* $1 \le r < q$.

Remark 1. An additional condition is needed to secure that this result is true for $r = q$ (see section 8).

Remark 2. The assertion of Theorem 2.8 is **not** true for $r = q$, as shown by the following examples.

Example 1. The function $f(n) = n^{\frac{1}{4}}$ if n is a square, and $f(n) = 0$ otherwise, has norm $\|f\|_q = 0$ as long as $q < 2$, and it is (trivially) in $\mathcal{B}^1$. All RAMANUJAN coefficients $a_r(f) = M(fc_r)/\varphi(r)$ vanish, but $\sum_{n\le x} |f(n)|^2 \sim \frac{1}{2}x$, and so $\|f\|_2^2 = M(|f|^2) = \frac{1}{2}$. But PARSEVAL's equation $M(|f|^2) = \sum \varphi(r)\cdot|a_r(f)|^2$ (see section 3) is violated, and f is not in $\mathcal{B}^2$. [This example is due to J.-L. MAUCLAIRE].

Similarly, the function $g(n) = \sqrt{\log n}$ if n is a prime, else $g(n) = 0$, has $\|g\|_1 = 0$, all $a_r(g) = 0$, $\|g\|_2^2 = 1$, and PARSEVAL's equation is violated again.

Example 2. (A. HILDEBRAND). Fix $q > 1$, and put $f(n) = 2^{k/q}$ if $n = 2^k$ is a power of 2, and $f(n) = 0$ otherwise. Then $\|f\|_r = 0$ if $1 \le r < q$, $\|f\|_q > 0$, but f is not in $\mathcal{B}^q$.

[The proof runs as follows: it is easy to calculate $x^{-1}\cdot\sum_{n\le x} f^q(n)$ and to show that $\lim_{x\to\infty} x^{-1}\cdot\sum_{n\le x} f^q(n)$ does not exist (for example, let $x \to \infty$ through the sequences 2^k and $2^{k+1}-1$); therefore the mean-value $M(f^q)$ does not exist, and so f^q is not in $\mathcal{A}^1$].

Proof of Theorem 2.8. Without loss of generality, let f be real-valued. Define the truncation f_K of f by

$$f_K(n) = \begin{cases} f(n), & \text{if } |f(n)| \le K, \\ K, & \text{if } f(n) > K, \\ -K, & \text{if } f(n) < -K. \end{cases}$$

$f \in \mathcal{B}^1$ implies that $f_K \in \mathcal{B}^1$, and - being bounded - the truncation f_K is in $\mathcal{B}^\ell$ for every $\ell \ge 1$. Define $s' > 1$ by $r \cdot s' = q$ and fix s by the equation $s^{-1}+s'^{-1} = 1$. Then, using HÖLDER's inequality,

$$\Delta(x) = x^{-1} \cdot \sum_{n \le x} |f(n) - f_K(n)|^r \le x^{-1} \cdot \sum_{n \le x, |f(n)|>K} |f(n)|^r$$

$$\le \left\{ x^{-1} \cdot \sum_{n \le x} |f(n)|^{rs'} \right\}^{q/(s'q)} \cdot \left\{ x^{-1} \cdot \sum_{n \le x, |f(n)|>K} 1 \right\}^{1/s}.$$

Next,

$$K^q \cdot \sum_{n \le x, |f(n)|>K} 1 \le \sum_{n \le x, |f(n)|>K} |f(n)|^q \le \left(2 \cdot \| f \|_q \right)^q \cdot x,$$

if x is large. Hence, we arrive at

$$\limsup_{x \to \infty} \Delta(x) \le \| f \|_q^{q/s'} \cdot \left(2 \| f \|_q / K \right)^{q/s} < \varepsilon,$$

if K is chosen large enough, and so f, being near $f_K \in \mathcal{B}^r$, is in $\mathcal{B}^r$. □

We state that for real-valued functions f in $\mathcal{B}^r$ the truncated function f_K tends to f in $\| . \|_r$, and that, for any $\varepsilon > 0$,

$$\limsup_{x \to \infty} x^{-1} \cdot \sum_{n \le x, |f(n)|>K} |f(n)|^r < \varepsilon \tag{2.10}$$

if K is sufficiently large.

Theorem 2.9 [DABOUSSI]. *Assume* $g \ge 0$, $\alpha \ge 1$, $\beta \ge 1$. *Then*

$$g^\alpha \in \mathcal{A}^\beta \textit{ if and only if } g \in \mathcal{A}^{\alpha \cdot \beta}.$$

The same result is true with $\mathcal{A}$ *replaced by* $\mathcal{D}$ *or* $\mathcal{B}$.

Corollary 2.10. *If* g *is non-negative and in* $\mathcal{A}^q$, *where* $q \ge 1$, *then* $g^{\frac{1}{2}q} \in \mathcal{A}^2$. *The same result is true with* $\mathcal{A}$ *replaced by* $\mathcal{D}$ *or* $\mathcal{B}$.

This corollary comes from Theorem 2.9 with $\beta = 2$, $\alpha = \frac{1}{2}q$ if $q \ge 2$. If $1 \le q < 2$, then put $f = g^{\frac{1}{2}q}$; then $f^{2/q} = g \in \mathcal{A}^q$, and so $f \in \mathcal{A}^{(2/q)\cdot q} = \mathcal{A}^2$.

Proof of Theorem 2.9, following DABOUSSI [1980].
(1) Assume that $g^\alpha \in \mathcal{A}^\beta$, $\varepsilon > 0$. Choose a trigonometric polynomial t in $\mathcal{A}$ such that $\| g^\alpha - t \|_\beta \le (\varepsilon/2)^\alpha$. According to the WEIERSTRASS Approximation Theorem (Theorem A.1.1) there is a polynomial Q with the property

$$| Q(u) - \{ \max(0,u) \}^{1/\alpha} | \le \tfrac{1}{2}\varepsilon \text{ in } |u| \le \|t\|_u.$$

Then the composition $Q \circ t$ is in $\mathcal{A}$ [resp. $\mathcal{B}$ or $\mathcal{D}$ in the other cases], and

$$\text{(2.11)} \qquad \| g - Q \circ t \|_{\alpha\beta} \le \varepsilon$$

gives the assertion $g \in \mathcal{A}^{\alpha\beta}$. In order to prove (2.11) we use the inequalities

(a) $| x-y |^\alpha \le | x^\alpha - y^\alpha |$ in $x \ge 0$, $y \ge 0$,
(b) $(x+y)^{\alpha\beta} \le 2^{\alpha\beta-1} \cdot \left(x^{\alpha\beta} + y^{\alpha\beta} \right)$ in $x \ge 0$, $y \ge 0$,
(c) $| \max(0,y) - \max(0,x) | \le |y-x|$, if x and y are real.

(b) follows from the convexity of $t \mapsto t^{\alpha\beta}$, (a) is proved utilizing the function $t \mapsto (y+t)^\alpha - t^\alpha$ (without loss of generality $x = y+t > y$). Therefore,

$$|g(n) - Q(t(n))|^{\alpha\beta} \le \Big\{ \big| g(n) - \{\max(0,t(n))\}^{1/\alpha} \big| + \big| \{ \max(0,t(n)) \}^{1/\alpha} - Q(t(n)) \big| \Big\}^{\alpha\beta},$$

and using (b), (a) and (c), this becomes

$$\le 2^{\alpha\beta-1} \cdot \Big\{ \big| g^\alpha(n) - t(n) \big|^\beta + \big| \{\max(0,t(n))\}^{1/\alpha} - Q(t(n)) \big|^{\alpha\beta} \Big\}.$$

Therefore

$$\| g - Q \circ t \|_{\alpha\beta} \le \varepsilon.$$

This is one part of the proof.

By Exercise 11 (or Corollary 2.5 (3)) $h \in \mathcal{A}^k$, $k \in \mathbb{N}$, implies $h^k \in \mathcal{A}^1$. In order to prove the other part, put

$$\gamma = ([\alpha \cdot \beta] + 1) \cdot (\alpha\beta)^{-1}.$$

Then $\gamma \geq 1$, and $\alpha \cdot \beta \cdot \gamma \in \mathbb{N}$. The function $h = g^{1/\gamma}$ satisfies $h^\gamma \in \mathcal{A}^{\alpha\beta}$, therefore [according to the first part of our proof] $h \in \mathcal{A}^{\alpha\beta\gamma}$. The number $\alpha\beta\gamma$ is an integer, and so $h^{\alpha\beta\gamma} \in \mathcal{A}^1$. Therefore, $g^{\alpha\beta} \in \mathcal{A}^1$ and the first part of the proof again gives $g^\alpha \in \mathcal{A}^\beta$. □

For the question of the existence of a limit distribution of real-valued functions the following result is useful, as it has already been shown for uniformly-almost-even functions.

Theorem 2.11. *Let* $q \geq 1$.

(1) *If* $f \in \mathcal{B}^q$ *is real-valued with values in some finite [or infinite] closed interval* $I = [a, b]$, *and if the function* $\Psi: I \to \mathbb{C}$ *is* LIPSCHITZ*-continuous (so that* $|\Psi(x) - \Psi(y)| \leq L \cdot |x - y|$ *for some constant* $L > 0$*), then the composed function* $\Psi \circ f$ *is in* $\mathcal{B}^q$ *again. The result remains true, if* $\mathcal{B}^q$ *is replaced by* $\mathcal{D}^q$ *or* $\mathcal{A}^q$.

(2) *If* $f \in \mathcal{B}^q$ *is complex-valued with values in some finite [or infinite] closed rectangle* R, *and if the function* $\Psi: R \to \mathbb{C}$ *is* LIPSCHITZ*-continuous, then the composed function* $\Psi \circ f$ *is in* $\mathcal{B}^q$ *again. The result remains true if* $\mathcal{B}^q$ *is replaced by* $\mathcal{D}^q$ *or* $\mathcal{A}^q$.

(3) *If* f *in* $\mathcal{B}^1$ [*or* $\mathcal{D}^1$ *or* $\mathcal{A}^1$] *is real-valued then the function* $n \mapsto \exp(it \cdot f(n))$ *is in* $\mathcal{B}^1$ [*or* $\mathcal{D}^1$ *or* $\mathcal{A}^1$] *for any real* t.

(4) *If* $q \geq 1$, $f \in \mathcal{B}^q$, f *is real-valued, and* $\inf_{n\in\mathbb{N}} |f(n)| = \delta > 0$, *then* $\frac{1}{f} \in \mathcal{B}^q$.

Proof.

(1) Let $\varepsilon > 0$. Choose a trigonometric polynomial t^* in $\mathcal{B}$ [resp. $\mathcal{D}$] near f, $\| f - t^* \|_q < \varepsilon$. The values of f are in I; t^* is real-valued, without loss of generality. If the values of t^* are not in the interval $I =]a, b[$, replace t^* by $t = \min\{b, \max(t^*, a)\}$ (with an obvious interpretation, if a or b are $\pm\infty$). t is $\| \ \|_q$ - nearer to f than t^*, therefore $\| f - t \|_q < \varepsilon$. Then $\Psi \circ t$ is even and so in $\mathcal{B}$ [resp. periodic and so is in $\mathcal{D}$], the values of f and t are in I, Ψ is LIPSCHITZ-continuous, and therefore

$$\begin{aligned} \| \Psi \circ f - \Psi \circ t \|_q^q &= \limsup_{x \to \infty} x^{-1} \cdot \sum_{n \leq x} \left| \Psi(f(n)) - \Psi(t(n)) \right|^q \\ &\leq \limsup_{x \to \infty} x^{-1} \cdot L^q \cdot \sum_{n \leq x} \left| f(n) - t(n) \right|^q \leq L^q \cdot \varepsilon^q. \end{aligned}$$

In the case where $f \in \mathcal{A}^q$ and $t \in \mathcal{A}$, the function $\Psi \circ t$ is in $\mathcal{A}^u$ by the WEIERSTRASS Theorem, and the proof works in this case, too.

(2) The complex case can be reduced to the real one. Assume that $R = [a_1, b_1] \times i \cdot [a_2, b_2]$. Then approximate Re f by an even function t_1 with values in $[a_1, b_1]$, and Im f by an even function t_2 with values in $[a_2, b_2]$. The even function $t = t_1 + i \cdot t_2$ has values in R, and $\| f - t \|_q \le \| \mathrm{Re}\, f - t_1 \|_q + \| \mathrm{Im}\, f - t_2 \|_q$. The rest may be concluded as in (1).

(3) and (4) are special cases: the functions $x \mapsto \exp(it \cdot x)$, defined on $\mathbb{R}$, where t is any real number, and $y \mapsto y^{-1}$, defined in $y \ge \delta$, are LIPSCHITZ-continuous. Thus $1/|f| \in \mathcal{B}^q$, and $f^{-1} = f \cdot |f|^{-2} \in \mathcal{B}^q$ by Theorem 2.6 (iii). □

Examples.

(a) If f is a bounded function in $\mathcal{A}^1$, and P a polynomial with complex coefficients, then the composed function $P \circ f$ is also in $\mathcal{A}^1$.
This follows from Theorem 2.11, but it could also be deduced from the fact that a bounded function in $\mathcal{A}^1$ is in $\mathcal{A}^q$ for every $q \ge 1$.

(b) If $f \in \mathcal{A}^q$ satisfies $b := \sup \mathrm{Re}(f(\mathbb{N})) < \infty$, then $\exp(f) \in \mathcal{A}^q$. The reason is that exp is LIPSCHITZ-continuous in the half-plane $\{z \in \mathbb{C};\ \mathrm{Re}\, z \le b\}$.

(c) If $f \in \mathcal{A}^q$ and $a := \inf(\mathrm{Re}\, f(\mathbb{N})) > 0$, then $\log f \in \mathcal{A}^q$, because the principal branch of the logarithm function is LIPSCHITZ-continuous in the half-plane $\{z \in \mathbb{C};\ \mathrm{Re}\, z \ge a\}$ with $L = a^{-1}$.

Remark. If P is an integer-valued polynomial with positive values, for example $P(n) = n^2+1$, then it is a difficult task to prove that $f \circ P$ is in $\mathcal{A}^1$ (or has a mean-value at least) if f is in some $\mathcal{A}^q$. The result is not known even for the function μ^2, if the degree of P is greater than two.

VI.3. PARSEVAL'S EQUATION

According to section 2 of this chapter the spaces $\mathcal{B}^2 \subset \mathcal{D}^2 \subset \mathcal{A}^2$ are complete vector-spaces with an "inner product"

$$\langle f, g \rangle = M(f \cdot \overline{g}).$$

This "inner product" is linear in the first argument; it satisfies

$\overline{\langle f, g \rangle} = \langle g, f \rangle$ and $\langle f, f \rangle \geq 0$, but $\langle f, f \rangle = 0$ is possible for functions $f \neq 0$. Thus, the quotient-spaces modulo null-functions, $B^2 \subset D^2 \subset A^2$, are HILBERT spaces.

Theorem 3.1 (PARSEVAL's equation).

(i) *If* f *is in* $\mathcal{B}^2$, *then*

$$\sum_{r=1}^{\infty} \varphi(r) \cdot |a_r(f)|^2 = \| f \|_2^2,$$

where the $a_r(f)$ *denote the* RAMANUJAN-FOURIER *coefficients*

$$a_r(f) = \frac{1}{\varphi(r)} \cdot M(f \cdot c_r),\ r = 1, 2, \ldots .$$

(ii) *If* f *is in* $\mathcal{D}^2$, *then*

$$\sum_{r=1}^{\infty} \sum_{1 \leq a \leq r, \gcd(a,r)=1} | M(f \cdot \overline{e_{a/r}}) |^2 = \| f \|_2^2 .$$

(iii) *If* f *is in* $\mathcal{A}^2$, *then*

$$\sum_{\alpha \in \mathbb{R}/\mathbb{Z}} | M(f \cdot \overline{e_\alpha}) |^2 = \| f \|_2^2 .$$

Corollary 3.2.

(i) *The set* $\{ (\varphi(r))^{-\frac{1}{2}} \cdot c_r,\ r = 1, 2, \ldots \}$ *is a complete orthonormal system in* $\mathcal{B}^2$. *If* f, g *are in* $\mathcal{B}^2$, *then*

$$\sum_{r=1}^{\infty} \varphi(r) \cdot a_r(f) \cdot \overline{a_r(g)} = M(f \cdot \overline{g}).$$

(ii) *The set* $\{ e_{a/r},\ r = 1, 2, \ldots, 1 \leq a \leq r, \gcd(a,r) = 1 \}$ *is a complete orthonormal system in* $\mathcal{D}^2$. *If* f, g *are in* $\mathcal{D}^2$, *then*

$$\sum_{r=1}^{\infty} \sum_{1 \leq a \leq r, \gcd(a,r)=1} M(f \cdot \overline{e_{a/r}}) \cdot \overline{M(g \cdot \overline{e_{a/r}})} = M(f \cdot \overline{g}).$$

(iii) *If* f, g *are in* $\mathcal{A}^2$, *then*

$$\sum_{\alpha \in \mathbb{R}/\mathbb{Z}} M(f \cdot \overline{e_\alpha}) \cdot \overline{M(g \cdot \overline{e_\alpha})} = M(f \cdot \overline{g}).$$

First Proof. The assertions of Corollary 3.2 come from the "Elementary Theory of HILBERT space", which is sketched in Appendix A.2. According to this theory, the validity of the PARSEVAL equation is equivalent to the denseness [with respect to $\|.\|_2$] of the sets $\mathcal{B}$, $\mathcal{D}$, $\mathcal{A}$ of linear combinations of RAMANUJAN sums [resp. exponential functions] in $\mathcal{B}^2$, $\mathcal{D}^2$, and $\mathcal{A}^2$; this is true by definition of these spaces. □

VI.4. A SECOND PROOF FOR PARSEVAL'S FORMULA

In this section we present a second proof for PARSEVAL's equation in the space $\mathcal{B}^2$. Some properties, perhaps of some interest, of arithmetical functions in $\mathcal{B}^2$ are exhibited, and these properties are used in the proof. Let r be a positive integer, and, for k dividing r, denote by χ_k the characteristic function of the set

$$A_k = \{n \in \mathbb{N}: \gcd(n,r) = k\}.$$

χ_k is a function in $\mathcal{B}_r \subset \mathcal{B}$ (with positive mean-value), therefore $f \cdot \chi_k \in \mathcal{B}^2$ for every f in $\mathcal{B}^2$. Consider the linear map

$$F_r: \mathcal{B}^2 \to \mathcal{B}_r, \; f \mapsto \sum_{k|r} \frac{M(f \cdot \chi_k)}{M(\chi_k)} \cdot \chi_k.$$

This function has the properties given in the following lemma.

Lemma 4.1.

(1) $F_r(f) = f$ *if and only if* $f \in \mathcal{B}_r$.

(2) *If* $f, g \in \mathcal{B}^2$, *then* $M(F_r(f) \cdot g) = M(f \cdot F_r(g))$.

(3) *If* $f \in \mathcal{B}^2$, *then* $F_r(f) = \sum_{k|r} a_k \cdot c_k$, *where* $a_k = \frac{1}{\varphi(k)} \cdot M(f \cdot c_k)$.

(4) *If* $f \in \mathcal{B}^2$, $g \in \mathcal{B}_r$, *then* $\| f - F_r(f) \|_2 \le \| f - g \|_2$. *So* $F_r(f)$ *is a "best" approximation in* $\mathcal{B}^r$.

(5) *For every* f *in* $\mathcal{B}^2$, *the sequence*

$$\Delta_R(f) = \| f - F_{R!}(f) \|_2, \; R = 1, 2, \ldots,$$

is monotonically decreasing to zero.

Proof. (1) A function f in $\mathcal{B}_r$ is constant on A_k, say equal to d_k, for every k dividing r. Therefore, $F_r(f) = \sum_{k|r} d_k \cdot \chi_k = f$.

(2) By definition of F_r and the linearity of the mean-value, we obtain

$$M(F_r(f) \cdot g) = \sum_{k|r} \frac{M(f \cdot \chi_k)}{M(\chi_k)} \cdot M(\chi_k \cdot g).$$

This expression is symmetric in f and g, and so also is equal to $M(F_r(g) \cdot f)$.

(3) By the orthogonality of the RAMANUJAN sums the coefficient a_k, using (2) and (1), equals

$$a_k = \{\varphi(k)\}^{-1} \cdot M(F_r(f) \cdot c_k) = \{\varphi(k)\}^{-1} \cdot M(F_r(c_k) \cdot f) = \{\varphi(k)\}^{-1} \cdot M(c_k \cdot f).$$

(4) Without loss of generality, we assume that f and g are real-valued. Let $x > 0$, $k|r$, and define the function

$$G_{x,k}: \mathbb{R} \to \mathbb{R},\ y \mapsto x^{-1} \cdot \sum_{n \le x, n \in A_k} (f(n) - y)^2.$$

This function has just one stationary point

$$m_x = \left(\sum_{n \le x} \chi_k(n)\right)^{-1} \cdot \left(\sum_{n \le x} f(n) \cdot \chi_k(n)\right) = \left(M(\chi_k)\right)^{-1} \cdot M(f \cdot \chi_k) + o(1)$$

as $x \to \infty$, and this point gives the absolute minimum of $G_{x,k}$. Therefore,

$$x^{-1} \cdot \sum_{n \le x, n \in A_k} \left(f(n) - F_r(f)(n)\right)^2 = x^{-1} \cdot \sum_{n \le x, n \in A_k} \left(f(n) - (M(\chi_k))^{-1} \cdot M(f \cdot \chi_k)\right)^2$$

$$= x^{-1} \cdot \sum_{n \le x, n \in A_k} (f(n) - m_x)^2 + o(1) \le x^{-1} \cdot \sum_{n \le x, n \in A_k} \left(f(n) - g(n)\right)^2 + o(1).$$

Summing over $k|r$, we find for $x \to \infty$

$$\| f - F_r(f) \|_2^2 \le \| f - g \|_2^2.$$

(5) At first $\Delta_{R+1}(f) \le \Delta_R(f)$, by (4). Now, given $\varepsilon > 0$, there exists an even function $g \in \mathcal{B}$ "near" f, $\| f - g \|_2 < \varepsilon$. Choose an integer R, for which g is in $\mathcal{B}_R$. Then

$$\| f - F_{R!}(f) \|_2 < \varepsilon,$$

again by (4). □

Now we are ready to prove PARSEVAL's equation

$$\sum_{r=1}^{\infty} \varphi(r) \cdot |a_r(f)|^2 = M(|f|^2) = \|f\|_2^2.$$

For every $f \in \mathcal{B}^2$, and for every integer R, a standard computation gives

$$\| f - \sum_{r|R!} a_r \cdot c_r \|_2^2 = \langle f - \sum_r a_r \cdot c_r , f - \sum_k a_k \cdot c_k \rangle$$

$$= \| f \|_2^2 - \sum_{r|R!} \varphi(r) \cdot |a_r|^2.$$

By (3), the left-hand side is $\| f - F_{R!}(f) \|_2^2$, which converges to zero by (5). □

VI.5. AN APPROXIMATION FOR FUNCTIONS IN $\mathcal{B}^1$

In the last section, the result

$$\| f - \sum_{r|R!} a_r(f) \cdot c_r \|_2 \to 0, \text{ as } R \to \infty,$$

was proved by [elementary] HILBERT-space methods. In this section, a similar result for arithmetical functions in $\mathcal{B}^1$ is given.

Theorem 5.1 (A. HILDEBRAND). *For every function* f *in* $\mathcal{B}^1$

(5.1) $$\lim_{R\to\infty} \| f - \sum_{r|R!} a_r(f) \cdot c_r \|_1 = 0,$$

where $a_r(f) = \{\varphi(r)\}^{-1} \cdot M(f \cdot c_r)$, $r = 1, 2, \ldots$ *denote the* RAMANUJAN *coefficients of the function* f.

The important feature of this result is that the coefficients of the even functions approximating f are *not changed* when R is increased. Note that the sequence $\{R!\}_{R=1,2,\ldots}$ may be substituted by every sequence $\{n_R\}_{R=1,2,\ldots}$ with the property $\lim_{R\to\infty} \gcd(n_R, r) = r$ for every integer r.

Remark. This theorem allows us to show [again] that the MÖBIUS function μ, with $\|\mu\|_1 = \frac{6}{\pi^2} > 0$, does not belong to $\mathcal{B}^1$. It is known from prime number theory that

$$M_d(\mu) = \lim_{x\to\infty} x^{-1} \cdot \sum_{n\le x, n\equiv 0 \bmod d} \mu(n) = 0$$

for every integer d: Therefore,

(5.2) $$\varphi(r) \cdot a_r(\mu) = \lim_{x\to\infty} x^{-1} \cdot \sum_{n\le x} \mu(n) \cdot c_r(n) = \sum_{d|r} d \cdot \mu(r/d) \cdot M_d(\mu) = 0.$$

First we collect some formulae, needed for the proof of Theorem 5.1, as follows.

Lemma 5.2. (1) *For every integer* k, r, R *satisfying* $r|R!$,

(5.3) $$\frac{1}{R!} \cdot \sum_{\substack{n\le R! \\ n\equiv 0\ (k)}} c_r(n) = \begin{cases} 0, & \text{if } r\nmid k, \\ \frac{\varphi(r)}{k}, & \text{if } r|k. \end{cases}$$

(2) *Denote by* χ_k *the characteristic function of the set* $A_k = \{ n \in \mathbb{N}: \gcd(n, R!) = k\}$, *where* $k|R!$ *is supposed. Put* $F_{R!} = \sum_{r|R!} a_r(f) \cdot c_r$, *for* $f \in \mathcal{B}^1$. *Then, for every* $k|R!$,

$$M(f \cdot \chi_k) = M(F_{R!} \cdot \chi_k). \tag{5.4}$$

Proof. (1) The RAMANUJAN sum c_r is $R!$ - even (and so $R!$-periodic) if $r | R!$, therefore the left-hand side of (5.3) is equal to

$$\begin{aligned}\lim_{x \to \infty} x^{-1} \cdot \sum_{n \le x, k|n} c_r(n) &= \sum_{d|r} d \cdot \mu(r/d) \cdot \lim_{x \to \infty} x^{-1} \sum_{n \le x, d|n, k|n} 1 \\ &= k^{-1} \cdot \sum_{d|r} \gcd(d,k) \cdot \mu(r/d) \\ &= k^{-1} \cdot \prod_{p^\ell \| r} \left(\gcd(p^\ell, k) - \gcd(p^{\ell-1}, k)\right),\end{aligned}$$

and this gives the right-hand side of (5.3).

(2) $F_{R!}$ is $R!$ - even, and so we obtain

$$\begin{aligned}\lim_{x \to \infty} x^{-1} \cdot \sum_{n \le x, k|n} F_{R!}(n) &= \frac{1}{R!} \cdot \sum_{\substack{n \le R! \\ k|n}} \sum_{r|R!} a_r(f) \cdot c_r(n) \\ &= \lim_{x \to \infty} x^{-1} \cdot \sum_{m \le x} f(m) \cdot \sum_{r|R!} \frac{c_r(m)}{\varphi(r)} \cdot \frac{1}{R!} \cdot \sum_{\substack{n \le R! \\ k|n}} c_r(n) \\ &= k^{-1} \cdot \lim_{x \to \infty} x^{-1} \cdot \sum_{m \le x} f(m) \cdot \sum_{r|k} c_r(m),\end{aligned}$$

using (1) and the fact that k divides $(R!)$. The inner sum $\sum_{r|k} c_r(m)$ is equal to

$$\prod_{p^\ell \| k} (1 + c_p(m) + \dots + c_{p^\ell}(m)) = \begin{cases} k, & \text{if } k|m, \\ 0, & \text{if } k \nmid m. \end{cases} \tag{5.5}$$

Therefore, the functions $F_{R!}$ and f have the same mean-value on the sets $M_k = \{n \in \mathbb{N}: k|n\}$, if $k|R!$, and so also on

$$A_k = M_k \setminus \bigcup_{\substack{\ell|R!, \ell \ne k \\ \ell \equiv 0 \bmod k}} M_\ell. \tag{5.6}$$ □

Proof of Theorem 5.1. We start by proving that for any real-valued function f in $\mathcal{B}^1$ and every real-valued k-even function g the estimate

(5.7) $$\|f - F_{R!}\|_1 \le 2 \cdot \|f-g\|_1, \text{ if } k|R!$$
holds.

$F_{R!}$ and g are R! - even and therefore constant on every set A_k if $k|R!$ (for the definition of A_k see Lemma 5.2 (2)). Denote the values, taken by $F_{R!}$ and g on A_k, by γ_k [resp. δ_k]. Fix k, and assume that $\gamma_k \ge \delta_k$. Then

$$\sum_{\substack{n\le x\\ n\in A_k}} |f(n) - F_{R!}(n)| = \sum_{\substack{n\le x\\ n\in A_k\\ f(n)\ge\gamma_k}} (f(n)-\gamma_k) + \sum_{\substack{n\le x\\ n\in A_k\\ f(n)<\gamma_k}} (\gamma_k - f(n))$$

$$= 2\cdot \sum_{\substack{n\le x\\ n\in A_k\\ f(n)\ge\gamma_k}} (f(n)-\gamma_k) - \sum_{\substack{n\le x\\ n\in A_k}} (f(n)-\gamma_k),$$

and, using Lemma 5.2 (2), this is

$$\le 2\cdot \sum_{\substack{n\le x\\ n\in A_k\\ f(n)\ge\gamma_k}} (f(n)-\delta_k) + o(x) \le 2\cdot \sum_{\substack{n\le x\\ n\in A_k}} |f(n)-g(n)| + o(x).$$

In the other case, $\gamma_k < \delta_k$, the same estimate is valid. The sets A_k, $k|R!$, are a partition of $\mathbb{N}$. Therefore, we obtain

$$x^{-1}\cdot\sum_{n\le x} |f(n) - F_{R!}(n)| \le 2\cdot x^{-1}\cdot \sum_{n\le x} |f(n)-g(n)| + o(1),$$

and (5.7) is proved.

To conclude the proof of Theorem 5.1, assume without loss of generality that $f \in \mathcal{B}^1$ is real-valued. Given $\varepsilon > 0$, choose a real-valued even function g near f, $\|f-g\|_1 < \varepsilon$. If g is k-even, then, according to (5.7), $\|f-F_{R!}\|_1 < 2\varepsilon$ for all $R \ge k$. □

VI.6. LIMIT DISTRIBUTIONS OF ARITHMETICAL FUNCTIONS

For a set $\mathcal{a}$ contained in $\mathbb{N}$ we define the counting function

$$A(N) = \sum_{n\le N, n\in\mathcal{a}} 1.$$

If f is a real-valued arithmetical function we put

$$F_{N,\mathfrak{a}}(t) = \{A(N)\}^{-1} \cdot \sum_{n \le N, n \in \mathfrak{a}, f(n) \le t} 1,$$

and in case $\mathfrak{a} = \mathbb{N}$ we write

$$F_N(t) = F_{N,\mathbb{N}}(t)$$

for short. As mentioned in Chapter IV, section 3, one says that f has a limit distribution with respect to $\mathfrak{a}$ if there is a distribution function[1) F such that

$$\lim_{N \to \infty} F_{N,\mathfrak{a}}(t) = F(t)$$

at every point of continuity of F. An additive function f has a limit distribution, according to P. ERDÖS and A. WINTNER's Theorem. [1939], on $\mathbb{N}$ if and only if the three series

$$\sum_{p, |f(p)| \le 1} p^{-1} \cdot f(p), \ \sum_{p, |f(p)| \le 1} p^{-1} \cdot |f(p)|^2, \ \sum_{p, |f(p)| > 1} p^{-1}$$

converge.

H. DABOUSSI [1981] gives examples of large classes of non-negative additive functions and sets $\mathfrak{a}$ having <u>no</u> limit distribution. He proved the following theorem.

Theorem 6.1. *Let* $\mathfrak{a} \subset \mathbb{N}$ *satisfy*

(i) $\lim_{N \to \infty} \{A(N)\}^{-1} \cdot \sum_{n \le N, n \in \mathfrak{a}, n \equiv 0 \bmod d} 1 = d^{-1} \cdot \omega(d)$

exists for every $d \in \mathbb{N}$, *where*

(ii) ω *is multiplicative and* $\sum_p \sum_{k \ge 2} p^{-k} \cdot \omega(p^k) < \infty$.

If f *is a non-negative additive function such that*

$$\sum_{0 \le f(p) \le 1} p^{-1} \cdot f(p) \cdot \omega(p) + \sum_{f(p) > 1} p^{-1} \cdot \omega(p) = +\infty, \tag{6.1}$$

then f *does* **not** *have a limit distribution on* $\mathfrak{a}$. *More precisely*

$$\lim_{N \to \infty} F_{N,\mathfrak{a}}(t) = 0 \text{ for every } t.$$

---

1) So $F \ge 0$, $F(-\infty) = 0$, $F(+\infty) = 1$, F is non-decreasing, and continuous from the right.

Corollary 6.2. *If* $\mathfrak{a}$ *is the set* $\{p+1\}$ *of translates of the primes, then* $\omega(d) = d/\varphi(d)$. *Thus, if* f *is additive and non-negative and satisfies* (6.1) *with* $\omega = \mathrm{id}/\varphi$, *then* f *does* **not** *have a limit distribution on* $\{p+1;\ p \text{ prime}\}$.

Proof of Theorem 6.1. The inequalities

$$(1-e^{-1})\cdot t \le 1-e^{-t} \le t \text{ in } 0 \le t \le 1,\ (1-e^{-1}) \le 1-e^{-t} \le 1 \text{ in } t \ge 1$$

(see Figure VI.3) imply

$$\sum_p p^{-1}\cdot\omega(p)\cdot(\ 1-e^{-f(p)}\ ) = \infty.$$

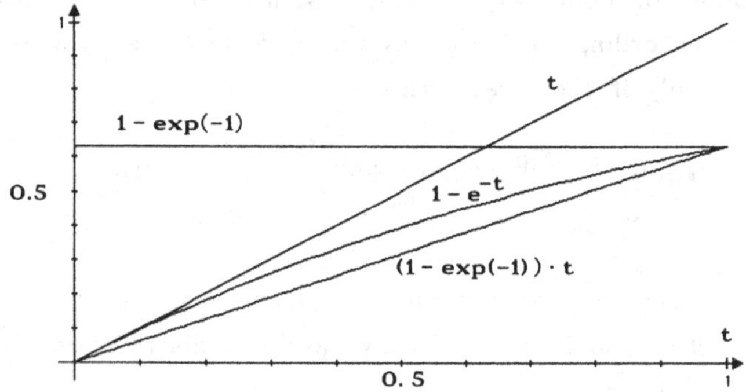

Figure VI.3

Define an additive function f_y by truncation: $f_y(p^k) = f(p^k)$ if $p^k \le y$, and $f_y(p^k) = 0$ otherwise. The convolution $g_y = \mu * \exp(-f_y)$ is multiplicative, $|g_y(p^k)| \le 2$, and $g_y(n) = 0$ except on a **finite** set S_y of integers, as is seen from the values of g_y at prime-powers, as well as the relation

$$G_y = \sum_n n^{-1}\cdot\omega(n)\cdot g_y(n) = \prod_{p\le y}\Big(1 - p^{-1}(1-e^{-f(p)})\cdot\omega(p) + \sum_{k\ge 2} p^{-k}g_y(p^k)\cdot\omega(p^k)\Big).$$

The convergence of the series $\sum_p\sum_{k\ge 2}p^{-k}g_y(p^k)\cdot\omega(p^k)$ and the divergence of $\sum_p p^{-1}(1-e^{-f(p)})\cdot\omega(p) = \infty$ imply $\lim_{y\to\infty} G_y = 0$. Next, $\exp(-f_y) = 1*g_y$, and so

$$\lim_{N\to\infty}\ \{A(N)\}^{-1}\cdot \sum_{n\le N, n\in\mathfrak{a}} \exp(\ -f_y(n)) = G_y.$$

Since $\exp\big(-f(n)\big) \le \exp\big(-f_y(n)\big)$, we obtain

$$\limsup_{N\to\infty}\ \{A(N)\}^{-1}\cdot \sum_{n\le N, n\in\mathfrak{a}} \exp(\ -f(n)) \le G_y \to 0, \text{ as } y\to\infty.$$

But

$$\{A(N)\}^{-1} \cdot \sum_{n \le N, n \in \mathcal{A}} \exp(-f(n)) = \int_0^\infty e^{-x} dF_{N,\mathcal{A}}(x) \ge e^{-t} \cdot F_{N,\mathcal{A}}(t),$$

and Theorem 6.1 is proved. □

[Positive] results on the existence of limit distributions for additive functions will be given in the next section (see Theorem 7.2).

VI.7. ARITHMETICAL APPLICATIONS

VI.7.A. Mean-Values, Limit Distributions.

In number theory the question of the existence of a mean-value is an important one. Some general mean-value theorems were proved in Chapter II, and a wealth of theorems are known on the existence of mean-values for special functions, in particular for multiplicative functions.

Functions in $\mathcal{A}^1$ do have a mean-value. Sections 1 and 2 of this chapter gave many results, producing new functions in $\mathcal{A}^1$ from "simpler" ones. So these results (Theorems 1.2, 2.3 [(3) - (6)], 2.4, 2.11, Corollaries 2.5, 2.7,) have consequences on the existence of mean-values, regardless of arithmetical properties such as multiplicativity or additivity. Of course, to make these results useful, it is necessary to provide criteria for functions f to be in $\mathcal{A}^1$. Some criteria of this kind will be given in Chapter VII - but then multiplicativity is relevant.

We formulate some examples of results on the existence of mean-values for functions in the spaces considered in this chapter. Proofs need not be given; they consist of the remark that the function in question is in $\mathcal{A}^1$, which follows from results given earlier in this chapter.[1)]

---

1) $\mathcal{A}^1$ is the largest of the spaces, defined in this chapter. So it is advisable to formulate conditions ensuring that some resulting function is in $\mathcal{A}^1$.

(1) If f is in $\mathcal{A}^1$, then the mean-values M(f), M(|f|), M(Re(f)), M(Im(f)) exist.

(2) If f is in $\mathcal{A}^1$ and g is in $\mathcal{A}^u$ [for example, g is a DIRICHLET character, a RAMANUJAN sum, an exponential function e_α, the characteristic function of a residue-class a mod k, where gcd(a,k) = 1, the characteristic function of the integers prime to [some fixed] k, etc., ...], then the mean-value $M(g \cdot f)$ exists.

(3) If f is in $\mathcal{A}^1$, then the mean-values of the [additively, resp. multiplicatively] shifted functions $f_{(a+)}: n \mapsto f(n+a)$, $f_{(b\cdot)}: n \mapsto f(b \cdot n)$, where $a \in \mathbb{Z}$, $b \in \mathbb{N}$, exist.
[As long as $n+a \le 0$, put $f(n+a) = 1$ for accuracy.]

(4) If $f, g \in \mathcal{A}^2$, then $M(f \cdot g_{(a+)})$ exists ($a \in \mathbb{Z}$); if $q, q' > 1$, $q^{-1} + q'^{-1} = 1$, f in $\mathcal{A}^q$, $g \in \mathcal{A}^{q'}$, then $M(f \cdot g_{(a+)})$ exists.

(5) If $f_1, \dots, f_k$ are in $\mathcal{A}^k$, and if $b_\varkappa > 0$, $a_\varkappa$ are integers ($\varkappa = 1, \dots, k$), then the function

$$F: n \mapsto f_1(b_1 n + a_1) \cdot \dots \cdot f_k(b_k n + a_k)$$

has a mean-value.

[This is a generalization of L. LUCHT's results; this author only dealt with multiplicative functions, but he obtained product formulae for the mean-values (see L. LUCHT [1979a, 1979b]). The continuity theorem for DIRICHLET series (see the Appendix) might be helpful in calculating the mean-value of the function F given in (5) in the case of multiplicative functions, but some additional conditions seem to be necessary to obtain "nice" results.]

(6) If f is real-valued and is in $\mathcal{A}^q$, where $q \ge 1$, and if the image $f(\mathbb{N})$ is contained in a closed interval $I \subset \mathbb{R}$, and if $\Psi: I \to \mathbb{C}$ is LIPSCHITZ-continuous, then the composed function $\Psi \circ f$ is in $\mathcal{A}^q$, and so it has a mean-value.

Examples for Ψ are the functions $z \mapsto z^{-1}$, $z \mapsto \exp(z)$, $z \mapsto \log(z)$, etc. Of course, one has to be careful about $f(\mathbb{N})$, and some assumptions on the values of f are necessary before (6) or other versions of Theorem 2.11 are applicable.

(a) Let $q \geq 1$. If $f \in \mathcal{A}^q$ is real-valued, and if $\inf_{n \in \mathbb{N}} |f(n)| = \delta > 0$, then $1/f \in \mathcal{A}^q$.

(b) If $f \in \mathcal{A}^q$ is complex-valued, and if $\sup_{n \in \mathbb{N}} \mathrm{Re}(f(n)) \leq K < \infty$, then $\exp(f) \in \mathcal{A}^q$.

(c) If $f \in \mathcal{A}^q$ is complex-valued, and if $\inf_{n \in \mathbb{N}} \mathrm{Re}(f(n)) \geq \delta > 0$, then $\log f \in \mathcal{A}^q$.

The calculation of the mean-value can (given appropriate circumstances) be dealt with by an application of the continuity theorem for DIRICHLET series:

Theorem 7.1. *If* $f\colon \mathbb{N} \to \mathbb{C}$ *has a mean-value* $M(f)$, *then*

$$M(f) = \lim_{\sigma \to 1+} \sum_{n=1}^{\infty} f(n) \cdot n^{-\sigma} \cdot \zeta^{-1}(\sigma). \tag{7.1}$$

In particular, if f is multiplicative, then the calculation of the limit (7.1) often is rather simple.

Proof. The existence of the limit M(f) implies

$$\sum_{n \leq x} f(n) = M(f) \cdot x + o(x), \text{ as } x \to \infty.$$

Partial summation gives, as long as $\sigma > 1$,

$$\sum_{n \leq x} f(n) \cdot n^{-\sigma} = \sum_{n \leq x} f(n) \cdot x^{-\sigma} - \int_1^x \sum_{n<u} f(n) \cdot (-\sigma) \cdot u^{-\sigma-1}\, du,$$

and so

$$\begin{aligned} \sum_{n=1}^{\infty} f(n) \cdot n^{-\sigma} &= M(f) \cdot \sigma \cdot (\sigma-1)^{-1} + \int_1^{\infty} o(u) \cdot u^{-\sigma-1}\, du \\ &= M(f) \cdot \sigma \cdot (\sigma-1)^{-1} + o((\sigma-1)^{-1}), \text{ as } \sigma \to 1+. \end{aligned}$$

The asymptotic relation $\zeta(\sigma) = (\sigma-1) + o((\sigma-1)\ )$, as $\sigma \to 1+$, gives the assertion. □

According to the continuity theorem for characteristic functions (see Chapter IV, section 3) the question of the existence of a limit distribution in the sense of probability theory is a problem of the existence (and continuity) of certain mean-values. We prove the following theorem.

Theorem 7.2. *If* g *is a real-valued arithmetical function in* $\mathcal{A}^1$, *then there is a* <u>*limit distribution*</u> *for* g; *this means that the limit*

$$\lim_{N \to \infty} N^{-1} \cdot \#\{n \leq N;\ g(n) \leq x\} = \Psi_g(x)$$

exists in the sense of probability theory.

For the **proof** it has to be shown that the mean-value

$$M_t = M\big(n \mapsto \exp\{ itg(n)\}\big) \tag{7.2}$$

exists for any real t, and that the function $t \mapsto M_t$ is continuous at $t = 0$. According to Theorem 2.11, the function $n \mapsto \exp\{ itg(n)\}$ is in $\mathcal{A}^1$; hence the mean-values M_t exist. The continuity of $t \mapsto M_t$ follows from the estimate

$$\Big| \lim_{x \to \infty} x^{-1} \cdot \sum_{n \le x} \Big(e^{itg(n)} - 1 \Big) \Big| \le \limsup_{x \to \infty} x^{-1} \cdot \sum_{n \le x} |t \cdot g(n)|$$

$$= |t| \cdot \|g\|_1. \qquad \square$$

VI.7.B. Applications to Power-Series with Multiplicative Coefficients.

Given an arithmetical function f, the region of analyticity for the generating power-series

$$F(z) = \sum_{n=1}^{\infty} f(n) \cdot z^n \tag{7.3}$$

may be of some interest. In some sense "most" power series with radius of convergence equal to 1 are non-continuable across the unit disc in the complex plane [see, for example, L. BIEBERBACH [1955]]. Of course, a number theorist would like to obtain an answer to the question of non-continuability of F(z) if the coefficients of this power series are arithmetical functions with some arithmetical property.

G. PÓLYA and G. SZEGÖ's Theorems [see BIEBERBACH [1955]] state: if the coefficients f(n) of the power series (7.3) with radius of convergence equal to one are integers [resp. assume at most finitely many distinct values], then either F represents a rational function or it is analytically non-continuable beyond the unit circle.

For *multiplicative* arithmetical functions L. LUCHT and F. TUTTAS [1979] proved the following result.

If f *is a multiplicative function with finite [semi-]norm* $\|f\|_2$, *and if the mean-value* M(f) *exists and is non-zero, then the power series* $F(z) = \sum_{n=1}^{\infty} f(n)\cdot z^n$ *is non-continuable beyond its circle of convergence if and only if*

$$\frac{f(p^{k-1})}{\varphi(p^k)} \neq \sum_{\nu\geq k} p^{-\nu}\cdot f(p^\nu)$$

for infinitely many prime-powers p^k. *Otherwise F(z) represents a rational function.*

This theorem relates special properties of the coefficients of the power series to the global behaviour of the function represented by this series. For example, the power series $\sum_{n=1}^{\infty} (n^{-1}\cdot\varphi(n))\cdot z^n$, and $\sum_{n=1}^{\infty} (n^{-1}\cdot\sigma(n))\cdot z^n$ are non-continuable. The LUCHT-TUTTAS condition is, in fact, a condition related with the RAMANUJAN coefficients of the arithmetical function f. We are going to show that the property *"multiplicativity"* does not play an essential role; more important is that the RAMANUJAN coefficients $a_r(f) = \{\varphi(r)\}^{-1}\cdot M(f\cdot c_r)$ do not vanish "too often".

Theorem 7.3. *Let* $f \in \mathcal{B}^2$.

(i) *If infinitely many of the* RAMANUJAN*-[*FOURIER*]-coefficients* $a_r(f) = \{\varphi(r)\}^{-1}\cdot M(f\cdot c_r)$ *are non-zero, then* $F(z) = \sum_{n=1}^{\infty} f(n)\cdot z^n$ *is non-continuable beyond the unit circle.*

(ii) *If only finitely many coefficients* $a_r(f)$ *are non-zero, and if* f *is represented [pointwise] by its* RAMANUJAN *expansion,*

$$f(n) = \sum a_r(f)\cdot c_r(n), \textit{ for } n = 1, 2, \ldots,$$

then the power-series $F(z) = \sum_{n=1}^{\infty} f(n)\cdot z^n$ *represents a rational function.*

Remark 1. By HILDEBRAND's Theorem (V, Theorem 1.2) the RAMANUJAN expansion is convergent to the correct values f(n) if f is in $\mathcal{B}^u$. Later we shall show that the same is true for multiplicative functions in $\mathcal{B}^2$, supposed that $M(f) \neq 0$ (see VIII, Theorem 5.1).

Remark 2. Using formulae for the RAMANUJAN coefficients, which will be deduced in Chapter VIII (see VIII, Theorem 4.4), it is easy to show that in case of multiplicative functions the non-vanishing condition of infinitely many RAMANUJAN coefficients is equivalent to LUCHT's condition given above.

Remark 3. The assumption $f \in \mathcal{B}^2$ may be replaced by $f \in \mathcal{B}^1$. The proof has to be changed in so far as PARSEVAL's equation has to be replaced by a result by A. HILDEBRAND, proved in section 5 (Theorem 5.1).

Remark 4. Differentiation does not destroy the property of being rational or non-continuable beyond the unit circle. Therefore, the result can easily be extended by replacing the assumption $f \in \mathcal{B}^2$ with:

There is some non-negative integer k *such that* $n \mapsto n^{-k}\cdot f(n)$ *is in* $\mathcal{B}^2$.

Example. Theorem 7.3 is no longer true if $f \in \mathcal{B}^2$ is replaced by $f \in \mathcal{D}^u$. This may be seen from the function f, defined by the uniformly convergent series

$$f(n) = \sum_{1\le k<\infty} k^{-2}\cdot \exp\left\{2\pi i\cdot(n/k)\right\}$$

This function is in $\mathcal{D}^u$; the power series

$$\sum_{n=1}^{\infty} f(n)\cdot z^n = \sum_{k=1}^{\infty} k^{-2}\cdot\left(1 - \exp(2\pi i/k)\cdot z\right)^{-1}\cdot \exp(2\pi i/k\cdot z)$$

is continuable beyond the unit circle, but it is *not* a rational function.

Proof of Theorem 7.3. PARSEVAL's equation gives

$$(7.4)\qquad \| f - \sum_{1\le r\le R} a_r(f)\cdot c_r \|_2^{\,2} = \sum_{r>R} \varphi(r)\cdot |a_r(f)|^2 < \varepsilon$$

if $R \ge R_o(\varepsilon)$ is sufficiently large. The generating power series for the function $\sum_{1\le r\le R} a_r(f)\cdot c_r$ is

$$\mathcal{R}(z) = \sum_{n=1}^{\infty} \left\{ \sum_{1\le r\le R} a_r(f)\cdot c_r(n) \right\}\cdot z^n$$

$$= \sum_{1\le r\le R} a_r(f)\cdot \sum_{\omega \bmod r} \omega\cdot z\cdot (1-\omega z)^{-1},$$

where ω runs through the primitive r^{th} roots of unity,

$$\omega = \omega_{a,r} = \exp(2\pi i\cdot(a/r)),\ \gcd(a,r) = 1.$$

This function $\mathcal{R}(z)$ is a rational function and is "near" $\sum_{n=1}^{\infty} f(n)\cdot z^n$ in the following sense:

$$\left| \sum_{n=1}^{\infty} f(n)\cdot z^n - \mathcal{R}(z) \right| < 2\varepsilon\cdot(1-|z|)^{-1}$$

if $|z| < 1$ is near 1. This may be seen from (7.4), using partial summation. Therefore, if $z = t \cdot \omega_{a,r}$, $0 < t < 1$, $t \to 1-$, and if $a_r(f) \neq 0$, then

$$\left| \sum_{n=1}^{\infty} f(n) \cdot z^n \right| \to \infty,$$

and so $\omega_{a,r}$ is a singular point for F(z). But if $a_r(f) \neq 0$ infinitely often, then the corresponding points $\omega_{a,r}$ are dense on the unit circle, and the non-continuability of F(z) is proved. [The asserted denseness of the points $\omega_{a,r}$ may be deduced from a Theorem from CH. HOOLEY [Acta Arithm. **8**, 1963], given as follows.

Theorem 7.4 (CH. HOOLEY). *Denote by* $1 = a_1 < a_2 < \ldots < a_{\varphi(r)} \le r$ *the integers in the interval* [1, r], *which are prime to* r. *Then*

$$\max_{1 \le i < \varphi(r)} (a_{i+1} - a_i) = o(r), \text{ as } r \to \infty. \;]$$

VI.7.C. Power Series Bounded on the Negative Real Axis

An interesting problem was posed by L. RUBEL and K. STOLARSKY [1980]. These authors sought the determination of all subsets $\mathbb{N}_1 \subset \mathbb{N}$ with the property that the power series

$$\sum_{n \in \mathbb{N}_1} x^n/n!$$

shares with the exponential function the property of being bounded on the negative real axis. The solution of this question by these two mathematicians shows that there are, besides $\emptyset$ and $\mathbb{N}$, exactly four subsets $\mathbb{N}_1$ with the property mentioned above.

Analogously, we seek the determination of all multiplicative or additive functions f in $\mathcal{B}$ (with the additional assumption $M(f) \neq 0$ in the multiplicative case) for which the function

$$E_f(z) = \sum_{n=1}^{\infty} \frac{f(n)}{n!} \cdot z^n$$

is bounded on the negative real axis. We prove the following theorem.

Theorem 7.5. (a) *Let* $f \in \mathcal{B}^2$ *be a multiplicative function with mean-value* $M(f) \neq 0$, *which is represented by its* RAMANUJAN *expansion, and assume that* $E_f(z)$ *is bounded on the negative real axis. Then*

$$f = 1 \textit{ or } f \textit{ is periodic with period } 4.$$

More exactly, with some complex parameter c, *the function* f *is given by*

$$f(n) = \begin{cases} 1, & \textit{if } n \equiv 1 \textit{ or } 3 \bmod 4, \\ 1-c, & \textit{if } n \equiv 2 \bmod 4, \\ 1+c, & \textit{if } n \equiv 0 \bmod 4. \end{cases}$$

The corresponding functions $E_f(z)$ *are*

$$e^x - 1 - c\cdot(\cos x + 1), \quad c \in \mathbb{C}.$$

(b) *If* $g \in \mathcal{B}^2$ *is additive, if g is represented by its* RAMANUJAN *expansion, and if* $E_g(z)$ *is bounded on the negative real axis, then*

$$g(n) = \begin{cases} c, & \textit{if } 4|n, \\ -c, & \textit{if } 2\|n, \\ 0, & \textit{if } 2\nmid n. \end{cases}$$

The corresponding functions E_g *are*

$$E_g(z) = c\cdot(\cos z - 1).$$

Remark. The assumption, that f (or g) is represented by its RAMANUJAN expansion, is automatically true for multiplicative or additive functions in $\mathcal{B}^2$ (see Chapter VIII).

For the **proof of Theorem** 7.5 consider the Laplace transform (this is called the "Borel transform" in the theory of entire functions)

$$\mathcal{L}_f(z) = \int_0^\infty E_f(-t)\cdot e^{-tz}\,dt = \sum_{n=1}^\infty (-1)^n\cdot f(n)\cdot z^{-n-1}.$$

The integral representation together with the boundedness of $E_f(-t)$ in $t \ge 0$ shows that the Laplace transform $\mathcal{L}_f(z)$ is holomorphic in $\mathrm{Re}\, z > 0$, and the second representation as a power series implies (using partial summation and $\sum_{n\le x} f(n) = \mathcal{O}(x)$ or the estimate $f(n) \ll \sqrt{n}$, which comes from $\|f\|_2 < \infty$) that $\mathcal{L}_f(z)$ is holomorphic in $|z| > 1$. Therefore, the power series

$$-\mathcal{L}_f(-z^{-1}) = z\cdot\sum_{n=1}^\infty f(n)\cdot z^n$$

is continuable beyond the unit circle; according to our results in the preceding section, this implies that at most finitely many of the RAMANUJAN coefficients of f are non-zero. The function f is represented by its RAMANUJAN expansion, therefore the rationality of the power series follows and f is in $\mathcal{B}$, and so

$$f(n) = \sum_{1\le r\le R} a_r \cdot c_r(n).$$

Inserting this formula into the power series expansion of $-\mathcal{L}_f(-z^{-1})$ we obtain, after a short calculation, the exponential polynomial

$$E_f(z) = \sum_{1\le r\le R} a_r \cdot \sum_{1\le a\le r, \gcd(a,r)=1} \left\{ \exp(\omega_{a,r} \cdot z) - 1 \right\}.$$

Boundedness on the negative real axis implies $a_r = 0$ for every exponent $\omega_{a,r} = \exp(2\pi i \cdot a/r)$ having real part $\mathrm{Re}(\omega_{a,r}) < 0$ [for the details of this argument see the paper by RUBEL & STOLARSKY [1980]].

For $r \neq 1, 4, 6$ there are primitive roots of unity ω with $\mathrm{Re}(\omega) < 0$[2], therefore $a_r = 0$ unless $r = 1$ or 4 or 6. We shall show in Chapter VIII, Theorem 4.4, that the function $r \mapsto \{M(f)\}^{-1} \cdot a_r(f)$ is multiplicative. Therefore, $a_6(f) = 0$ because $a_2(f) = 0$. Next,

$$1 = f(1) = a_1 c_1(1) + a_4 c_4(1) = a_1,$$

and all the possible solutions of our problem are

$$f(n) = 1 + a_4 c_4(n);$$

these are indeed solutions, and the multiplicative case is settled.

In the additive case we know (this will be shown later) that $a_r = 0$, if r is not a power of a prime, therefore in this case $a_6 = 0$, too. The value f(1) is zero, and so $a_1 = f(1) = 0$, and all the solutions are given by

$$f(n) = a_4 \cdot c_4(n). \qquad \square$$

---

2) If $r \ge 3$ is odd, then $\gcd(\frac{1}{2}(r+1), r) = 1$, and $\omega = \exp(2\pi i \cdot \frac{1}{2}(r+1)/r)$ will do. If $4|r$ and $r \ge 8$, then take $\omega = \exp(2\pi i \cdot (\frac{1}{2} r + 1)/r)$. If $2\|r$, $r \ge 10$, then take $\omega = \exp(2\pi i \cdot (\frac{1}{2} r + 2)/r)$.

VI.8. A $\mathcal{B}^q$-CRITERION

The condition $f \in \mathcal{B}^q$, where $q \geq 1$, implies $f \in \mathcal{B}^1$ and $\|f\|_q < \infty$. Is the reverse assertion true? It follows from Theorem 2.8 that $f \in \mathcal{B}^1$ and $\|f\|_q < \infty$ for some q lead to $f \in \mathcal{B}^r$ for $1 \leq r < q$, but $f \in \mathcal{B}^q$ is not true in general. In his dissertation (1988, Frankfurt; parts ot this dissertation are published as [1989a, b, c]) P. KUNTH proved the following theorem.

Theorem 8.1. *For every* $q > 1$ *the conditions* (8.1) *and* (8.2) *are equivalent.*

(8.1) $$f \in \mathcal{B}^q,$$

(8.2) $$\begin{cases} \text{(a)} & f \in \mathcal{B}^1, \\ \text{(b)} & \|f\|_q < \infty, \\ \text{(c)} & \lim_{r \to q-} \|f\|_r = \|f\|_q. \end{cases}$$

The same theorem holds (with the same proof) for the spaces $\mathcal{A}^q$ *and* $\mathcal{D}^q$ *instead of* $\mathcal{B}^q$.

Remarks.

(1) In his proof, P. KUNTH used tools from functional analysis centering around the concept of uniform convexity. The proof given here uses standard approximating techniques.

(2) For every arithmetical function f, the function

$$r \mapsto \|f\|_r, \ [1, \infty[\ \to [0, \infty],$$

is non-decreasing. (8.2) (c) means that this function is semi-continuous (from the left) at the point $r = q$.

(3) Condition (8.2) (c) is clearly equivalent with

(8.2) (c^*) $$\lim_{r \to q-} \|f\|_r^r = \|f\|_q^q.$$

If (c) or (c^*) is violated, then $\|f\|_s = \infty$ for any $s > q$ because $f \in \mathcal{B}^q$ is not true.

Proposition 8.2. *For* $q \ge 1$ *and every arithmetical function* f

$$f \in \mathcal{B}^q \text{ if and only if } f \in \mathcal{B}^1 \text{ and } |f| \in \mathcal{B}^q.$$

The same assertion is true for the spaces $\mathcal{D}^q$, $\mathcal{A}^q$.

Proof. The implication from left to right is contained in Theorem 2.3. So, let $f \in \mathcal{B}^1$, $|f| \in \mathcal{B}^q$ be given. We factorize f:

$$f = g \cdot \left(\frac{f}{g} \right), \text{ where } g = \max\{1, |f|\}.$$

The first factor g is in $\mathcal{B}^q$ for $1 \in \mathcal{B}^q$, $|f| \in \mathcal{B}^q$. Since $g \ge 1$, by Theorem 2.11 we obtain $(1/g) \in \mathcal{B}^1$. Therefore, the second factor (f/g) is in $\mathcal{B}^1$; it is bounded, and Theorem 2.6 gives $f \in \mathcal{B}^q$. □

Proof of the easy implication (8.1) ⇒ (8.2). It suffices to prove (c). This will be performed in two steps.

(8.3) *Every bounded function* $f \in \mathcal{B}^1$ *has property* (c^*).

Proof of (8.3). Let $q > r \ge 1$, $\delta = q - r > 0$, $K = \sup_{n \in \mathbb{N}} |f(n)|$. Then, if $x > 0$,

$$\Delta = \sum_{n \le x} \left| |f(n)|^q - |f(n)|^r \right| = \sum_{\substack{n \le x \\ 0 < |f(n)| \le 1}} |f(n)|^r \cdot \left| |f(n)|^\delta - 1 \right| + \sum_{\substack{n \le x \\ 1 < |f(n)| \le K}} |f(n)|^r \cdot \left| |f(n)|^\delta - 1 \right|.$$

Using

$$|y^\delta - 1| \le \begin{cases} \delta \cdot |\log y|, & \text{if } 0 < y \le 1, \\ \delta \cdot |\log y| \cdot y^\delta, & \text{if } y \ge 1, \end{cases}$$

we obtain

$$\Delta \le \sum_{\substack{n \le x \\ 0 < |f(n)| \le 1}} |f(n)|^r \cdot \left| \log |f(n)| \right| \cdot \delta + \sum_{\substack{n \le x \\ 1 < |f(n)| \le K}} |f(n)|^q \cdot \log |f(n)| \cdot \delta$$

$$\le \left(\sup_{0 < y \le 1} y^r \cdot |\log y| + K^q \cdot \log K \right) \cdot \delta x.$$

Hence,

$$\|f\|_q^q \le c \cdot \delta + \|f\|_r^r, \text{ and } \|f\|_r^r \le c \cdot \delta + \|f\|_q^q.$$

Therefore, $\left| \|f\|_q^q - \|f\|_r^r \right| \le c \cdot \delta = c \cdot (q - r)$, and (8.3), and so (8.2) for bounded functions is proved.

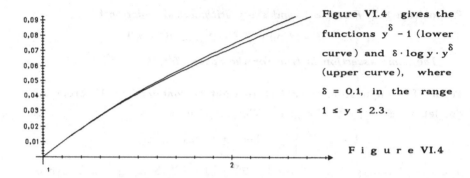

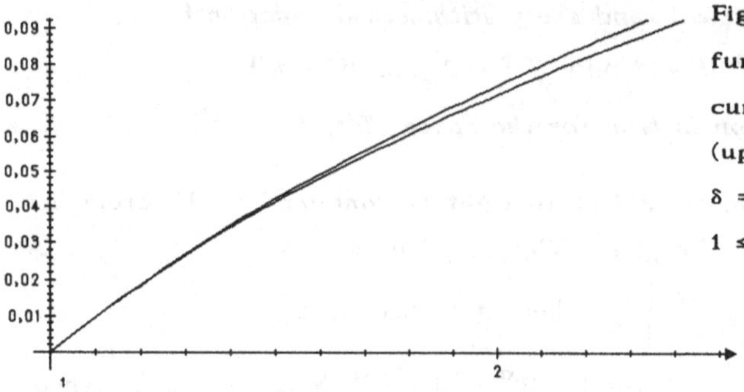

Figure VI.4 gives the functions $y^{\delta} - 1$ (lower curve) and $\delta \cdot \log y \cdot y^{\delta}$ (upper curve), where $\delta = 0.1$, in the range $1 \le y \le 2.3$.

Figure VI.4

In the second step we show (8.2) (c), if $f \in \mathcal{B}^q$. Without loss of generality, we assume that $f \ge 0$ ($f \in \mathcal{B}^q$ implies $|f| \in \mathcal{B}^q$), and use the truncated function

$$f_K = \min \{f, K\}.$$

From section 2 (see (2.10)) we know that $f_K \in \mathcal{B}^q$ for any $K > 0$, and that $\lim_{K\to\infty} \| f - f_K \|_q = 0$. If $1 \le r \le q$, the inequality

$$0 \le \| f \|_q - \| f \|_r \le (\| f \|_q - \| f_K\|_q) + (\| f_K\|_q - \| f_K\|_r) + (\| f_K\|_r - \| f \|_r)$$

$$\le 2 \cdot \| f - f_K \|_q + (\| f_K\|_q - \| f_K\|_r)$$

holds. Given $\varepsilon > 0$, we find $K > 0$ such that $2 \cdot \| f - f_K \|_q < \frac{1}{2} \cdot \varepsilon$, and (by (8.3)) a real number $r_0 \in [1, q[$ with the property $\|f_K\|_q - \|f_K\|_r < \frac{1}{2} \cdot \varepsilon$ for every r in $[r_0, q]$. For these r we obtain $\| f \|_q - \| f \|_r < \varepsilon$, and (c) is proved.

Proof of the implication (8.2) ⇒ (8.1) in three steps. Let $q > 1$.

(8.4) *For every $\varepsilon > 0$ there exists a real number $r_0 \in [1, q[$ with the property: for every sequence $(a_n)_{n=1,2,\dots}$ with $0 \le a_n \le 1$ the inequality*

$$0 \le \sum_{n \le x} (a_n^r - a_n^q) < \varepsilon \cdot x$$

holds for all $x > 0$, $r \in [r_0, q[$.

In order to show (8.4), put $\delta = q - r$, where $q > r$. The function $h(x) = x^r - x^q$ takes a maximal value in the intervall $[0,1]$ at the point $x_0 = \left(\frac{r}{q}\right)^{1/\delta}$, and the maximal value is $h(x_0) = \left(\frac{r}{q}\right)^{r/\delta} \cdot \left(\frac{\delta}{q}\right) \le \delta \cdot \left(\frac{1}{q}\right)$.

Therefore, there exists a $\delta_0 > 0$ such that the maximal value of h is less than ε for every $\delta \in]0,\delta_0[$. The desired inequality is correct for $r = r_0 := q - \delta_0$, and by monotonicity for every r in $[r_0, q[$.

(8.5) *For any* $f \in \mathcal{B}^1$, $f \geq 0$, *with* $\|f\|_q < \infty$ *and* $\varepsilon > 0$ *there exists a real number* $r_0 \in [1,q[$ *such that*

$$0 \leq \|f - f_K\|_q^q - \|f - f_K\|_r^r \leq \|f\|_q^q - \|f\|_r^r + \varepsilon$$

for all $K > 0$ *and all* r *in* $[r_0, q[$.

Proof. The difference $f(n) - f_K(n) = 0$ if $f(n) \leq K$, and $= f(n) - K$ if $f(n) > K$. Using (8.4) and the monotonicity of $x \mapsto x^q - x^r$ in $[1,\infty[$, we calculate for $K > 0$, $x \geq 1$,

$$\sum_{n\leq x} \left(|f(n) - f_K(n)|^q - |f(n) - f_K(n|^r \right) = \Sigma_1 + \Sigma_2,$$

where

$$0 \leq -\Sigma_1 = \sum_{\substack{n\leq x \\ K\leq f(n)\leq K+1}} \left((f(n)-K)^r - (f(n)-K)^q \right) \leq \tfrac{1}{2}\varepsilon\cdot x$$

for all $r \in [r_0, q[$, and

$$\Sigma_2 = \sum_{n\leq x, f(n)>K+1} \Big((f(n)-K)^q - (f(n)-K)^r \Big) \leq \sum_{n\leq x, f(n)>K+1} \Big(f^q(n) - f^r(n) \Big)$$

$$\leq \sum_{n\leq x, f(n)\geq 1} \Big(f^q(n) - f^r(n) \Big) = \sum_{n\leq x} \Big(f^q(n) - f^r(n) \Big) + R,$$

where

$$0 \leq R = \sum_{n\leq x, 0<f(n)<1} \left(f^r(n) - f^q(n) \right) < \tfrac{1}{2}\varepsilon\cdot x,$$

by (8.4), for every $r \in [r_0, q[$. So, for all $x \geq 1$,

$$0 \leq x^{-1}\cdot\Sigma_2 \leq x^{-1}\cdot \Sigma_{n\leq x} f^q(n) - x^{-1}\cdot\Sigma_{n\leq x} f^r(n) + \tfrac{1}{2}\varepsilon.$$

Since $f \in \mathcal{B}^r$ (where $1 \leq r < q$) and $f \geq 0$, Theorem 2.9 gives $f^r \in \mathcal{B}^1$, and the mean-value of f^r exists; we obtain

$$\limsup_{x\to\infty} x^{-1}\cdot\Sigma_2 \leq \|f\|_q^q - \|f\|_r^r + \tfrac{1}{2}\varepsilon.$$

Collating our estimates, we obtain the estimate

$$x^{-1}\,\Sigma_{n\leq x} |f(n) - f_K(n)|^q \leq x^{-1}\,\Sigma_{n\leq x} |f(n) - f_K(n)|^r + x^{-1}\cdot|\Sigma_1| + x^{-1}\cdot\Sigma_2$$

for all $K > 0$, $r \in [r_o, q[$, $x \geq 1$. For $x \to \infty$ this implies

$$\| f - f_K \|_q^q \leq \| f - f_K \|_r^r + \| f \|_q^q - \| f \|_r^r + \varepsilon,$$

and (8.5) is proved.

Now the missing implication in Theorem 8.1 is easily proved. Given $f \in \mathcal{B}^1$ with $\| f \|_q < \infty$, and $\lim_{r \to q-} \| f \|_r^r = \| f \|_q^q$, we assume that $f \geq 0$ without loss of generality because of Proposition 8.2. Having chosen $\varepsilon > 0$, we find a real number $r_o \in [1, q[$ by (8.5) such that

$$\| f - f_K \|_q^q \leq \| f - f_K \|_r^r + \| f \|_q^q - \| f \|_r^r + \tfrac{1}{3}\varepsilon$$

for all $K > 0$ and $r \in [r_o, q[$. Choose r such that $\| f \|_q^q - \| f \|_r^r < \frac{1}{3}\varepsilon$, and then $K > 0$ so that $\| f - f_K \|_r^r < \frac{1}{3}\varepsilon$. Then $\| f - f_K \|_q^q < \varepsilon$. The function f_K is in $\mathcal{B}^q$, and so Theorem 8.1 is proved. □

Remark. We give a second proof for the more difficult implication (8.1) ⇒ (8.2) of Theorem 8.1, using DABOUSSI's Theorem 2.9. Given $f \in \mathcal{B}^1$, with (8.1) (c^*) and $\| f \|_q < \infty$, we assume $f \geq 0$ without loss of generality because of Proposition 8.2. For every r, $1 \leq r < q$, and $x > 0$, we see that

$$\sum_{n \leq x} \left(f^{\frac{1}{2}q}(n) - f^{\frac{1}{2}r}(n) \right)^2 = \sum_{n \leq x} f^q(n) + \sum_{n \leq x} f^r(n) - 2 \cdot \sum_{n \leq x} f^{\frac{1}{2}(q+r)}(n).$$

By Theorem 2.8, f is in $\mathcal{B}^r$; DABOUSSI's Theorem gives $f^r \in \mathcal{B}^1$, and so the mean-value $M(f^r) = \| f \|_r^r$ exists. The same argument applies to $M(f^{\frac{1}{2}(q+r)})$. Therefore,

$$\| f^{\frac{1}{2}q} - f^{\frac{1}{2}r} \|_2^2 = \| f \|_q^q + \| f \|_r^r - 2 \cdot \| f \|_{\frac{1}{2}(q+r)}^{\frac{1}{2}(q+r)}.$$

Making use of (c^*) we obtain

$$\lim_{r \to q-} \| f^{\frac{1}{2}q} - f^{\frac{1}{2}r} \|_2^2 = 0.$$

f is in $\mathcal{B}^r$, therefore $f^{\frac{1}{2}r} \in \mathcal{B}^2$ (again by Theorem 2.9). Being approximated by functions in $\mathcal{B}^2$, the function $f^{\frac{1}{2}q}$ itself is in $\mathcal{B}^2$. Using Theorem 2.9 once more, the function f is in $\mathcal{B}^q$. □

VI.9. EXERCISES

1) Give a [simple] direct proof for the fact that arithmetical functions in $\mathcal{A}^1$ have a mean-value.

2) If $f: \mathbb{N} \to \mathbb{R}$ is an integer-valued function in $\mathcal{B}^u$, then f is in $\mathcal{B}$. Give an integer-valued function in $\mathcal{B}^1$ which is not in $\mathcal{B}$.

3) Denote by ADD resp. ADD^s the set of additive [resp. strongly additive] functions. Prove that these are subspaces of $\mathbb{C}^{\mathbb{N}}$, and that the $\|.\|_1$- completion of ($\text{ADD} \cap \mathcal{B}^1$) [resp. of ($\text{ADD}^s \cap \mathcal{B}^1$)] is a subspace of $\mathcal{B}^1$.

4) Assume that $f \in \mathcal{D}^1$ is a non-negative multiplicative arithmetical function. Denote by $M_p(f)$ the limit $\lim_{x \to \infty} x^{-1} \cdot \sum_{n \le x, p|n} f(n)$. Prove that for every prime $\sum_{k \ge 1} p^{-k} \cdot f(p^k) < \infty$ if and only if $M_p(f) \neq M(f)$.

5) Let f be a multiplicative function in $\mathcal{D}^1$. For every prime power p^k, prove

(a) $\lim_{x \to \infty} x^{-1} \cdot \sum_{n \le x, p^k \| n} f(n) = p^{-k} \cdot f(p^k) \cdot (M(f) - M_p(f))$,

(b) $\lim_{x \to \infty} x^{-1} \cdot \sum_{n \le x, p^k | n} f(n) = \sum_{\ell \ge k} p^{-\ell} \cdot f(p^\ell) \cdot (M(f) - M_p(f))$,

if the series on the right-hand side converges absolutely.

6) Prove Theorem 2.11 (3) directly.

7) Let $\gamma > 0$ be an irrational number. Denote by g(n) the number of positive integers m with the property $[\gamma \cdot m] = n$. Prove:

(a) g is in $\mathcal{A}^2$.

(b) Put $\delta = \gamma^{-1} - [\gamma^{-1}]$. Then the FOURIER coefficients of the function g are $\hat{g}(\alpha) = \gamma^{-1}$, if $\alpha = 0$,

$$\hat{g}(\alpha) = (2\pi i \alpha \gamma)^{-1} \cdot \left(e^{2\pi i \alpha \gamma (\delta-1)} - 1\right), \text{ if } \alpha \in \gamma^{-1} \cdot \mathbb{Z}, \alpha \neq 0,$$

and $\hat{g}(\alpha) = 0$ otherwise.

(c) What does PARSEVAL's equation mean?

Answer: $\sum_{n=1}^{\infty} n^{-2} \cdot \sin^2(\pi\delta n) = \frac{1}{6}\pi^2 \cdot (\delta - \delta^2)$, where $0 \le \delta < 1$.

8) Give a proof for PARSEVAL's equation in $\mathcal{D}^2$, using methods similar to those used in section 4.
Hint: $A_k = \{n \in \mathbb{N};\ n \equiv k \bmod r\}$, $F_r(f) = r \cdot \sum_{1 \le k \le r} M(f \cdot \chi_k) \cdot \chi_k$.

9) If f is in $\mathcal{B}^1$ and $\delta > 0$, then the function h,

$$h(n) = \begin{cases} \dfrac{f(n)}{|f(n)|}, & \text{if } |f(n)| > \delta, \\ \delta^{-1} \cdot f(n), & \text{if } |f(n)| \le \delta, \end{cases}$$

belongs to $\mathcal{B}^1$.

10) For every function $f \in \mathcal{D}^1$ and every residue-class s mod r the mean-value $\lim_{x \to \infty} x^{-1} \cdot \sum_{n \le x,\, n \equiv s \bmod r} f(n)$ exists. Prove this result for coprime r, s, using the formula

$$\sum_{n \le x,\, n \equiv s \bmod r} f(n) = \{\varphi(r)\}^{-1} \cdot \sum_{\chi \bmod r} \left(\overline{\chi(s)} \cdot \sum_{n \le x} \chi(n) f(n) \right).$$

11) If $q_1 > 1, \ldots, q_k > 1$, $q_1^{-1} + \ldots + q_k^{-1} = 1$, and $f_1 \in \mathcal{A}^{q_1}, \ldots, f_k \in \mathcal{A}^{q_k}$, then prove that the product $f_1 \cdot f_2 \cdot \ldots \cdot f_k$ is in $\mathcal{A}^1$.

12) Let $k|r$, where k and r are positive integers. Calculate the mean-value of the indicator-function of the set $\{n \in \mathbb{N};\ \gcd(n, r) = k\}$.

E. WIRSING

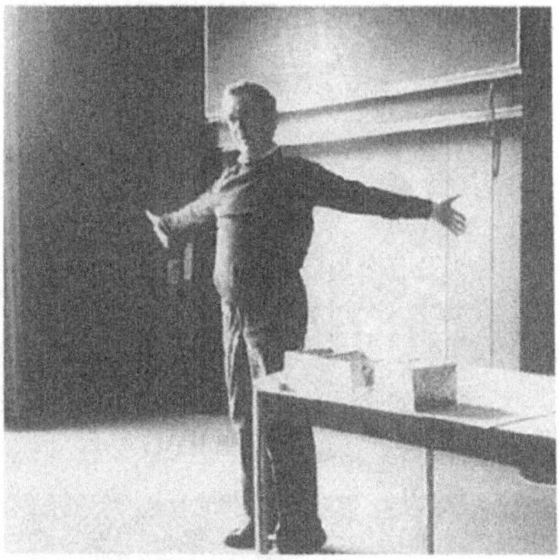

P. D. T. A. ELLIOTT

H. DABOUSSI

H. DELANGE

R. RANKIN

A. RÉNYI
(1921–1970)

P. Erdös

A. Selberg

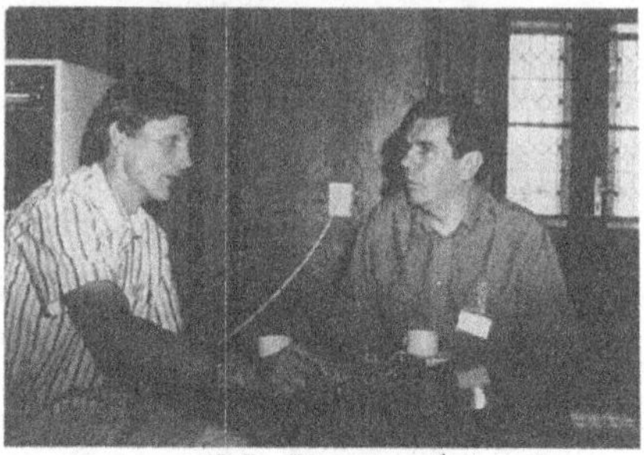
M. Jutila &
M. N. Huxley

H. E. Richert

A. Karacuba
A. Ivić

C. L. Siegel
(1896–1981)

J.-L. Mauclaire

M. Nair

Chapter VII

The Theorems of ELLIOTT and DABOUSSI

ABSTRACT. *This chapter deals with multiplicative arithmetical functions* f*, and relations between the values of these functions taken at prime powers, and the almost periodic behaviour of* f. *More exactly, we prove that the convergence of four series, summing the values of* f *at primes, respectively prime powers [with appropriate weights], implies that* f *is in* $\mathcal{B}^q$, *and (if in addition the mean-value* M(f) *is supposed to be non-zero) vice versa. For this part of the proof we use an approach due to H. DELANGE and H. DABOUSSI [1976] in the special case where* q = 2; *the general case is reduced to this special case using the properties of spaces of almost-periodic functions obtained in Chapter VI. Finally, DABOUSSI's characterization of multiplicative functions in* $\mathcal{A}^q$ *with non-empty spectrum is deduced.*

VII.1. INTRODUCTION

As shown in the preceding chapter, q-almost-even and q-almost-periodic functions have nice and interesting properties; for example, there are mean-value results for these functions (see VI.7) results concerning the existence of limit distributions and some results on the global behaviour of power series with almost-even coefficients. These results seem to provide sufficient motivation in the search for a, hopefully, rather simple characterization of functions belonging to the spaces $\mathcal{A}^q \supset \mathcal{D}^q \supset \mathcal{B}^q$ of almost-periodic functions, defined in VI.1. Of course, in number theory we look for functions having some distinguishing arithmetical properties, and the most common of these properties are additivity and multiplicativity.

According to the heuristics outlined in Chapter III.1, conditions characterizing membership of an arithmetical function to, say, $\mathcal{B}^q$, ought to be formulated using the values of f at primes and prime powers.

Historically, theorems of this kind were given for the first time in connection with the problem of the characterization of multiplicative functions with a non-zero mean-value. The E. WIRSING Theorem, proved in II.4, is an example of the fact that assumptions about the behaviour in the mean of values of a multiplicative function, taken at primes, imply asymptotic formulae for the sum $\sum_{n\le x} f(n)$. But these results do not characterize multiplicative functions with a non-zero mean-value. In 1961, H. DELANGE proved the following theorem.

Theorem 1.1. *Let* $f\colon \mathbb{N} \to \mathbb{C}$ *be a multiplicative function satisfying* $|f| \le 1$. *Then the following conditions are equivalent:*

(1.1) *The mean-value* $M(f) = \lim_{x \to \infty} x^{-1} \cdot \sum_{n \le x} f(n)$ *exists and is non-zero.*

(1.2)
(i) *The series* $S_1(f) = \sum_p p^{-1} \cdot (f(p) - 1)$ *is convergent,*
(ii) $\sum_{0 \le k < \infty} p^{-k} \cdot f(p^k) \neq 0$ *for all primes* p.

Remark. The assumption $|f| \le 1$ implies that $\left|\sum_{0\le k<\infty} p^{-k}\cdot f(p^k)\right| \ge \frac{1}{2}$ for every prime $p \ge 3$. Therefore, as did DELANGE, the validity of (1.2ii) is to be assumed only for $p = 2$, and it may be substituted by the DELANGE condition

$$f(2^k) \neq -1 \text{ for some } k \ge 1.$$

In 1965 A. RÉNYI gave a simple proof of the implication ((1.2) $\Rightarrow$ (1.1)), using the TURÁN-KUBILIUS inequality (see I.4). This method of proof will be the basis of the more general result, given as Proposition 3.2 in this chapter. The condition $|f| \le 1$ was removed by P. D. T. A. ELLIOTT in 1975, who replaced this severe restriction by the assumption $\| f \|_q < \infty$, with the semi-norm $\| f \|_q$ defined in VI, (1.3). We define the [$\mathcal{E}$LLIOTT-] set $\mathcal{E}_q$ of multiplicative functions $f: \mathbb{N} \to \mathbb{C}$ by the following conditions:

Definition 1.2. $f \in \mathcal{E}_q$ *if and only if*

(i) *the* DELANGE *series* $S_1(f) = \sum_p p^{-1} \cdot (f(p) - 1)$ *is [conditionally] convergent,*

(ii) *the series*

$$S_2'(f) = \sum_{p,\ |f(p)| \le 5/4} p^{-1} \cdot | f(p) - 1 |^2,$$

and

$$S_{2,q}''(f) = \sum_{p,\ |f(p)| > 5/4} p^{-1} \cdot |f(p)|^q$$

are convergent,

(iii) *the series*

$$S_{3,q}(f) = \sum_p \sum_{k\ge 2} p^{-k} |f(p^k)|^q$$

is convergent.

Remarks. 1) The series $S_1(f)$ is conditionally convergent, the primes being ordered canonically according to their size. The other series are absolutely convergent.

2) In the special case where $q = 2$, condition (ii) is equivalent to the convergence of the series

(ii') $$S_2(f) = \sum_p p^{-1} \cdot | f(p) - 1 |^2.$$

Using this notation, P. D. T. A. ELLIOTT [1975] proved the following theorem.

Theorem 1.3. *Assume that* $f: \mathbb{N} \to \mathbb{C}$ *is a multiplicative function, and assume that* $q > 1$. *Then the following conditions are equivalent.*

(1.3) $\quad \|f\|_q < \infty$ *and the mean-value* $M(f)$ *exists and is non-zero.*

(1.4) $\quad f$ *is in* $\mathcal{E}_q$ *and condition* (1.2ii) *is satisfied.*

In this chapter we are going to show that the convergence of the series in Definition 1.2 implies, in fact, that the multiplicative function f is in $\mathcal{B}^q$ (**Theorem** 4.1). Furthermore, following DABOUSSI and DELANGE, we prove (**Theorem 5.1**) that for any multiplicative function f with mean-value $M(f) \neq 0$ the following properties are equivalent:

$$\begin{cases} f \in \mathcal{E}_q, \\ f \in \mathcal{B}^q, \\ f \in \mathcal{D}^q, \\ f \in \mathcal{A}^q. \end{cases}$$

Finally, we characterize multiplicative functions in $\mathcal{A}^q$, possessing a non-void spectrum (see **Theorem 6.1**).

We begin with some rather simple consequences of the condition $\|f\|_q < \infty$.

If $\|f\|_q < \infty$, then there exists some positive constant c such that

(1.5) $$|f(n)| \le c \cdot n^{1/q} \text{ for every } n \in \mathbb{N},$$

and [by partial summation from $\sum_{n \le x} |f(n)| \le C \cdot x$]

(1.6) $$\sum_{n=1}^{\infty} n^{-s} \cdot |f(n)|^q < \infty, \text{ if } \operatorname{Re} s > 1.$$

Lemma 1.4. *If* $\|f\|_q < \infty$ *for some* $q > 1$, *then,*

$$\sum_p \left|\frac{f(p)}{p}\right|^2 < \infty, \text{ and}$$

$$\sum_p \sum_{k \ge 2} p^{-k} \cdot |f(p^k)|^r < \infty \text{ for every } r \text{ in } 1 \le r < q.$$

In particular, using the notation of Chapter III, Section 1, a multiplicative arithmetical function f, *satisfying* $\|f\|_q < \infty$, *belongs to the set*

$$\mathcal{G} = \Big\{ f: \mathbb{N} \to \mathbb{C},\ f \text{ multiplicative},\ \sum_p \Big|\frac{f(p)}{p}\Big|^2 < \infty,\ \sum_p \sum_{k\ge 2} p^{-k}\cdot |f(p^k)| < \infty \Big\}.$$

Proof. Choose an $\varepsilon > 0$ such that $1 + 2\cdot\varepsilon < q$. HÖLDER's inequality and (1.5) imply

$$\sum_{p\le x} \Big|\frac{f(p)}{p}\Big|^2 \le c\cdot \sum_{p\le x} \frac{|f(p)|}{p^{(1+\varepsilon)/q}} \cdot \frac{1}{p^{2-(2+\varepsilon)/q}}$$

$$\le c\cdot \Big(\sum_{p\le x} \frac{|f(p)|^q}{p^{(1+\varepsilon)}} \Big)^{1/q} \cdot \Big(\sum_{p\le x} \Big\{ \frac{1}{p^{2-(2+\varepsilon)/q}} \Big\}^{q'} \Big)^{1/q'}.$$

By (1.6) and the choice of ε, both series on the right converge for $x \to \infty$. Similarly, with $\varepsilon > 0$, $1 + 2\varepsilon < \frac{q}{r}$, the estimate

$$\sum_p \sum_{\substack{k\ge 2\\ p^k\le x}} p^{-k}\cdot|f(p^k)|^r \le \Big(\sum_p \sum_{\substack{k\ge 2\\ p^k\le x}} p^{-k(1+\varepsilon)}\cdot |f(p^k)|^q \Big)^{r/q} \cdot \Big(\sum_p \sum_{\substack{k\ge 2\\ p^k\le x}} p^{-k(1-\varepsilon\frac{r}{q-r})} \Big)^{1-\frac{r}{q}}$$

proves the convergence of the second series. □

Example. The following example shows that an extension of Lemma 1.4 to $r = q$ is not possible. Define a multiplicative function f by $f(p^k) = 0$ if $p > 2$ or k is odd, and

$$f(2^k) = (\ell^{-1}\cdot 2^{2\ell})^{1/q}, \text{ if } k = 2\cdot\ell \text{ is even.}$$

Then $\|f\|_q = 0$, but $\sum_p \sum_{k\ge 2} p^{-k}\cdot |f(p^k)|^q = \sum_{k\ge 2} 2^{-k}\cdot f^q(2^k) = \infty$.

Lemma 1.5. *Let* $q > 1$, $f: \mathbb{N} \to \mathbb{C}$ *be multiplicative*, $\|f\|_q < \infty$, *and assume that the mean-value* $M(f)$ *exists and is non-zero. Then there exists a prime* p_0 *with the properties*

(1) $$M_{(p)}(f) = \lim_{x\to\infty} x^{-1}\cdot \sum_{n\le x,\, p\nmid n} f(n) = M(f)\cdot \{\varphi_f(p,1)\}^{-1}$$

for every prime $p \ge p_0$, *and*

(2) $$M_{(d)}(f) = \lim_{x\to\infty} x^{-1}\cdot \sum_{n\le x,\,(n,d)=1} f(n) = M(f)\cdot \prod_{p|d} \{\varphi_f(p,1)\}^{-1}$$

for every positive integer d *which consists only of primes* $p \ge p_0$.

(3) $$M_p(f) = \lim_{x\to\infty} x^{-1}\cdot \sum_{n\le x,\ n\equiv 0 \bmod p} f(n) = M(f)\cdot\{\varphi_f(p,1)-1\}\cdot\{\varphi_f(p,1)\}^{-1}.$$

Remark. If f is 2-multiplicative, so that $f(p^k) = 0$ for every $k \ge 2$, then the mean-values in question are given by

$$M_{(p)}(f) = M(f)\cdot\left\{1+p^{-1}\cdot f(p)\right\}^{-1},\ M_{(d)}(f) = M(f)\cdot\prod_{p|d}\left\{1+p^{-1}\cdot f(p)\right\}^{-1},$$

$$M_p(f) = M(f)\cdot\frac{f(p)}{p}\cdot\left\{1+p^{-1}\cdot f(p)\right\}^{-1}.$$

Proof. In Re $s \geq 1$, (1.5) implies

$$\left|\sum_{k\geq 1} p^{-ks}\cdot f(p^k)\right| \leq c\cdot\left\{p^{1-1/q}-1\right\}^{-1},$$

therefore there is some p_0 such that [recalling the abbreviation $\varphi_f(p,s) = 1 + p^{-s}\cdot f(p) + p^{-2s}\cdot f(p^2) + \dots$], for every prime $p \geq p_0$,

$$\left|\varphi_f(p,s)\right| \geq 1 - c\cdot\left\{p^{1-1/q}-1\right\}^{-1} \geq 1 - c\cdot\left\{p_0^{\,1-1/q}-1\right\}^{-1} \geq \tfrac{1}{2}$$

in Re $s \geq 1$. Let p^* be a fixed prime greater than or equal to p_0. Define a multiplicative function g by

$$g(p^k) = \begin{cases} f(p^k), & \text{if } p \neq p^*, \\ 0, & \text{if } p = p^*. \end{cases}$$

The functions f and g are related, f is in $\mathcal{G}$, $|g| \leq |f|$, therefore $g \in \mathcal{G}$. For every prime $p \geq p_0$ the factor $\varphi_f(p,s) \neq 0$ in Re $s \geq 1$, and for primes $p < p_0$ the values $f(p^k)$ and $g(p^k)$ are equal. Therefore III, Theorem 4.1 (Remark) gives the existence of

$$M_{(p^*)}(f) = M(g) = M(f)\cdot\{\varphi_f(p^*,1)\}^{-1}.$$

Next,

$$x^{-1}\cdot\sum_{n\leq x,\,p^*|n} f(n) = x^{-1}\cdot\sum_{n\leq x} f(n) - x^{-1}\cdot\sum_{n\leq x,\,p^*\nmid n} f(n),$$

and so

$$M_{p^*}(f) = M(f) - M_{(p^*)}(f) = M(f)\cdot\left\{\varphi_f(p^*,1)-1\right\}\cdot\left\{\varphi_f(p^*,1)\right\}^{-1}.$$

The remaining assertion of Lemma 1.5 is left as Exercise 1. □

Lemma 1.6. *If* $q \geq 1$, $f \in \mathcal{E}^q$ *is multiplicative, and* $M(f) \neq 0$, *then the mean-value has the product representation*

$$M(f) = \prod_p (1-p^{-1})\cdot\varphi_f(p,1).$$

In particular, for every prime, $\varphi_f(p,1) \neq 0$.

Proof. Partial summation (see I) gives the convergence of the DIRICHLET series $\sum_{n=1}^{\infty} n^{-\sigma} \cdot f(n)$ in $\sigma > 1$. By the continuity theorem,

$$M(f) = \lim_{\sigma \to 1+} \zeta^{-1}(\sigma) \cdot \sum_{n=1}^{\infty} n^{-\sigma} \cdot f(n) = \lim_{\sigma \to 1+} \prod_p (1 - p^{-\sigma}) \cdot \varphi_f(p,\sigma).$$

Using results on infinite products (see Appendix A.7) and the assumption $f \in \mathcal{E}^q$, the assertion is obtained. □

VII.2. MULTIPLICATIVE FUNCTIONS WITH MEAN-VALUE $M(f) \neq 0$, SATISFYING $\|f\|_2 < \infty$.

In this section, in the special case where $q = 2$, we prove one of the two implications of Theorem 1.3.

Proposition 2.1. *Assume that* f *is multiplicative,* $\|f\|_2 < \infty$, *and the mean-value* $M(f)$ *exists and is non-zero. Then the series (see Definition 1.2)*

$$S_1(f) = \sum_p p^{-1} \cdot (f(p) - 1),$$

$$S_2(f) = \sum_p p^{-1} \cdot |f(p) - 1|^2,$$

and

$$S_{3,2}(f) = \sum_p \sum_{k \geq 2} p^{-k} \, |f(p^k)|^2$$

are convergent, and so $f \in \mathcal{E}_2$.

Proof. First we prove $S_2(f) < \infty$ in the following way: we calculate the RAMANUJAN coefficients of a slightly changed, related function g and utilize BESSEL's inequality. In order to obtain the convergence of the other series $S_1(f)$ and $S_3(f)$, we use H. DELANGE and H. DABOUSSI's method [1976].

1) Take p_1 so large that $|\varphi_f(p,s)| \geq \frac{1}{2}$ for every prime $p \geq p_1$, and every

s in Re $s \geq 1$. Then $f \in \mathcal{G}$ (see Lemma 1.4). Define a multiplicative function g by

$$g(p^k) = \begin{cases} f(p^k), & \text{if } p < p_1, \\ f(p), & \text{if } p \geq p_1,\ k = 1, \\ 0, & \text{if } p \geq p_1,\ k \geq 2. \end{cases}$$

The functions f and g are related, and - as before in the proof of Lemma 1.5 - the Relationship Theorem III.4.1 is applicable. According to this result the mean-value M(g) exists, and

$$M(g) = M(f) \cdot \prod_{p \geq p_1} \left(1 + \frac{f(p)}{p}\right) \cdot \left(\varphi_f(p,1)\right)^{-1} \neq 0.$$

Using the representation $c_p(n) = \sum_{d|(p,n)} d \cdot \mu\left(\frac{p}{d}\right)$ for the RAMANUJAN sum, we obtain (for $p \geq p_1$)

$$a_p(g) = \frac{1}{\varphi(p)} \cdot M(g \cdot c_p) = \frac{1}{p-1} \cdot \left(- M(g) + p \cdot M_p(g) \right)$$

$$= \frac{M(g)}{p-1} \cdot \left(g(p) - 1 - \frac{g(p)}{\sqrt{p}} \cdot \frac{p^{-\frac{1}{2}} \cdot g(p)}{1 + p^{-1} \cdot g(p)} \right).$$

Therefore

$$\left| \frac{M(g)}{p-1} \cdot \left(g(p) - 1 \right) \right| \leq |a_p(g)| + \mathcal{O}\left(\frac{|g(p)|}{p^{3/2}} \right),$$

and so

$$(p-1) \cdot \left| \frac{M(g)}{p-1} \cdot \left(g(p) - 1 \right) \right|^2 \leq 2 \cdot \varphi(p) \cdot |a_p(g)|^2 + \mathcal{O}\left(\frac{|g(p)|}{p^2} \right)^2.$$

Summing over the primes $p \geq p_1$, we obtain from BESSEL's inequality

$$|M(g)|^2 \cdot \sum_{p \geq p_1} \frac{|g(p) - 1|^2}{p} \leq 2 \cdot \|g\|_2^2 + \mathcal{O}(1) = \mathcal{O}(1).$$

The mean-value M(g) is non-zero, therefore $S_2(f) = \sum p^{-1} \cdot |g(p) - 1|^2$ is convergent.

Next we follow H. DABOUSSI and H. DELANGE in order to conclude the proof of Proposition 2.1. We have to show that the series

$$S_1(f) = \sum_p p^{-1} \cdot (f(p) - 1) \quad \text{and} \quad S_{3,2}(f) = \sum_p \sum_{k \geq 2} p^{-k} \cdot |f(p^k)|^2$$

are convergent. Denote the partial sums of the series $S_1(f)$ by

$$\alpha(u) = \sum_{p \leq \exp(u)} p^{-1} \cdot (f(p) - 1).$$

For s > 0, partial summation gives the relation

$$\sum_{p\le x} p^{-(1+s)}\cdot(f(p)-1) = x^{-s}\cdot\sum_{p\le x} p^{-1}\cdot(f(p)-1) + \int_0^{s\cdot\log x} \alpha\left(\frac{t}{s}\right)\cdot e^{-t}\,dt.$$

Using the convergence of the series $S_2(f)$, the CAUCHY-SCHWARZ inequality, applied to $x^{-s}\cdot\sum_{p\le x} p^{-\frac12}\cdot p^{-\frac12}(f(p)-1)$, proves that $x^{-s}\cdot\sum_{p\le x} p^{-1}\cdot(f(p)-1)\to 0$ for any s > 0, as $x\to\infty$. Therefore,

$$(2.1)\qquad \alpha(s^{-1}) - \sum_p p^{-1-s}\cdot(f(p)-1) = \int_0^\infty e^{-t}\cdot\left(\alpha(\tfrac{1}{s}) - \alpha(\tfrac{t}{s})\right)dt.$$

Having proved that the two limits

$$(2.2)\qquad \lim_{s\to 0+}\int_0^\infty e^{-t}\cdot\left(\alpha(\tfrac{1}{s}) - \alpha(\tfrac{t}{s})\right)dt = 0,$$

and

$$(2.3)\qquad \lim_{s\to 0+}\sum_p p^{-1-s}\cdot(f(p)-1) = \alpha$$

exist, relation (2.1) gives the existence of $\lim_{s\to 0+}\alpha(s^{-1})$, so that the series $S_1(f)$ is convergent.

For a proof of (2.2) we apply LEBESGUE's Dominated Convergence Theorem. In order to be able to do so, we have to estimate the difference $\alpha(\frac{1}{s}) - \alpha(\frac{t}{s})$ by an integrable function of t, uniformly in s. In $0 < y < z$, the CAUCHY-SCHWARZ inequality yields

$$|\alpha(z)-\alpha(y)|^2 = \left|\sum_{\exp(y)<p\le\exp(z)} p^{-1}\cdot(f(p)-1)\right|^2$$

$$\le \left(\sum_p p^{-1}\cdot|f(p)-1|^2\right)\cdot\left(\sum_{\exp(y)<p\le\exp(z)} p^{-1}\right).$$

The first series is convergent, the second sum is less than $\log(z/y) + C$; we know from elementary prime number theory (see I, section 6) that

$$\sum_{\exp(y)<p\le\exp(\lambda y)} p^{-1}\to\log\lambda,\ \text{as } y\to\infty.$$

Therefore, assuming $t\ge 1$ without loss of generality,

$$|\alpha(\tfrac{1}{s}) - \alpha(\tfrac{t}{s})|^2 \le C'\cdot\left(\log t + C\right).$$

The difference

$$|\alpha(\tfrac{1}{s}) - \alpha(\tfrac{t}{s})|^2 \le \left(\sum_{p>\exp(1/s)} p^{-1}\cdot|f(p)-1|^2\right)\cdot\log t$$

tends to zero as $s \to 0+$. LEBESGUE's Dominated Convergence Theorem gives assertion (2.2).

For (2.3), the existence of the mean-value M(f) implies $\sum_{n=1}^{\infty} n^{-s} \cdot f(n) \sim M(f) \cdot (s-1)^{-1}$ for $s \to 1+$ by partial summation (see VI, Theorem 7.1), and so [as $s \to 1+$]

$$\prod_p \left(1 + \frac{f(p)-1}{p^s} + \frac{f(p^2)-f(p)}{p^{2s}} + \dots \right) = \zeta^{-1}(s) \cdot \sum_{n=1}^{\infty} \frac{f(n)}{n^s} \sim M(f).$$

In particular, no one of the factors $\left(1 + \frac{f(p)-1}{p^s} + \frac{f(p^2)-f(p)}{p^{2s}} + \dots \right)$ is zero. The product over the primes is split into a finite product $\prod_{p \le L}(\dots)$, the product $\prod_{p>L} \left(1 + \frac{f(p)-1}{p^s} \right)$ and the product

$$\prod_{p>L} \left(1 + \frac{f(p)-1}{p^s} \right)^{-1} \cdot \left(1 + \frac{f(p)-1}{p^s} + \frac{f(p^2)-f(p)}{p^{2s}} + \dots \right).$$

If L is chosen large enough $\left[\text{so that } |p^{-1} \cdot (f(p)-1)| \le \frac{1}{2}\right]$, then the last product is absolutely convergent in $\mathrm{Re}(s) \ge 1$. Therefore

$$\lim_{s \to 1+} \prod_{p>L} \left(1 + \frac{f(p)-1}{p^s} \right) = \beta \neq 0$$

exists. Taking logarithms and using the absolute convergence of the series

$$\sum_p \left\{ \log \left(1 + \frac{f(p)-1}{p^s} \right) - \frac{f(p)-1}{p^s} \right\} \text{ in } \mathrm{Re}\, s \ge 1,$$

one sees that $\lim_{s \to 1+} \sum_{p>L} \frac{f(p)-1}{p^s}$ exists. Thus (2.3) is true.

3) For the convergence of $S_3(f) = \sum_p \sum_{k \ge 2} p^{-k} \cdot \left|f(p^k)\right|^2$, one starts, assuming $1 < s \le 2$, with

$$(2.4) \quad \zeta^{-1}(s) \cdot \sum_{n=1}^{\infty} n^{-s} \cdot |f(n)|^2 = \prod_p \left(1 - p^{-s}\right) \cdot \left(1 + \sum_{k=1}^{\infty} p^{-ks} \cdot |f(p^k)|^2 \right).$$

The finiteness of $\|f\|_2$ implies the boundedness of the left-hand side in $1 < s \le 2$; hence any partial product of the right-hand side is $\le c_1$, say. Let $f^*(p) = \min\{|f(p)|, \frac{5}{4}\}$. Then $1 + p^{-s} \cdot f^*(p)^2 \le c_2^{-1}$, where $c_2 = \frac{32}{57}$. We use

$$1 + x \ge \exp(x - x^2) \text{ in } x \ge -\tfrac{1}{2}.$$

For every factor of (2.4) and for every $K \ge 2$ we obtain

$$\left(1 - p^{-s}\right) \cdot \left(1 + \sum_{k=1}^{\infty} \frac{|f(p^k)|^2}{p^{ks}} \right) \ge \left(1 - p^{-s}\right) \cdot \left(1 + \frac{f^*(p)^2}{p^s}\right) \cdot \left(1 + c_2 \cdot \sum_{k=2}^{K} \frac{|f(p^k)|^2}{p^{ks}}\right)$$

$$\geq \left(1 + c_2 \cdot \sum_{k=2}^{K} \frac{|f(p^k)|^2}{p^{ks}}\right) \cdot \exp\left(\frac{f^*(p)^2 - 1}{p^s} - \frac{f^*(p)^4 + 1}{p^{2s}}\right).$$

Using $\prod_{p \le y} (1 + x_p) \ge \sum_{p \le y} x_p$ for $x_p \ge 0$, and letting s tend to 1+, we obtain

$$\sum_{p \le y} \sum_{k=2}^{K} \frac{|f(p^k)|^2}{p^k} \le \frac{c_1}{c_2} \cdot \exp\left(\sum_{p \le y} \frac{f^*(p)^4 + 1}{p^2} - \sum_{p \le y} \frac{f^*(p)^2 - 1}{p}\right),$$

for every $y \ge 2$ and $K \ge 2$. The series on the right-hand side are dominated by $\sum_p p^{-2} = \mathcal{O}(1)$, resp. by

$$\sum_{|f(p)| \le 5/4} \frac{|f(p)|^2 - 1}{p} + \sum_{|f(p)| > 5/4} \frac{1}{p} = S_1(f) + \overline{S_1(f)} + \mathcal{O}(S_2(f)) = \mathcal{O}(1).$$

Therefore, the partial sums of $\sum_p \sum_{k \ge 2} \frac{|f(p^k)|^2}{p^k}$ are bounded and $S_{3,2}(f)$ is convergent. This concludes the proof of Proposition 2.1. □

VII.3. CRITERIA FOR MULTIPLICATIVE FUNCTIONS TO BELONG TO $\mathcal{B}^1$

In this section we give another partial answer to the problem of characterizing multiplicative functions in $\mathcal{B}^q$. We show that the condition $f \in \mathcal{E}_q$ implies that f is in $\mathcal{B}^1$, and $\|f\|_q < \infty$. First a rather special result is proved.

Lemma 3.1. *Assume that* $f: \mathbb{P} \to \mathbb{C}$, *and, for every prime* p, $|f(p) - 1| \le \frac{1}{4}$. *Write the values* f(p) *in polar coordinates,*

$$f(p) = r(p) \cdot \exp\{i \cdot \vartheta(p)\}, \ -\pi < \vartheta \le \pi.$$

If the two series

$$S_1(f) = \sum_p p^{-1} \cdot \left(f(p) - 1\right), \ S_2(f) = \sum_p p^{-1} \cdot \left|f(p) - 1\right|^2$$

are convergent, the following five series converge:

$$\Sigma_I = \sum_p p^{-1} \cdot \vartheta(p),$$

$$\Sigma_{II} = \sum_p p^{-1} \cdot \vartheta^2(p),$$

$$\Sigma_{III} = \sum_p p^{-1} \cdot \log r(p),$$

$$\Sigma_{IV} = \sum_p p^{-1} \cdot \left| \log r(p) \right|^2,$$

$$\Sigma_V = \sum_p p^{-1} \cdot \left(r^q(p) - 1 \right), \textit{ for any } q \geq 1.$$

Proof. Clearly, $\frac{3}{4} \leq r(p) \leq \frac{5}{4}$, and $\cos(\vartheta(p)) \geq \frac{1}{2}\sqrt{3}$, and so $-\frac{1}{6}\pi < \vartheta(p) < \frac{1}{6}\pi$. Taking real and imaginary parts, the convergence of the two series $S_1(f)$ and $S_2(f)$ implies the convergence of the four series $\Sigma_1, \ldots, \Sigma_4$, where

$$\Sigma_1 = \sum_p p^{-1} \cdot \left\{ r(p) \cdot \cos \vartheta(p) - 1 \right\},$$

$$\Sigma_2 = \sum_p p^{-1} \cdot \left\{ r(p) \cdot \sin \vartheta(p) \right\},$$

$$\Sigma_3 = \sum_p p^{-1} \cdot \left\{ r(p) \cdot \cos \vartheta(p) - 1 \right\}^2 =$$

$$= \sum_p p^{-1} \cdot \left\{ r^2(p) \cdot \cos^2 \vartheta(p) - 2 \cdot (r(p) \cdot \cos \vartheta(p) - 1) - 1 \right\},$$

and

$$\Sigma_4 = \sum_p p^{-1} \cdot \left\{ r(p) \cdot \sin \vartheta(p) \right\}^2.$$

The inequality $r^2(p) \geq 9/16$ implies

$$\Sigma_5 = \sum_p p^{-1} \cdot \left\{ 1 - \cos^2(\vartheta(p)) \right\} < \infty.$$

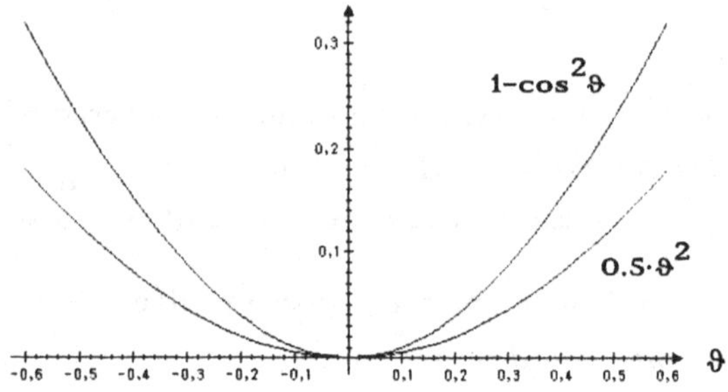

Figure VII.1

Throughout the interval $-\frac{1}{6}\pi < \vartheta(p) < \frac{1}{6}\pi$ the relation $1 - \cos^2(\vartheta(p)) \geq \gamma \cdot \vartheta^2(p)$ holds with a suitable positive constant γ. This implies the convergence of the series

$$\Sigma_{II} = \sum_p p^{-1} \cdot \vartheta^2(p).$$

The relation $|1 - \cos \vartheta| \leq \gamma' \cdot \vartheta^2$ and the convergence of $\Sigma_{II} = \sum_p \frac{\vartheta^2(p)}{p}$ show that

$$\sum_p p^{-1} \cdot r(p) \cdot (1 - \cos \vartheta(p)) < \infty. \tag{3.1}$$

The sum of this series and of Σ_1 is $\sum_p p^{-1}\cdot(r(p)-1)$, and so it is convergent. Similarly, starting with (3.1) and utilizing the convergence of Σ_3 and Σ_4, we find that

$$\sum_p p^{-1}\cdot(r(p)-1)^2 < \infty.$$

Since

$$r^q - 1 = q\cdot(r-1) + \mathcal{O}\left((r-1)^2\right)$$

in $\frac{3}{4} \le r \le 1+\frac{1}{4}$, the series Σ_V is convergent; the approximations

$$\log r = (r-1) + \mathcal{O}\left((r-1)^2\right),\ \log^2 r = \mathcal{O}\left((r-1)^2\right),\ \tfrac{3}{4} \le r \le 1+\tfrac{1}{4},$$

imply the convergence of Σ_{III} and Σ_{IV}. Finally, $\sin\vartheta = \vartheta + \mathcal{O}(\vartheta^2)$ gives the convergence of $\sum_p p^{-1}\cdot r(p)\cdot\vartheta(p)$. Together with the CAUCHY-SCHWARZ estimate

$$\left(\sum_p p^{-1}\cdot|r(p)-1|\cdot|\vartheta(p)|\right)^2 \le \sum_p p^{-1}\cdot(r(p)-1)^2\cdot\sum_p p^{-1}\cdot\vartheta^2(p) < \infty$$

we obtain that

$$\Sigma_I = \sum_p p^{-1}\cdot r(p)\cdot\vartheta(p) - \sum_p p^{-1}\cdot(r(p)-1)\cdot\vartheta(p)$$

is convergent, and the lemma is established. □

Proposition 3.2. *Assume that* f *is a strongly multiplicative arithmetical function, for which the two series*

$$S_1(f) = \sum_p p^{-1}\cdot(f(p)-1)$$

and

$$S_2(f) = \sum_p p^{-1}\cdot|f(p)-1|^2$$

are convergent. Assume, furthermore, that for all primes p *the condition*

$$|f(p)-1| \le \tfrac{1}{4}$$

is satisfied. Then $f \in \mathcal{B}^1$ *and* $\|f\|_q < \infty$ *for any* $q \ge 1$.

Proof. 1) First we obtain $\|f\|_q < \infty$, using RANKIN's trick (II.3). Recall that f is strongly multiplicative and satisfies $|f(p)| \le \frac{5}{4}$. Therefore,

$$\log x \cdot \sum_{\sqrt{x} < n \le x} |f(n)|^q \le 2 \cdot \sum_{\sqrt{x} < n \le x} |f(n)|^q \cdot \log n$$

$$\le 2 \cdot \sum_{n \le x} |f(n)|^q \cdot \sum_{p^k \| n} \log p^k = 2 \sum_{p^k \le x} \log p^k \cdot \sum_{m \le x/p^k, p \nmid m} |f(m \cdot p^k)|^q$$

$$\le 2^{1+q} \cdot \sum_{m \le x} |f(m)|^q \cdot \sum_{p^k \le x/m} \log p^k \ll x \cdot \sum_{m \le x} \frac{|f(m)|^q}{m}$$

by TCHEBYCHEFF's results (see Chapter I). Obviously,

$$\sum_{m \le x} \frac{|f(m)|^q}{m} \le \prod_{p \le x} \left\{ 1 + \frac{|f(p)|^q}{p} + \dots \right\} \le \exp\left\{ \sum_{p \le x} \left(\frac{|f(p)|^q}{p} + \mathcal{O}\left(\frac{1}{p^2}\right)\right)\right\}.$$

The sum $\Sigma_V = \sum_p p^{-1} \cdot (|f(p)|^q - 1)$ is convergent (see Lemma 3.1), and $\Sigma_{p \le x} p^{-1} = \log \log x + \mathcal{O}(1)$. Therefore,

$$\sum_{m \le x} \frac{|f(m)|^q}{m} = \mathcal{O} (\log x),$$

and $\sum_{\sqrt{x} < n \le x} |f(n)|^q = \mathcal{O}(x)$. Combined with the trivial estimate (use II, Theorem 3.1, for example)

$$\sum_{n \le \sqrt{x}} |f(n)|^q \le \sum_{n \le \sqrt{x}} 2^{q \cdot \omega(n)} = \mathcal{O}(x^{\frac{1}{2}+\varepsilon}),$$

we obtain the estimate $\| f \|_q < \infty$.

2) Define a strongly multiplicative function f^* by

$$f^*(p) = f(p), \text{ if } p \le K, \text{ and } f^*(p) = 1, \text{ if } p > K,$$

where K is a [large] constant, which is to be fixed later (depending on the convergence properties of the series appearing in Lemma 3.1). The function f^* just defined is obviously periodic, with period

$$\mathcal{P} = \prod_{p \le K} p.$$

Moreover, f^* is *even* mod $\mathcal{P}$: if $n = \prod_{p|n} p^{\nu(p)}$, then

$$f^*((n,\mathcal{P})) = \prod_{p|(n,\mathcal{P})} f^*(p) = \prod_{p|n, p \le K} f(p) = \prod_{p|n} f^*(p) = f^*(n).$$

Therefore, f^* is in $\mathcal{B}$. We are going to prove that

$$\lim_{K \to \infty} \| f - f^* \|_1 = 0.$$

Let N be sufficiently large. Since $f(n) = f^*(n)\cdot\prod_{p|n,p>K} f(p)$, we obtain

$$\Delta_N := N^{-1}\cdot\sum_{n\le N}\left|f(n) - f^*(n)\right| \le N^{-1}\cdot\sum_{n\le N}\left|f^*(n)\right|\cdot\left|\prod_{p|n,p>K} f(p) - 1\right|.$$

For $|f(p)-1| \le \frac{1}{4}$, the values $\log f(p)$ are well-defined. Next, a strongly *additive* function $w\colon \mathbb{N}\to\mathbb{C}$, is introduced by

$$w(n) = \sum_{p|n,\ p>K} \log f(p).$$

Then

$$w(p) = \log f(p) = \log r(p) + i\cdot\vartheta(p), \text{ if } p > K,$$

in the notation of Lemma 3.1. Making use of the inequality

$$\left| e^z - 1 \right| = \left| \int_0^z e^\zeta\, d\zeta \right| \le |z|\cdot\max\left\{1, e^{\mathrm{Re}\, z}\right\} \le |z|\cdot\left(1+|e^z|\right),$$

we obtain

$$\left|\prod_{p|n,p>K} f(p) - 1\right| = \left|e^{w(n)} - 1\right| \le \left|w(n)\right|\cdot\left(1+\left|e^{w(n)}\right|\right).$$

Starting with $\Delta_N \le N^{-1}\cdot\sum_{n\le N}|w(n)|\cdot(|f^*(n)| + |f^*(n)\cdot e^{w(n)}|)$, the CAUCHY-SCHWARZ inequality gives

$$\Delta_N \le \left\{\frac{1}{N}\cdot\sum_{n\le N}|w(n)|^2\right\}^{\frac12}\cdot 2\left\{\left(\frac{1}{N}\cdot\sum_{n\le N}|f^*(n)|^2\right)^{\frac12} + \left(\frac{1}{N}\cdot\sum_{n\le N}\left|f^*(n)\cdot e^{w(n)}\right|^2\right)^{\frac12}\right\}$$

$$= \{\Delta_N^{(1)}\}^{\frac12}\cdot 2\left\{\left(\Delta_N^{(2)}\right)^{\frac12} + \left(\Delta_N^{(3)}\right)^{\frac12}\right\}.$$

First it will be proved that $\limsup_{N\to\infty} \Delta_N^{(2)}$ is bounded uniformly in K. Using the $\mathcal{P}$-evenness of f^*, we obtain

$$\Delta_N^{(2)} = N^{-1}\cdot\sum_{n\le N}|f(\gcd(n,\mathcal{P}))|^2 = N^{-1}\cdot\sum_{d|\mathcal{P}}|f(d)|^2\cdot\sum_{m\le N/d,(m,\mathcal{P}/d)=1} 1$$

$$= N^{-1}\cdot\sum_{d|\mathcal{P}}|f(d)|^2\cdot\left\{\varphi\left(\frac{\mathcal{P}}{d}\right)\cdot\frac{N}{\mathcal{P}} + \Theta\cdot\frac{\mathcal{P}}{d}\right\},$$

where $|\Theta| \le 1$. The error term is

$$\frac{\mathcal{P}}{N}\sum_{d|\mathcal{P}}\frac{1}{d}\cdot|f(d)|^2 = \frac{\mathcal{P}}{N}\cdot\prod_{p\le K}\left\{1+\frac{|f(p)|^2}{p}\right\} \le \frac{\mathcal{P}}{N}\cdot 2^{\omega(\mathcal{P})}.$$

The main term is

$$\Delta_N^{(21)} = \mathcal{P}^{-1}\cdot\{|f|^2 * \varphi\}(\mathcal{P}) = \prod_{p\le K} p^{-1}\cdot\{\varphi(p) + |f(p)|^2\} = \prod_{p\le K}\left\{1+\frac{|f(p)|^2-1}{p}\right\}.$$

[The star * denotes convolution, see Chapter I, Section 1]. The inequality $1 + x \le e^x$, valid in $-\infty < x < \infty$, and the convergence of $\Sigma_V = \sum_p p^{-1} \cdot \{ |f(p)|^2 - 1 \}$ (from Lemma 3.1) imply

$$\Delta_N^{(21)} \le \exp\Big(\sum_{p \le K} p^{-1} \cdot \{ |f(p)|^2 - 1 \} \Big) \le C_3 < \infty,$$

where the bound C_3 can be chosen independently of K. Thus $\limsup_{N\to\infty} \Delta_N^{(2)}$ is bounded.

Observing that $f^*(n) \cdot e^{w(n)} = f(n)$, and using the CAUCHY-SCHWARZ inequality, one immediately obtains

$$\limsup_{N \to \infty} \Delta_N^{(3)} = \| f \|_2^2 < \infty$$

(by part 1). The proof of Proposition 3.2 will be concluded by showing $\limsup_{N\to\infty} \Delta_N^{(1)} \to 0$ as $K \to \infty$. First

$$\Delta_N^{(1)} \le \frac{2}{N} \sum_{n \le N} \Big| w(n) - \sum_{p \le N} \frac{w(p)}{p} \Big|^2 + 2 \Big| \sum_{p \le N} \frac{w(p)}{p} \Big|^2 = \Delta_N^{(11)} + \Delta_N^{(12)}.$$

The TURÁN-KUBILIUS inequality (I, 4) immediately gives, for every $N > K$,

$$\Delta_N^{(11)} \le 2\, C_1 \cdot \sum_{K < p \le N} p^{-1} \cdot \left\{ \log^2 r(p) + \vartheta^2(p) \right\},$$

and

$$\Delta_N^{(12)} = 2 \cdot \Big| \sum_{K < p \le N} p^{-1} \cdot \log r(p) \Big|^2 + 2 \cdot \Big| \sum_{K < p \le N} p^{-1} \cdot \vartheta(p) \Big|^2.$$

The four series appearing in these estimates are convergent, by Lemma 3.1. Thus

$$\limsup_{N\to\infty} \Delta_N^{(1)} \to 0, \text{ if } K \to \infty. \qquad \square$$

Proposition 3.3. *Let* $q \ge 1$, *and let* $f \in \mathcal{E}_q$ *be multiplicative. Then* $f \in \mathcal{B}^1$ *and* $\| f \|_q < \infty$.

Remark. From VI. Theorem 2.8, the finiteness of $\| f \|_q$ and the fact that $f \in \mathcal{B}^1$ imply $f \in \mathcal{B}^r$ for every r in $1 \le r < q$. In fact, for multiplicative functions f with mean-value $M(f) \neq 0$, the stronger conclusion $f \in \mathcal{B}^q$ is true. This will be shown in this chapter, Theorem 5.2.

Corollary 3.4 [H. DELANGE]. *If* f *is multiplicative, if the series*

$$S_1(f) = \sum p^{-1} \cdot \left\{ f(p) - 1 \right\}$$

converges, and if $|f| \le 1$, *then the mean-value* $M(f)$ *exists and* f *is in* $\mathcal{B}^1$. *In fact,* $\|f\|_q < \infty$ *for every* $q \ge 1$, *and so* $f \in \mathcal{B}^q$ *for any* $q \ge 1$.

This follows immediately from Proposition 3.3. The estimate

$$p^{-1}\cdot\left|f(p)-1\right|^2 \le p^{-1}\cdot\left|f(p)-1\right|^2 + p^{-1}\cdot\left(1-|f(p)|^2\right) = -\frac{f(p)-1}{p} - \frac{\overline{f(p)}-1}{p}$$

and the convergence of $S_1(f)$ imply the boundedness, hence convergence of $S_2(f)$. Therefore, $f \in \mathcal{E}_2$, hence $f \in \mathcal{B}^1$. The finiteness of $\|f\|_q$ is obvious from the estimate $|f| \le 1$. □

The **Proof of Proposition 3.3** is achieved by an application of the Relationship Theorem of Chapter III, which enables us to reduce the assertions of Proposition 3.3 to Proposition 3.2.

1) Let f satisfy the assumptions of Proposition 3.3. The convergence of the series $\sum p^{-1}\cdot(f(p)-1)$ implies the existence of a constant $L \ge 3$ with the property

$$|f(p)| < \tfrac{1}{2}\cdot(p-1), \text{ if } p \ge L.$$

Define a strongly multiplicative function f^* by

$$f^*(p) = \begin{cases} f(p), \text{ if } |f(p)-1| \le \frac{1}{4} \text{ and } p \ge L, \\ 1 \quad \text{otherwise.}\end{cases}$$

The functions f and f^* are related. In fact more is the case: for $j = 1, 2$ one has

$$\begin{aligned}\sum p^{-1}\cdot|f(p)-f^*(p)|^j &\le \sum_{|f(p)-1|>\frac14} p^{-1}\cdot|f(p)-1|^j + \sum_{p\le L} p^{-1}\cdot|f(p)-1|^j \\ &\le 4\cdot\sum_p p^{-1}\cdot|f(p)-1|^2 + \sum_{p\le L} p^{-1}\cdot|f(p)-1|^j \\ &\le \gamma(f,L) < \infty.\end{aligned}$$

Moreover,

$$\sum p^{-1}\cdot\{f^*(p)-1\} = \sum p^{-1}\cdot\{f(p)-1\} + \sum p^{-1}\cdot\{f^*(p)-f(p)\}$$

is convergent. The convergence of

$$S_2(f^*) = \sum p^{-1} \cdot | f^*(p) - 1 |^2$$

follows from the inequality

$$| f^*(p) - 1 |^2 \le 2 \, | f^*(p) - f(p) |^2 + 2 \, | f(p) - 1 |^2.$$

Proposition 3.2 gives $f^* \in \mathcal{B}^1$ and $\| f^* \|_2 < \infty$, and $\| f^* \|_q < \infty$. The Relationship Theorem is used to transform these results into corresponding results about f. The property $f^* \in \mathcal{G}$ is trivial, and it is easy to check that f belongs to

$$\mathcal{G} = \{ f: \mathbb{N} \to \mathbb{C}, \text{ multiplicative}, \sum |p^{-1} \cdot f(p)|^2 < \infty, \sum_p \sum_{k \ge 2} |p^{-k} \cdot f(p^k)| < \infty \}.$$

For example,

$$\sum |p^{-1} \cdot f(p)|^2 \le 2 \sum |p^{-1} \cdot (f(p)-1)|^2 + 2 \sum p^{-2} < \infty,$$

and, if $q > 1$ $\left[\text{define } q' \text{ in the usual way by } q^{-1} + q'^{-1} = 1\right]$,

$$\sum_p \sum_{k \ge 2} p^{-k} \cdot |f(p^k)| \le \sum_p \Big(\sum_{k \ge 2} p^{-k} \cdot |f(p^k)|^q \Big)^{1/q} \cdot \Big(\sum_{k \ge 2} p^{-k} \Big)^{1/q'}$$

$$\ll \Big\{ \sum_p \Big(\sum_{k \ge 2} p^{-k} \cdot |f(p^k)|^q \Big) \Big\}^{1/q} \cdot \Big\{ \sum_p \Big(\sum_{k \ge 2} p^{-k} \Big) \Big\}^{1/q'}.$$

Finally, the arithmetical function f^* is in $\mathcal{G}^*$. By the choice of L, for any prime $p \ge L$, the factors of the EULER product satisfy

$$|\varphi_{f^*}(p,s)| = \left| 1 + p^{-s} \cdot f^*(p) + \ldots \right| = \left| 1 + (p^s - 1)^{-1} \cdot f^*(p) \right|$$

$$\ge 1 - (p-1)^{-1} \cdot |f^*(p)| > 0 \text{ in } \operatorname{Re} s \ge 1.$$

For primes $p \le L$,

$$\varphi_{f^*}(p,s) = 1 + (p^s - 1)^{-1} \ne 0 \text{ in } \operatorname{Re} s \ge 1.$$

The Relationship Theorem (III.7.1) implies that $f \in \mathcal{B}^1$.

2) The finiteness of the [semi-]norm $\| f \|_q$ can be shown by similar arguments (or, by an application of II, Theorem 3.1). Choose $L \ge 3$ so large that

$$|f(p)|^q < \tfrac{1}{2} \cdot (p-1), \text{ if } p \ge L.$$

Then $|f|^q \in \mathcal{G}$, $|f^*|^q \in \mathcal{G}^*$ and these two functions f, f^* are related. Since $\| |f^*|^q \|_1 < \infty$, the Relationship Theorem gives $\| f \|_q^q = \| |f|^q \|_1 < \infty$. □

VII.4. CRITERIA FOR MULTIPLICATIVE FUNCTIONS TO BELONG TO $\mathcal{B}^q$

Up to now the assumption $f \in \mathcal{E}_q$ for multiplicative functions only leads to $f \in \mathcal{B}^1$. We expect that $f \in \mathcal{B}^q$ is true. This is proved in the following theorems.

Theorem 4.1. *If* f *is multiplicative, and* $f \in \mathcal{E}_q$, *where* $q \geq 1$, *then* $f \in \mathcal{B}^q$.

Theorem 4.2. *If* f *is multiplicative,* $f \in \mathcal{E}_1$, $\|f\|_q < \infty$, *and* $M(f) \neq 0$, *then* $f \in \mathcal{B}^q$.

We give two proofs for Theorem 4.1: a direct proof in subsection VII.4.A, and a second proof by an application of BESSEL's inequality in VII.4.B for the special case where $q = 2$, $f \geq 0$, $M(f) \neq 0$, and f is completely multiplicative. Theorem 4.2 is proved in VII.4.C.

4.A. First Proof of Theorem 4.1.

Let $K \geq 2$ be a fixed [large] constant. We define the *multiplicative* function f^* by truncation of f, $f^*(p^k) = f(p^k)$ if $|f(p^k)| \leq K$, and $f^*(p^k) = 1$ if $|f(p^k)| > K$; so

$$f^*(n) = \prod_{p^k \| n,\ |f(p^k)| \leq K} f(p^k).$$

The assertion $f \in \mathcal{B}^q$ follows from the following two statements.

(i) For every K the function f^* is in $\mathcal{B}^q$.
(ii) If K is large, then f^* is near f with respect to $\|\,.\,\|_q$.

(i) It is easy to check that f^* is in $\mathcal{E}_{q+1}$:
The series $S_1(f^*) = S_1(f) - \sum_{|f(p)|>K} p^{-1} \cdot (f(p)-1)$ is convergent because $S''_{2,q}(f)$ is a majorant for the last series. Next, $S'_2(f^*) = S'_2(f) < \infty$, $S''_{2,q+1}(f^*) \ll S''_{2,q}(f) = \mathcal{O}(1)$, and $S_{3,q+1}(f^*) = \mathcal{O}(\sum_p \sum_{k\geq 2} p^{-k}) = \mathcal{O}(1)$.

Therefore, by Proposition 3.3, f^* is in $\mathcal{B}^1$, and $\|f^*\|_{q+1} < \infty$, hence $f^* \in \mathcal{B}^q$ by VI. Theorem 2.8.

(ii) Consider the sum

$$\Delta(x) = \sum_{n\le x} |f(n) - f^{*}(n)|^q = \sum_{n\le x} |f^{*}(n)|^q \cdot \Big| \prod_{p^k \| n, |f(p^k)|>K} f(p^k) - 1 \Big|^q.$$

The product is equal to one if there is no prime-power $p^k \| n$, for which $|f(p^k)| > K$. Using the inequality $|\alpha - 1|^q \le 2^q \cdot |\alpha|^q$, valid for $|\alpha| \ge 1$, we obtain

$$\Delta(x) \le 2^q \cdot \sum_{n\le x}^{(*)} |f(n)|^q,$$

where the condition (*) means that there exists at least one prime-power $p^k \| n$ for which $|f(p^k)| > K$. So, isolating prime-powers p^k, for which $|f(p^k)|$ is large, we obtain

$$\sum_{n\le x}^{(*)} |f(n)|^q \le \sum_p \sum_{\substack{k\ge 1\\ |f(p^k)|>K}} \sum_{\substack{m\le x/p^k\\ p\nmid m}} |f(m)|^q \cdot |f(p^k)|^q$$

$$\le C\cdot x \cdot \Big\{ \sum_{\substack{p\\ |f(p)|>K}} p^{-1} \cdot |f(p)|^q + \sum_p \sum_{\substack{k\ge 2\\ |f(p^k)|>K}} p^{-k} \cdot |f(p^k)|^q \Big\}.$$

Recall that the series $S''_{2,q}(f)$ and $S_{3,q}(f)$ without the conditions $|f(p)| > K$, $|f(p^k)| > K$, are convergent; therefore the summands tend to zero if $K \to \infty$. Thus $\lim_{K\to\infty} \|f - f^{*}\|_q = 0$, and the theorem is proved. □

4.B. A second proof for Theorem 4.1.

In the **special case**

$$q = 2,\ f \ge 0,\ f \text{ completely multiplicative},\ M(f) \ne 0,$$

a simple proof for Theorem 4.1 is available. We calculate the RAMANUJAN coefficients of f and use BESSEL's inequality.

Theorem 4.3. *If* $f \in \mathcal{E}_1$ *is completely multiplicative and has a mean-value* M(f), *then:*

1) *For every* $r \in \mathbb{N}$

$$a_r(f) = \{\varphi(r)\}^{-1} \cdot (f * \mu)(r) \cdot M(f).$$

2) *If in addition* $M(f) \ne 0$, *then the map* $r \mapsto \{M(f)\}^{-1} \cdot a_r(f)$ *is multiplicative and*

$$\{M(f)\}^{-1} \cdot a_{p^k}(f) = \{p-1\}^{-1} \cdot \{f(p) - 1\} \cdot \left(\frac{f(p)}{p}\right)^{k-1}.$$

3) *If* $M(f) \neq 0$ *and* $\|f\|_2 < \infty$, *then*

$$\sum_{r=1}^{\infty} |a_r(f)|^2 \cdot \varphi(r) = \prod_p \frac{p-1}{p-|f(p)|^2} \le \|f\|_2^2.$$

Proof. 1) Using the representation of the RAMANUJAN sum c_r as a sum over the common divisors of n and r, we easily obtain

$$M(f \cdot c_r) = (f*\mu)(r) \cdot M(f).$$

2) The function $r \mapsto a_r(f)/M(f) = (f*\mu)(r)/\varphi(r)$ is obviously multiplicative, and the values of the RAMANUJAN coefficients at prime-powers, given in the assertion of the theorem, are easily checked.

3) BESSEL's inequality implies that

$$A(f) = |M(f)|^2 \cdot \sum_{r=1}^{\infty} \{\varphi(r)\}^{-1} \cdot |(f*\mu)(r)|^2 = \sum_{r=1}^{\infty} |a_r(f)|^2 \cdot \varphi(r) \le \|f\|_2^2$$

is finite. From Lemma 1.6 we know that $M(f) = \prod_p \{p-1\} \cdot \{p-f(p)\}^{-1}$. Therefore,

$$A(f) = \prod_p \frac{|p-1|^2}{|p-f(p)|^2} \cdot \left(1 + \frac{p}{p-1} \cdot \frac{|f(p)-1|^2}{p-|f(p)|^2}\right) = \prod_p \frac{p-1}{p-|f(p)|^2}. \qquad \square$$

We are now going to prove the following special case of Theorem 4.1.

Theorem 4.1'. *If* $f \in \mathcal{E}_1$ *is completely multiplicative and non-negative with mean-value* $M(f) \neq 0$, *and if* $\|f\|_2 < \infty$, *then* $f \in \mathcal{B}^2$.

Proof. First we show $f^2 \in \mathcal{E}_1$: the series $S_1(f^2)$ converges, since $f^2(p) - 1 = 2\cdot\big(f(p)-1\big) + \big(f(p)-1\big)^2$, and $S_1(f)$ and $S_2(f)$ do converge, and $S_2'(f^2) = \mathcal{O}(S_2'(f)) = \mathcal{O}(1)$. The convergence of the other two series follows from BESSEL's inequality.

$$\sum_p \sum_{k\ge1} \left|a_{p^k}(f)\cdot\{M(f)\}^{-1}\right|^2 \cdot \varphi(p^k) \le \|f\|_2^2 \cdot |M(f)|^{-2} < \infty.$$

According to Theorem 4.3, every summand has the form

$$\frac{|f(p)-1|^2}{p-1}\,\frac{|f(p^{k-1})|^2}{p^{k-1}}.$$

Using $x^2 \le 5\cdot(x-1)^2$, if $|x| \ge \frac{1}{2}\cdot\sqrt{5}$, we obtain

$$\sum_{|f^2(p)|\ge\frac{5}{4}} \sum_{k\ge1} \frac{|f(p^k)|^2}{p^k} < \infty.$$

This series is a majorant of $S''_{2,1}(f^2)$ and of $S_{3,1}(f^2)$, so these series are also convergent.

From $f^2 \in \mathcal{E}_1$ we conclude $f^2 \in \mathcal{B}^1$, by Theorem 3.3; and $f \in \mathcal{B}^2$ follows from VI, Theorem 2.9. □

4.C. Proof of Theorem 4.2.

The multiplicative function f is factorized as $f = |f| \cdot h$. We prove a criterion, ensuring that $|f|$ is in $\mathcal{B}^q$ and $h \in \mathcal{B}^q$ (Theorems 4.5 and 4.7). In the case where $M(f) \neq 0$ these results give a criterion for the function f to belong to $\mathcal{B}^q$.

Lemma 4.4. *Let* $f \in \mathcal{A}^1$, *and assume that* $M(|f|^r)$ *exists for some* $r \geq 0$. *If* $M(f) \neq 0$, *then* $M(|f|^r) \neq 0$.

Proof. First $|f| \in \mathcal{A}^1$, $M(|f|)$ exists and $\gamma = M(|f|) \geq |M(f)| > 0$. Choose a trigonometric polynomial $t \in \mathcal{A}$ near $|f|$, so that $\| \, |f| - t \, \|_1 < \frac{1}{8} \cdot \gamma$. Then, for $x \geq x_1$,

$$S = x^{-1} \sum_{n \leq x, |f(n)| \geq \frac{1}{4}\gamma} |f(n)| = x^{-1} \cdot \sum_{n \leq x} |f(n)| - x^{-1} \cdot \sum_{n \leq x, |f(n)| < \frac{1}{4}\gamma} |f(n)|$$

$$\geq \tfrac{1}{2}\gamma - \tfrac{1}{4}\gamma = \tfrac{1}{4}\gamma > 0.$$

On the other hand, for $x \geq x_2$,

$$S \leq x^{-1} \cdot \sum_{n \leq x} \Big| |f(n)| - t(n) \Big| + x^{-1} \cdot \sum_{n \leq x, |f(n)| \geq \frac{1}{4}\gamma} |t(n)|$$

$$\leq \frac{1}{8} \cdot \gamma + \|t\|_u \cdot x^{-1} \cdot \sum_{n \leq x, |f(n)| \geq \frac{1}{4}\gamma} 1.$$

So, for every $x \geq \max\{x_1, x_2\}$

$$x^{-1} \cdot \sum_{n \leq x, |f(n)| \geq \frac{1}{4}\gamma} 1 \geq \frac{1}{8} \cdot \gamma \cdot \|t\|_u^{-1},$$

hence

$$x^{-1} \cdot \sum_{n \leq x} |f(n)|^r \geq (\tfrac{1}{4}\gamma)^r \cdot \frac{1}{8} \cdot \gamma \cdot \|t\|_u^{-1} > 0.$$

This proves Lemma 4.4. □

Theorem 4.5. *If* $f \in \mathcal{B}^1$ *is multiplicative with non-zero mean-value* $M(f)$, *and if* $\|f\|_q < \infty$ *for some* $q \geq 1$, *then* $|f| \in \mathcal{B}^q$.

Proof. The function $g = |f|^{\frac{1}{2}q}$ is non-negative and multiplicative; its 2-norm is $\|g\|_2 = (\|f\|_q)^{\frac{1}{2}q} < \infty$. VI, Theorem 2.8, gives $|f| \in \mathcal{B}^r$, if $1 \le r < q$. Hence:

- if $q \ge 2$, then $|f| \in \mathcal{B}^{\frac{1}{2}q}$ and $g \in \mathcal{B}^1$ (see VI, Theorem 2.9);
- if $1 \le q < 2$, then $|f| \in \mathcal{B}^1$, and so $g \in \mathcal{B}^{2/q} \subset \mathcal{B}^1$, by VI, Theorem 2.9.

In accordance with Lemma 4.4 the mean-value $M(g)$ is non-zero. Therefore, Proposition 2.1 gives $g \in \mathcal{E}_2$. This easily implies $g^2 \in \mathcal{E}_1$, since

$$S_1(g^2) = 2 \cdot S_1(g) + S_2(g),\ S_2'(g^2) = \mathcal{O}(\ S_2(g)),$$
$$S''_{2,1}(g^2) = \mathcal{O}(S_2(g)),\ S_{3,1}(g^2) = S_{3,2}(g),$$

using the notation of Definition 1.2. By Proposition 3.3 the function $|f|^q = g^2$ is in $\mathcal{B}^1$, hence $|f| \in \mathcal{B}^q$. □

Now, for an arithmetical function f, define a function h by

$$h(n) = \begin{cases} \dfrac{f(n)}{|f(n)|}, & \text{if } f(n) \neq 0, \\ 0\ , & \text{if } f(n) = 0. \end{cases}$$

If f is real-valued, then h is the sign-function. A first result on this "generalized sign-function" h is the following proposition.

Proposition 4.6. *Let* f *be an element of* $\mathcal{B}^1$. *If there exists a constant* $\delta > 0$, *for which the upper density*

$$\overline{\text{dens}}\{\ n;\ |f(n)| < \delta\ \} = \limsup_{x \to \infty} x^{-1} \cdot \#\{\ n \le x,\ |f(n)| < \delta\ \} = 0,$$

then the function h *is in* $\mathcal{B}^1$ *again.*

Proof. The function $s_\delta : \mathbb{C} \to \mathbb{C}$ will be defined by

$$s_\delta(z) = \begin{cases} \dfrac{z}{|z|}, & \text{if } |z| \ge \delta, \\ \dfrac{z}{\delta}, & \text{if } |z| < \delta. \end{cases}$$

Then

$$|\,s_\delta(z_1) - s_\delta(z_2)\,| \le 2 \cdot \delta^{-1} \cdot |z_1 - z_2|, \text{ for every } z_1, z_2 \in \mathbb{C},$$

and so s_δ is LIPSCHITZ-continuous; by Theorem VI.2.11 (2) the composition $s_\delta \circ f$ is in $\mathcal{B}^1$, and h is $\|.\|_1$-near $s_\delta \circ f$ on behalf of

$$\sum_{n\le x} \left| h(n) - s_\delta(f(n)) \right| = \sum_{n\le x, |f(n)|<\delta} \left| h(n) - s_\delta(f(n)) \right| = o(x).$$

The last line follows from the assumption $\overline{\text{dens}}\{n;\ |f(n)| < \delta\} = 0$ and the boundedness of $|h(n) - s_\delta(f(n))|$. □

Theorem 4.7. *Suppose* $f \in \mathcal{E}_1$ *is multiplicative. If the mean-value* M(f) *exists and is non-zero, the [multiplicative] function* h *is in* $\mathcal{B}^1$.

Remarks.

(a) The function h is bounded, therefore h is in $\mathcal{B}^q$ by Theorem VI.2.8.

(b) In Theorem 4.7, the condition $M(f) \neq 0$ cannot be omitted, as is shown by the multiplicative function f: $n \mapsto n^{-1}\cdot\mu(n)$; obviously $M(f) = \|f\|_1 = 0$, so $f \in \mathcal{B}^1$, but $h(n) = \mu(n)$ is not in $\mathcal{B}^1$. (The idea of the proof is: the assumption $\mu \in \mathcal{B}^1$ implies $\mu \in \mathcal{B}^2$, and this contradicts the PARSEVAL equation.)

Proof of Theorem 4.7. Obviously, $|h| \le 1$. According to Corollary 3.4 it is sufficient to prove the convergence of

$$S_1(h) = \sum p^{-1} \cdot \left\{ h(p) - 1 \right\}.$$

The function f being in $\mathcal{E}_1$, the series $S_2'(f) = \sum_{|f(p)|\le 5/4} p^{-1}\cdot|f(p)-1|^2$ and $S_{2,1}''(f) = \sum_{|f(p)|>5/4} p^{-1}\cdot|f(p)|$ are convergent. Therefore,

$$(4.1) \qquad \sum_{|f(p)|\le 3/4} p^{-1} \quad \text{and} \quad \sum_{|f(p)|>5/4} p^{-1}$$

are convergent. So it is sufficient to prove the convergence of

$$S_1^*(h) = \sum_p{}^{*}\, p^{-1} \cdot \left\{ h(p) - 1 \right\},$$

where the condition * means $\frac{3}{4} \le |f(p)| \le \frac{5}{4}$. Using

$$r^{-1} = 1 - (r-1) + \mathcal{O}\left(|r-1|^2\right), \text{ as long as } |r-1| \le \tfrac{1}{4},$$

we obtain

$$h(p) - 1 = f(p)\cdot \left\{ 1 - \left(|f(p)| - 1\right) + \mathcal{O}\left(\left(|f(p)|-1\right)^2\right)\right\} - 1,$$

and so

$$S_1^*(h) = \sum_p{}^* \frac{f(p)-1}{p} - \sum_p{}^* \frac{f(p)-1}{p}\cdot\big(|f(p)|-1\big) - \sum_p{}^* \frac{|f(p)|-1}{p} + O\Big(\sum_p{}^* \frac{(|f(p)|-1)^2}{p}\Big).$$

The first series on the right-hand side of this equation converges because of the convergence of $S_1(f)$ and of (4.1). The second and the fourth series have the convergent majorant $S_2'(f)$. Thus it remains to prove that the third series is convergent, and because of (4.1) it suffices to show the convergence of $S_1(|f|)$:

> This series equals $S_1(|f|) = 2\, S_1(|f|^{\frac12}) + S_2(|f|^{\frac12})$. According to VI, Theorem 2.9, the result $|f| \in \mathcal{B}^1$ implies $|f|^{\frac12} \in \mathcal{B}^2$. Lemma 4.4 gives $M(|f|^{\frac12}) > 0$, and Proposition 2.1 yields $|f|^{\frac12} \in \mathcal{E}_2$. □

Proof of Theorem 4.2. Write $f = h\cdot|f|$. First of all, by Proposition 3.3 and Theorem 4.5, the function $|f|$ is in $\mathcal{B}^q$. Next, the function h is in $\mathcal{B}^1$ according to Theorem 4.7. The function h is bounded; therefore VI, Theorem 2.6 (iii) gives $h\cdot|f| \in \mathcal{B}^q$. □

VII.5. MULTIPLICATIVE FUNCTIONS IN $\mathcal{A}^q$ WITH MEAN-VALUE $M(f) \neq 0$

The aim of this section is to prove the following theorem.

Theorem 5.1. *Let* f *be a multiplicative arithmetical function with mean-value* $M(f) \neq 0$. *Assume that* $q \geq 1$. *Then the following four statements are equivalent:*

(1) $f \in \mathcal{E}_q$,

(2) $f \in \mathcal{B}^q$,

(3) $f \in \mathcal{D}^q$,

(4) $f \in \mathcal{A}^q$.

If $q \geq 2$, *then* (1) *to* (4) *are also equivalent to*

(5) $\|f\|_q < \infty$.

Furthermore, in any case and for every prime p,

$$\varphi_f(p,1) = 1 + p^{-1} \cdot f(p) + p^{-2} \cdot f(p^2) + \ldots \neq 0.$$

Remark. Without the assumption $M(f) \neq 0$, the implication (4) ⇒ (1) is wrong, as may be seen from the example following Lemma 1.4.

Proof. The implication (1) ⇒ (2) is contained in Theorem 4.1. The implications (2) ⇒ (3) ⇒ (4) are trivial. The implication (4) ⇒ (1) will be proved in two steps: first the convergence of $S''_{2,q}(f)$ and $S_{3,q}(f)$ will be shown, then that of $S_1(f)$ and of $S_2'(f)$ by relationship arguments.

1) We consider the function $g = |f|^{\frac{1}{2}q}$ and use Proposition 2.1. Since $|f| \in \mathcal{A}^q$, Corollary VI.2.10 gives $g \in \mathcal{A}^2$. The mean-value $M(g)$ is non-zero for $M(|f|) \geq |M(f)| > 0$ and Lemma 4.4. Proposition 2.1 shows that the three series

$$S_1(g) = \sum_p p^{-1} \cdot \big(g(p) - 1\big),$$

$$S_2(g) = \sum_p p^{-1} \cdot |g(p) - 1|^2,$$

$$S_{3,2}(g) = \sum_p \sum_{k\geq 2} p^{-k} \, |g(p^k)|^2$$

are convergent. So $S_{3,q}(f) = S_{3,2}(g)$ is convergent.

From the convergence of $S_2(|f|^{\frac{1}{2}q})$ we obtain

(5.1) $$\sum_{p,|f(p)|>5/4} p^{-1} < \infty, \quad \text{and} \quad \sum_{p,|f(p)|<3/4} p^{-1} < \infty.$$

Using $|z|^q = \mathcal{O}\big((|z|^{\frac{1}{2}q} - 1)^2\big)$ in $|z| \geq \frac{5}{4}$, we obtain

$$S''_{2,q}(f) = \sum_{p,|f(p)|>5/4} p^{-1} \cdot |f(p)|^q = \mathcal{O}\big(S_2(g)\big) = \mathcal{O}(1).$$

2) The convergence of $S''_{2,q}(f)$ and $S_{3,q}(f)$ shows the existence of a prime p_0 with the property

$$\sum_{k\geq 1} \frac{|f(p^k)|}{p^k} \leq \tfrac{1}{4} \text{ for every prime } p \geq p_0.$$

Define a multiplicative function F by

$$F(p^k) = \begin{cases} f(p^k), & \text{if } p < p_0, \\ f(p), & \text{if } p \geq p_0, \ |f(p)| \leq \frac{5}{4} \text{ and } k = 1, \\ 0 & \text{otherwise.} \end{cases}$$

Then f and F are related, f and F are in $\mathcal{G}$, and

$$|\varphi_f(p,s)| \geq 1 - \sum_{k\geq 1} \frac{|f(p^k)|}{p^k} \geq \tfrac{3}{4} \text{ for all primes } p \geq p_0, \text{ and for } \mathrm{Re}\, s \geq 1.$$

Theorem III, 7.1 allows the conclusion $F \in \mathcal{A}^1$, and gives for the mean-value

$$M(F) = M(f) \cdot \prod_p \frac{\varphi_F(p,1)}{\varphi_f(p,1)} = M(f) \cdot \prod_{p\geq p_0} \frac{1 + p^{-1} \cdot F(p)}{1 + p^{-1} \cdot f(p) + p^{-2} \cdot f(p^2) + \dots} \neq 0.$$

The values of f must be changed for a second time. Let $K \geq 2$ be an integer, and denote by μ_K the characteristic function of the set of K-free integers:

> $\mu_K(n) = 0$, *if there exists a prime* p *with* p^K *dividing* n, *and*
> $\mu_K(n) = 1$ *otherwise.*

The function μ_K is related to $1 \in \mathcal{B}^1$, and so μ_K is in $\mathcal{B}^1$. μ_K is bounded, F is in $\mathcal{A}^1$, therefore the pointwise product

$$F_K = \mu_K \cdot F \text{ is in } \mathcal{A}^1.$$

The mean-value $M(F_K)$ exists and $M(F_K) \neq 0$ if K is chosen large enough (another possibility of showing $M(F_K) \neq 0$ would be to use the representation of the mean-values $M(F)$ and $M(F_K)$ as infinite products):

The map $M(.)\colon \mathcal{A}^1 \to \mathbb{C}$, $f \mapsto M(f)$, is continuous; $M(F) \neq 0$, therefore $M(g)$ is non-zero for any $g \in \mathcal{A}^1$ near F. So we have to show that $\|F - \mu_K \cdot F\|_1$ is small if K is sufficiently large. First we calculate

$$x^{-1} \cdot \sum_{n\leq x} |F(n)| \cdot |1-\mu_K(n)| = x^{-1} \cdot \sum_{n\leq x, \exists p<p_0 : p^K | n} |F(n)|$$

$$\leq x^{-1} \cdot \sum_{p<p_0} \sum_{n\leq x,\, n\equiv 0(p^K)} |F(n)|$$

$$\leq x^{-1} \cdot \sum_{p<p_0} \sum_{k\geq K} \sum_{n\leq x,\, p^k \| n} |F(n)|$$

$$\ll \sum_{p<p_0} \sum_{k\geq K} p^{-k} \cdot |F(p^k)|.$$

The convergence of the series $S_{3,1}(f)$ shows that this last sum is small as soon as K is chosen sufficiently large.

In order to show $\|F_K\|_2 < \infty$, Theorem II.3.1 is applied to $|F_K|^2$. The values $F_K(p^k)$ are uniformly bounded, so the assumption of this theorem is fulfilled. Using

$$\sum_{p < p_0} \sum_{2 \le k \le K} p^{-k} \cdot |F_K(p^k)|^2 = \mathcal{O}(1),$$

we obtain the upper estimate

$$x^{-1} \cdot \sum_{n \le x} |F_K(n)|^2 = \mathcal{O}\Big(\exp \Big(\sum_{p \le x} p^{-1} \cdot \big\{ |F_K(p)|^2 - 1 \big\} \Big) \Big).$$

Therefore, by (5.1), the result $\|F_K\|_2 < \infty$ is obtained as soon as the convergence of

$$\sum_{p,\ 3/4 < |f(p)| < 5/4} p^{-1} \cdot \big(|f(p)|^2 - 1 \big)$$

is proved. The inequality

$$\big| x^\beta - 1 - \beta \cdot (x-1) \big| \le \tfrac{1}{2} \cdot \big| \beta \cdot (\beta - 1) \big| \cdot \big| x-1 \big|^2 \cdot \max \{1, x^{\beta - 2}\}$$

is valid in $x > 0$, $\beta > 0$. Therefore, if $3/4 < |f(p)| < 5/4$,

$$|f(p)|^2 - 1 = 4 \cdot q^{-1} \cdot \big\{ |f(p)|^{\frac{1}{2}q} - 1 \big\} + \mathcal{O}\big(\big\{ |f(p)|^{\frac{1}{2}q} - 1 \big\}^2 \big),$$

and the convergence of $S_1(|f(p)|^{\frac{1}{2}q})$ and $S_2(|f(p)|^{\frac{1}{2}q})$ shows that $\|F_K\|_2 < \infty$.

Proposition 2.1, the finiteness of $\| F_K \|_2$, and $M(F_K) \neq 0$ imply the convergence of the series

$$S_2(F_K) = \sum_p p^{-1} \cdot | F_K(p) - 1 |^2, \text{ and } S_1(F_K) = \sum_p p^{-1} \cdot \{ F_K(p) - 1 \},$$

and so $S_1(f)$ and $S_2'(f)$ are convergent, and f is in $\mathcal{E}_q$. This concludes the proof of the equivalence of (1) - (4).

Lemma 1.6 yields $\varphi_f(p,1) \neq 0$ for every prime p.

Finally, we discuss (5) in the case where $q \ge 2$. Assume that $\| f \|_q < \infty$, where $q \ge 2$. Then $f \in \mathcal{E}_2$ by Proposition 2.1, and f is in $\mathcal{B}^2$ according to Theorem 4.1. The function $g = |f|^{\frac{1}{2}q}$ has a non-zero mean-value, as is seen from Lemma 4.4.

Since $\|g\|_2 = \| f \|_q^{\frac{1}{2}q} < \infty$, Proposition 2.1 and Theorem 4.1 give $g \in \mathcal{B}^2$, therefore $|f| \in \mathcal{B}^q$ by DABOUSSI's Theorem VI.2.9. Proposition VI, 8.2 shows that $f \in \mathcal{B}^q$. Therefore, condition (5) implies (2). (2) $\Rightarrow$ (5) is trivial, and we are done. □

As a corollary of Theorem 5.1 and 4.2 we state the following result.

Theorem 5.2. *Let* $q \geq 1$, *and let* f *be multiplicative with mean-value* $M(f) \neq 0$. *If* $f \in \mathcal{B}^1$ *and* $\|f\|_q < \infty$, *then* $f \in \mathcal{B}^q$.

Remark. The assumption $M(f) \neq 0$ can be weakened to $\|f\|_1 > 0$ (see Exercise 5).

VII.6. MULTIPLICATIVE FUNCTIONS IN $\mathcal{A}^q$ WITH NON–VOID SPECTRUM

If f is an arithmetical function in $\mathcal{A}^1$, its FOURIER-coefficients $\hat{f}(\beta) = M(f\cdot\overline{e}_\beta)$ exist for every $\beta \in \mathbb{R}$, and its FOURIER-BOHR spectrum is defined as

$$\mathrm{spec}(f) = \left\{\beta \in \mathbb{R}/\mathbb{Z} : \limsup_{x\to\infty} \left| x^{-1}\cdot \sum_{n\leq x} f(n)\cdot e_{-\beta}(n) \right| > 0\right\}.$$

Remarks. 1) For functions in $\mathcal{A}^1$ the *lim sup* in the definition of spec(f) can be replaced by $\lim_{x\to\infty}$ - this limit exists.
2) For every arithmetical function f the condition $M(f) \neq 0$ implies $\mathrm{spec}(f) \neq \emptyset$, and this implies $\|f\|_1 > 0$.

H. DABOUSSI proved the following theorem in 1980.

Theorem 6.1. *Let* f *be a multiplicative arithmetical function, and assume that* $q \geq 1$.

(I) *If*

(D.1) $$f \in \mathcal{A}^q,$$

and

(D.2) $$\mathrm{spec}(f) \neq \emptyset,$$

then there exists a DIRICHLET *character* χ, *such that the four series*

(D.3) $$S_1(\chi\cdot f),\quad S_2'(\chi\cdot f),\quad S''_{2,q}(f),\quad \textit{and}\quad S_{3,q}(f)$$

are convergent.

(II) *Conversely, if the series* (D.3) *are convergent for some* DIRICHLET *character* χ, *then* $f \in \mathcal{D}^q$.

Corollary 6.2. *If* f *is a multiplicative function with* $\mathrm{spec}(f) \neq \emptyset$, *and if* $q \geq 1$, *then the following three statements are equivalent.*

(1) *There exists a* DIRICHLET *character* χ *such that the four series given in* (D3) *are convergent.*

(2) $f \in \mathcal{A}^q$.

(3) $f \in \mathcal{D}^q$.

First we give a variation of DABOUSSI's result II.6.2.

Lemma 6.3. *If* $f \in \mathcal{A}^1$ *is multiplicative, then, for every* <u>*irrational*</u> β *the* FOURIER *coefficient*

$$\hat{f}(\beta) = M(f \cdot e_{-\beta})$$

is zero.

Proof. Without loss of generality, assume that $M(|f|) > 0$; otherwise $|\hat{f}(\beta)| = 0$ because of $|\hat{f}(\beta)| \leq M(|f|)$. Theorem 5.1 shows that $|f|$ is in $\mathcal{E}_1$. Choose a prime p_0 so large that for all primes $p > p_0$

$$p^{-1} \cdot |f(p)| \leq \tfrac{1}{4}, \text{ and } \sum_{k\geq 1} p^{-k} \cdot |f(p^k)| \leq \tfrac{1}{2}.$$

Define a multiplicative function F by

$$F(p^k) = \begin{cases} 1, & \text{if } \left(p \leq p_0 \text{ or } |f(p)| > \frac{5}{4}\right), \text{ and } k = 1, \\ f(p), & \text{if } p > p_0, |f(p)| \leq \frac{5}{4}, \text{ and } k = 1, \\ 0, & \text{if } k \geq 2. \end{cases}$$

The functions f and F are related:

$$\sum_p p^{-1} \cdot |f(p) - F(p)| = \mathcal{O}\left(S''_{2,1}(|f|)\right) = \mathcal{O}(1).$$

Both of the functions f, F are in $\mathcal{G}$ since

$$\sum_p \left|p^{-1} \cdot f(p)\right|^2 = \mathcal{O}(1) + \mathcal{O}\left(S''_{2,1}(|f|)\right) = \mathcal{O}(1),$$

$$\sum_p \sum_{k\geq 2} p^{-k} \cdot |f(p^k)| = S_{3,1}(|f|) = \mathcal{O}(1).$$

The estimate $|\varphi_F(p,1) - 1| = p^{-1} \cdot |F(p)| \leq \frac{5}{4} \cdot p^{-1} < 1$ shows that F is in $\mathcal{G}^*$. The Relationship Theorem III.7.1 (or III, Exercise 9) gives, with a multiplicative function h, satisfying $f = h * F$ and $\sum n^{-1} \cdot |h(n)| < \infty$,

$$(6.1) \qquad \hat{f}(\beta) = \sum_{n=1}^{\infty} n^{-1} \cdot h(n) \cdot \hat{F}(n \cdot \beta)$$

for every real β. The values $|F(p)|$ are bounded, therefore DABOUSSI's Theorem II.6.2 gives $\hat{F}(\alpha) = 0$ for every irrational α, and thus equation (6.1) gives the statement of Lemma 6.1 as soon as $\|F\|_2 < \infty$ is proved. By II, Theorem 3.1, (3.3)

$$x^{-1} \cdot \sum_{n \le x} |F(n)|^2 = \mathcal{O}\Big(\exp\Big\{ \sum_{p \le x, |f(p)| \le 5/4} p^{-1} \cdot (|f(p)|^2 - 1) \Big\} \Big).$$

The last series is seen to be convergent, using $x^2 - 1 = (x-1)^2 + 2(x-1)$, and

$$\sum_{|f(p)| \le 5/4} p^{-1} \cdot (|f(p)| - 1)^2 = S_2'(|f|) = \mathcal{O}(1),$$

$$\sum_{|f(p)| \le 5/4} p^{-1} \cdot (|f(p)| - 1) = S_1(|f|) - \sum_{|f(p)| > 5/4} p^{-1} \cdot (|f(p)| - 1),$$

$$\sum_{|f(p)| > 5/4} p^{-1} \cdot (|f(p)| - 1) \le \sum_{|f(p)| > 5/4} p^{-1} \cdot |f(p)| = S''_{2,1}(|f|) = \mathcal{O}(1). \square$$

Next we prove Theorem 6.1 (I) in the special case where f is *completely* multiplicative.

Lemma 6.4. *Suppose* f *is completely multiplicative and* $q \ge 1$. *If*

(D.1) $$f \in \mathcal{A}^1,$$

and

(D.2) $$\operatorname{spec}(f) \ne \emptyset,$$

then there is a DIRICHLET *character* χ *for which* $M(\chi \cdot f) \ne 0$.

Proof. Assume that $M(\chi \cdot f) = 0$ for all DIRICHLET characters χ. The calculation of the FOURIER coefficient $\hat{f}(\beta)$, where $\beta = \frac{a}{r}$, $r \ge 1$, is rational, is achieved in the following way:

$$x^{-1} \cdot \sum_{n \le x} f(n) \cdot e_{-a/r}(n) = \sum_{1 \le \rho \le r} e_{-a/r}(\rho) \cdot x^{-1} \cdot \sum_{n \le x,\ n \equiv \rho \bmod r} f(n)$$

$$= \sum_{d | r} \sum_{\substack{1 \le \rho \le r \\ \gcd(\rho, r) = d}} \Big(e_{-a/r}(\rho) \cdot \frac{1}{x} \cdot \sum_{\substack{n \le x \\ n \equiv \rho \bmod r}} f(n) \Big)$$

$$= \sum_{d | r} \frac{f(d)}{d} \cdot \sum_{\substack{1 \le \rho' \le r' \\ \gcd(\rho', r') = 1}} \Big(e_{-a/r'}(\rho') \frac{1}{x/d} \cdot \sum_{\substack{m \le x/d \\ m \equiv \rho' \bmod r'}} f(m) \Big),$$

with the abbreviations $\rho' = \rho/d$, $r' = r/d$. The orthogonality relations for the characters χ mod r' [in case $\gcd(r', \rho') = 1$] imply

$$\frac{1}{x} \cdot \sum_{\substack{m \le x \\ m \equiv \rho' \bmod r'}} f(m) = \frac{1}{\varphi(r')} \cdot \sum_{\chi \bmod r'} \overline{\chi(\rho')} \cdot \frac{1}{x} \cdot \sum_{m \le x} \chi(m) \cdot f(m) = o(1),$$

by our assumption $M(\chi \cdot f) = 0$ for every character χ. Thus we obtain $\hat{f}(\beta) = 0$ for every rational number β. Since $\mathrm{spec}(f) \subset \mathbb{Q}$ (see Lemma 6.3), we have a contradiction to (D.2), and Lemma 6.4 is proved. □

Proof of Theorem 6.1.

I) Our goal is to show that, given a multiplicative function f in $\mathcal{A}^q$, where $q \ge 1$, with non-void spectrum spec(f), there exists a DIRICHLET character χ such that $M(\chi \cdot f) \neq 0$.

Given $f \in \mathcal{A}^q$ with non-void spectrum, then $|f| \in \mathcal{A}^q$ and $M(|f|) > 0$. Therefore, we deduce $|f| \in \mathcal{E}_q$, using Theorem 5.1. In particular, the series

$$S''_{2,q}(|f|) = S''_{2,q}(f), \text{ and } S_{3,q}(|f|) = S_{3,q}(f)$$

are convergent. The convergence of these series enables us to choose an integer P_0 with the properties

$$|p^{-1} \cdot f(p)| < \tfrac{1}{4}, \text{ and } \sum_{k \ge 1} p^{-k} \cdot |f(p^k)| < \tfrac{1}{2} \text{ for any } p \ge P_0.$$

Define the "nearly-completely" multiplicative function f^* through

$$f^*(p^k) = \begin{cases} f(p^k), & \text{if } p \le P_0, \\ \{f(p)\}^k, & \text{if } p > P_0. \end{cases}$$

Then f and f^* are in $\mathcal{G}$, and $\varphi_f(p,s) \neq 0$ in Re $s \ge 1$ for every prime $p > P_0$. Furthermore, f and f^* are related, and Theorem III, 7.1 admits the conclusion

$$f^* = f * h, \text{ where } \sum_{n=1}^{\infty} n^{-1} \cdot |h(n)| < \infty;$$

therefore f^* is in $\mathcal{A}^1$. Next, define

$$f^{**}(n) = \eta(n) \cdot f^*(n),$$

where η is the multiplicative function defined by

$$\eta(n) = \begin{cases} 0, & \text{if } \gcd\left(n, \prod_{p \le P_0} p\right) \ne 1, \\ 1 & \text{otherwise.} \end{cases}$$

η is periodic, therefore f^{**} is in $\mathcal{D} \cdot \mathcal{A}^1 \subset \mathcal{A}^1$. Moreover, f^{**} is in $\mathcal{G}^*$, f and f^{**} are related, therefore

$$f = f^{**} * H, \text{ where } \sum_{n=1}^{\infty} n^{-1} \cdot |H(n)| < \infty.$$

Clearly, $\text{spec}(f^{**}) \ne \emptyset$ [otherwise it follows that $\text{spec}(f) = \emptyset$, too], and Lemma 6.4 can be applied to the completely multiplicative function f^{**}. We obtain a DIRICHLET character χ with the property $M(\chi \cdot f^{**}) \ne 0$.

Theorem 5.1 is now applied twice. Since $\chi \cdot f^{**} \in \mathcal{D} \cdot \mathcal{A}^1 \subset \mathcal{A}^1$, we obtain the convergence of the series

$$S_1(\chi \cdot f^{**}) = \sum p^{-1} \cdot \{ f^{**}(p) \cdot \chi(p) - 1 \} \text{ and } S_2'(\chi \cdot f^{**}) = \sum p^{-1} \cdot \left| f^{**}(p) \cdot \chi(p) - 1 \right|^2.$$

The values $f^{**}(p) = f(p)$ are equal, except for a finite number of primes, therefore the series

$$S_1(\chi \cdot f) \text{ and } S_2'(\chi \cdot f)$$

are convergent. The arithmetical function $|f|$ is in $\mathcal{A}^q$, and its mean-value is $M(|f|) \ge |\hat{f}(\beta)|$, which is > 0 for some real β. Therefore, again using Theorem 5.1, we find

$$S''_{2,q}(f) = S''_{2,q}(|f|) < \infty, \text{ and } S_{3,q}(f) = S_{3,q}(|f|) < \infty,$$

and (D3) is proved.

II) Let χ be a DIRICHLET character, for which the four series (D3) are convergent. The function $\chi \cdot f$ is in $\mathcal{B}^q$, utilizing Theorem 4.1 and $|\chi f| \le |f|$, $\chi \cdot f \in \mathcal{E}_q$.

Denote by d the module of the character χ and define multiplicative functions η, ν by

$$\eta(p^k) = \begin{cases} 1, & \text{if } p \nmid d, \\ 0, & \text{if } p \mid d, \end{cases} \qquad \nu(p^k) = \begin{cases} 0, & \text{if } p \nmid d, \\ 1, & \text{if } p \mid d. \end{cases}$$

Then

$$\eta \cdot f = \overline{\chi}(\chi f) \in \mathcal{D} \cdot \mathcal{B}^q \subset \mathcal{D}^q,$$

therefore $\eta \cdot |f| \in \mathcal{D}^q$, and $\eta \cdot |f|^q \in \mathcal{D}^1$ (see Theorem VI.2.9). Since

$$(\eta \cdot |f|^q) * (\nu \cdot |f|^q) = |f|^q,$$

and, by the convergence of $S_{3,q}(f)$,

$$\sum_{n=1}^{\infty} n^{-1} \cdot \nu(n) \cdot |f(n)|^q = \prod_{p|d} \sum_{k \geq 0} p^{-k} \cdot |f(p^k)|^q < \infty,$$

we obtain $|f|^q \in \mathcal{D}^1$, therefore $|f| \in \mathcal{D}^q$. Starting with $\eta \cdot f \in \mathcal{D}^1$, the same argument leads to $f \in \mathcal{D}^1$. Proposition 8.2 finally gives $f \in \mathcal{D}^q$. □

VII.7. EXERCISES

1) Let f be a multiplicative function and assume that $\| f \|_q < \infty$ for some $q > 1$. Then prove: the existence of the mean-value $M(f) \neq 0$ implies the existence of

$$M_{(d)}(f) = \lim_{x \to \infty} x^{-1} \cdot \sum_{n \leq x, \gcd(n,d)=1} f(n)$$

if d is composed solely from sufficiently large primes, and

$$M_{(d)}(f) = M(f) \cdot \prod_{p|d} \{\varphi_f(p,1)\}^{-1}.$$

2) Define an arithmetical function $\chi_{r,d}$ in the following way:

$$\chi_{r,d}(n) = \begin{cases} 1, & \text{if } \gcd(n,r) = d, \\ 0, & \text{if } \gcd(n,r) \neq d. \end{cases}$$

Prove for any function $f \in \mathcal{B}^1$:

(a) $\|f\|_1 > 0$ if and only if there exist positive integers r, d such that $M(f\chi_{r,d}) \neq 0$.

(b) If f is <u>multiplicative</u> in addition, then $\|f\|_1 > 0$ if and only if there exists a DIRICHLET character $\chi = \chi_{r,1}$ for which $M(f\chi) \neq 0$.

3) If f is multiplicative, $q \geq 1$ and $\| f \|_q > 0$, then prove that

$$f \in \mathcal{E}_q \Leftrightarrow f \in \mathcal{B}^q.$$

4) If $f \in \mathcal{B}^1$ is multiplicative and $\|f\|_1 > 0$, then the function

$$h(n) = \frac{f(n)}{|f(n)|}, \text{ if } f(n) \neq 0, \; h(n) = 0, \text{ if } f(n) = 0,$$

is in $\mathcal{B}^1$.

5) Let f be multiplicative, and assume that $\| f \|_1 > 0$. If f is in $\mathcal{B}^1$, $q \geq 1$, and $\| f \|_q < \infty$, then $f \in \mathcal{B}^q$.

6) Given $d \in \mathbb{N}$, define the function η by $\eta(n) = 1$ if $\gcd(n, d) = 1$, and $\eta(n) = 0$ otherwise. Assume that f is multiplicative, $q \geq 1$, $\|\eta \cdot f \|_q < \infty$, and - for every prime p - the sum $\sum_{k \geq 0} p^{-k} \cdot |f(p^k)|^q < \infty$. Then prove that $\| f \|_q < \infty$.

Chapter VIII

Ramanujan Expansions

Abstract. *In this chapter, for given classes of arithmetical functions, mean-values and* RAMANUJAN *coefficients* $a_r(f) = \{\varphi(r)\}^{-1} \cdot M(f \cdot c_r)$ *are calculated, and the convergence properties of* RAMANUJAN *expansions are studied. To achieve this, it is advisable to deal with mean-values* $M_d(f) = \lim_{x \to \infty} \sum_{n \le x, n \equiv 0 \bmod d} f(n)$ *of arithmetical functions in residue-classes. Rather simple criteria use the* ERATOSTHENES-MÖBIUS *transform* $f' = f * \mu$. *Better results are obtained when the results of Chapter VII are used to obtain information on mean-values* $M_d(f)$ *and* RAMANUJAN *coefficients* $a_r(f)$. *For multiplicative functions in* $\mathcal{A}^2$ *the* RAMANUJAN *expansion* $\sum a_r(f) \cdot c_r(n)$ *is pointwise convergent. Finally, still another proof of* PARSEVAL*'s equation is given for multiplicative functions in* $\mathcal{A}^2$.

VIII.1. INTRODUCTION

The RAMANUJAN sums c_r, r = 1, 2, ..., were defined in I, §3. In this chapter we shall utilize both of the representations

$$c_r(n) = \sum_{d|\gcd(r,n)} d \cdot \mu(r/d) = \sum_{1 \le a \le r, \gcd(a,r)=1} \exp\left(2\pi i \frac{a}{r} n\right),$$

and the multiplicativity of the map $r \mapsto c_r(n)$. Due to the orthogonality relations for RAMANUJAN sums,

$$M(c_r \cdot c_s) = \varphi(r), \text{ if } r = s, \text{ and } M(c_r \cdot c_s) = 0 \text{ otherwise}$$

(see I, Theorem 3.1), for an arithmetical function f we expect a RAMANUJAN expansion

(1.1) $$f \sim \sum_r a_r \cdot c_r,$$

where the coefficients $a_r = a_r(f)$, in the case of the existence of the limits involved, are given by

(1.2) $$a_r(f) = (\varphi(r))^{-1} \cdot M(f \cdot c_r) = (\varphi(r))^{-1} \cdot \langle f, c_r \rangle,$$

using the inner product notation $\langle f,g \rangle = M(f \cdot \bar{g})$. There are many examples of arithmetical functions possessing a [convergent or not convergent] RAMANUJAN expansion (1.1): *the coefficients* (1.2) *do exist, for example, for all functions in* $\mathcal{A}^1$. There are different concepts of "convergence" of the RAMANUJAN expansion. In VI.4 for functions $f \in \mathcal{B}^2$ the relation

$$\lim_{R \to \infty} \| f - \sum_{r|R!} a_r(f) \cdot c_r \|_2 = 0$$

was proved (a still better result is provided by PARSEVAL's equation), and in VI.5 we proved, analogously, that

$$\lim_{R \to \infty} \| f - \sum_{r|R!} a_r(f) \cdot c_r \|_1 = 0$$

for functions f in $\mathcal{B}^1$. A rather trivial example of the convergence of series with RAMANUJAN sums, but with "wrong" coefficients, was given in Chapter V, Theorem 1.1. The difficult question of *pointwise convergence* of expansion (1.1) for a "large" class of arithmetical functions was dealt with in Chapter V: Following A. HILDEBRAND, it was shown (V, Theorem 1.1) that the RAMANUJAN expansion of any function f in $\mathcal{B}^u$ is pointwise convergent. Many special examples of functions with pointwise convergent RAMANUJAN expansions are given in HARDY's paper [1921].

R. BELLMAN [1950] suggested the deduction of asymptotic results for such sums as $\sum_{n\le x} f(P(n))$, where P is an integer-valued polynomial, by using the [convergent] RAMANUJAN expansion of the arithmetical function f to be investigated. However, in order to obtain good results, one has to have intimate knowledge of the convergence properties of the RAMANUJAN expansion, and so this approach may not be very promising. Unfortunately, this method does not work for $f = \mu^2$, for example.

VIII.2. WINTNER'S CRITERION

A first general and simple result is due to A. WINTNER. It has the advantage of being valid for every arithmetical function, satisfying condition (2.1), which unfortunately is rather restrictive. On the other hand, the assumption of multiplicativity is not needed.

For any arithmetical function f the function $f' = \mu * f$ is called its ERATOSTHENES-MÖBIUS transform.

Theorem 2.1. *Assume that the ERATOSTHENES transform* $f' = \mu * f$ *of an arithmetical function* f *satisfies the WINTNER condition*

$$\sum_{n=1}^{\infty} n^{-1} \cdot |f'(n)| < \infty, \tag{2.1}$$

then:

(i) *The function* f *is in* $\mathcal{B}^1$.

(ii) *Its RAMANUJAN coefficients* (1.2) *exist and are equal to*

$$a_r(f) = \sum_{1\le d<\infty, d\equiv 0 \bmod r} d^{-1} \cdot f'(d). \tag{2.2}$$

(iii) *The* FOURIER *coefficients*

$$\hat{f}(\tfrac{a}{r}) = M(\, f \cdot e_{-a/r}\,)$$

exist, and $\hat{f}(\tfrac{a}{r}) = a_r(f)$ *if* $\gcd(a,r) = 1$. *So the* FOURIER *coefficients do not depend on* a.

(iv) *If, moreover, the series*

$$\sum_{n=1}^{\infty} 2^{\omega(n)} \cdot n^{-1} \cdot |f'(n)| \tag{2.3}$$

is convergent, then the RAMANUJAN *expansion*

$$\sum_{r=1}^{\infty} a_r \cdot c_r(n) = f(n)$$

is pointwise convergent, and its sum has the "correct" value $f(n)$:

$$\sum_{r=1}^{\infty} a_r \cdot c_r(n) = f(n) \quad \textit{for every } n \in \mathbb{N}.$$

Remark. In fact, the method of proof of (iii) gives a stronger result: if $M(|f * \mu|) = 0$, and $f \in \mathcal{A}^1$, then $\hat{f}(\tfrac{a}{r})$ exists and equals

$$\hat{f}(\tfrac{a}{r}) = \sum_{n \equiv 0 \bmod r} n^{-1} \cdot (f * \mu)(n), \text{ if } \gcd(a,r) = 1.$$

The condition $M(|f * \mu|) = 0$ is satisfied, for example, if f is in $\mathcal{E}_q$ for some $q > 1$ (Theorem 3.5).

Proof of Theorem 2.1. (i) and (ii). The function $f_K(n) = \sum_{d|n, d \le K} f'(d)$ is even mod K!, and so is in $\mathcal{B}$. We expect that f_K is "near" $f: n \mapsto \sum_{d|n} f'(d)$, if K is large. Using (2.1), the norm estimate

$$\|f - f_K\|_1 \le \limsup_{x \to \infty} x^{-1} \cdot \sum_{n \le x} \sum_{d|n, d > K} |f'(d)| \le \sum_{d > K} |f'(d)| \cdot \frac{1}{d} \to 0, \text{ as } K \to \infty,$$

shows that $f \in \mathcal{B}^1$. Therefore, the RAMANUJAN coefficients $a_r(f)$ exist. Next,

$$\varphi(r) \cdot a_r(f_K) = \lim_{x \to \infty} x^{-1} \cdot \sum_{n \le x} \sum_{d|n, d \le K} f'(d) \cdot c_r(n)$$

$$= \sum_{d \le K} f'(d) \cdot \lim_{x \to \infty} x^{-1} \cdot \sum_{n \le x, n \equiv 0 \bmod d} c_r(n)$$

$$= \sum_{d \le K, d \equiv 0 \bmod r} f'(d) \cdot d^{-1} \cdot \varphi(r),$$

as is easily shown using the representation of $c_r(n)$ as an exponential sum (see Exercise 2). The estimate

$$|a_r(f) - a_r(f_K)| \le \{\varphi(r)\}^{-1} \cdot \| f - f_K \|_1 \cdot \sup_{n \in \mathbb{N}} |c_r(n)| \le \| f - f_K \|_1$$

yields, by letting $K \to \infty$, the truth of (2.2).

(iii) Assume that $\gcd(a,r) = 1$. Put $f = 1 * f'$. Then, with $\mathcal{O}$-constants depending only on r,

$$x^{-1} \cdot \sum_{n \le x} f(n) \cdot e_{-a/r}(n) = x^{-1} \cdot \sum_{n \le x} e_{-a/r}(n) \cdot \sum_{d|n} f'(d)$$

$$= \sum_{d \le x} f'(d) \cdot x^{-1} \cdot \sum_{m \le x/d} e\left(-\frac{ad}{r} \cdot m \right)$$

$$= \sum_{d \le x, d \equiv 0 \bmod r} f'(d) \cdot x^{-1} \cdot \left(\frac{x}{d} + \mathcal{O}(1) \right) + x^{-1} \cdot \sum_{d \le x} |f'(d)| \cdot \mathcal{O}(1).$$

The absolute convergence of $\sum_{d \le x} d^{-1} \cdot f'(d)$ gives $\sum_{d \le x} |f'(d)| = o(x)$ [by partial summation], and the formula for the FOURIER coefficients follows.

If f is in $\mathcal{A}^1$, then the FOURIER coefficients exist. Therefore, if $M(|f'|) = 0$ is assumed, the last displayed equation gives that

$$\lim_{x \to \infty} \sum_{d \le x, d \equiv 0(r)} f'(d) \cdot d^{-1} \text{ exists and equals } \hat{f}(a/r).$$

(iv) Using (2.2), and

$$\sum_{r|d} c_r(n) = d, \text{ if } d|n, \ \sum_{r|d} c_r(n) = 0 \text{ otherwise}$$

(see Exercise 3), we estimate the difference

$$\Delta_R = f(n) - \sum_{r \le R} a_r(f) \cdot c_r(n) = \sum_d d^{-1} \cdot f'(d) \cdot \left(d - \sum_{r|d, r \le R} c_r(n) \right)$$

$$= \sum_d d^{-1} \cdot f'(d) \cdot \left(\sum_{r|d} c_r(n) - \sum_{r|d, r \le R} c_r(n) \right) = \sum_d d^{-1} \cdot f'(d) \cdot \sum_{r|d, r > R} c_r(n).$$

Thus we obtain

$$|\Delta_R| \le \sum_{d > R} d^{-1} \cdot |f'(d)| \cdot \sum_{r|d} |c_r(n)|.$$

The map $d \mapsto \sum_{r|d} |c_r(n)|$ is multiplicative, and so, if $p^m \| n$,

$$\sum_{r|d} |c_r(n)| = \prod_{p^\ell \| d} \sum_{0 \le k \le \ell} |c_{p^k}(n)|$$

$$\le \prod_{p|d} \left(1 + \varphi(p) + \dots + \varphi(p^m) + p^m \right) \le 2^{\omega(d)} \cdot n.$$

This gives

$$|\Delta_R| \le n \cdot \sum_{d>R} d^{-1} \cdot |f'(d)| \cdot 2^{\omega(d)} \to 0,$$

as $R \to \infty$, and the convergence of the RAMANUJAN expansion to the value $f(n)$ is proved. □

Examples. We mention the [absolutely convergent] RAMANUJAN expansions

$$n^{-1} \cdot \sigma(n) = \frac{1}{6}\, \pi^2 \cdot \sum_{r=1}^{\infty} r^{-2} \cdot c_r(n),$$

$$n^{-1} \cdot \varphi(n) = 6\, \pi^{-2} \cdot \sum_{r=1}^{\infty} \mu(r) \cdot \{\varphi_2(r)\}^{-1} \cdot c_r(n),$$

where $\varphi_2(r) = r^2 \cdot \prod_{p|r} \left(1 - p^{-2}\right)$ (Exercise 4). Several approximations of $n^{-1} \cdot \varphi(n)$, by partial sums of its RAMANUJAN expansion, are given in Figure 1-6. Abbreviate $6\, \pi^{-2} \cdot \sum_{r \le R} \mu(r) \cdot \{\varphi_2(r)\}^{-1} \cdot c_r(n)$ by $S_R(n)$.

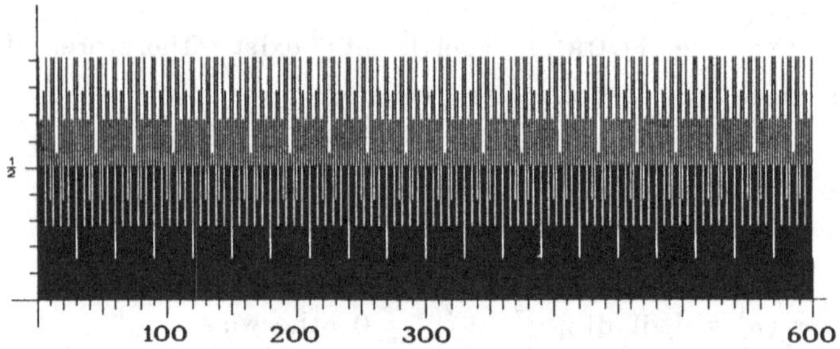

Figure VIII.1

Values of $S_5(n)$ in the range $1 \le n \le 600$.

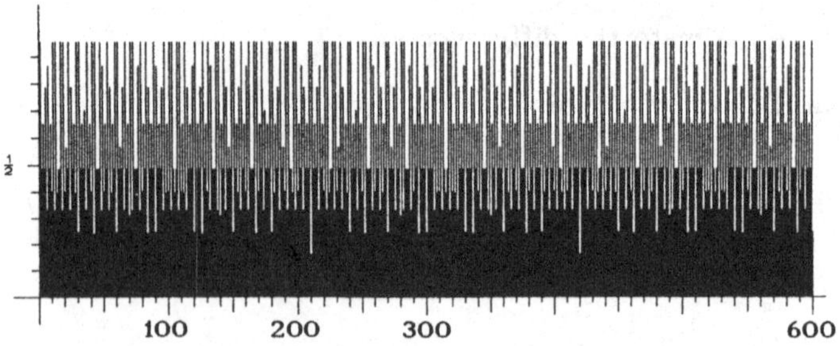

Figure VIII.2

Values of $S_{10}(n)$ in the range $1 \le n \le 600$.

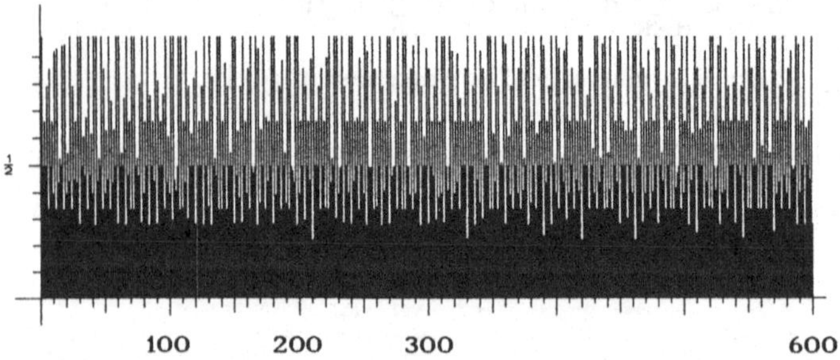

Figure VIII.3

Values of $S_{20}(n)$ in the range $1 \le n \le 600$.

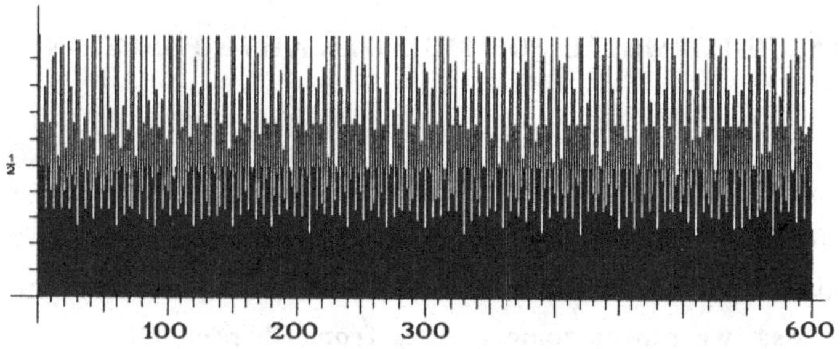

Figure VIII.4

Values of $S_{40}(n)$ in the range $1 \le n \le 600$.

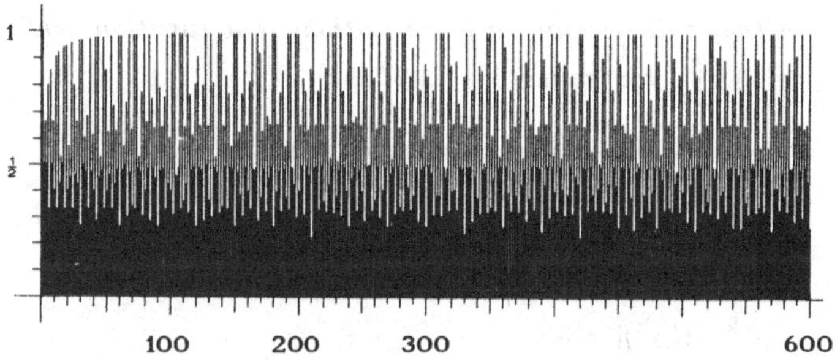

Figure VIII.5

Values of $n^{-1} \cdot \varphi(n)$ in the range $1 \le n \le 600$.

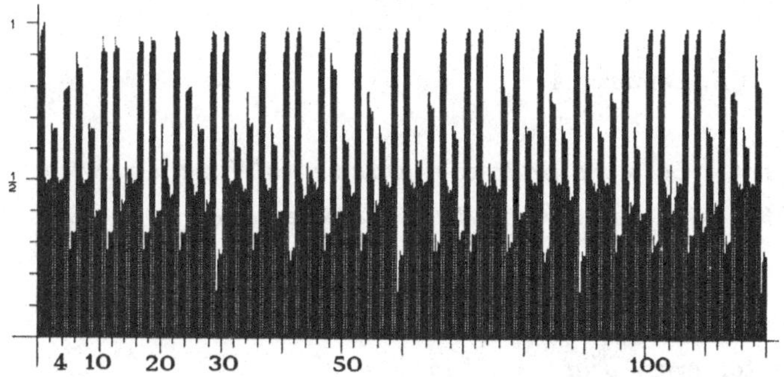

Figure VIII.6
Values of S_i ($i = 5, 10, 20, 40$), and of $n \mapsto n^{-1} \cdot \varphi(n)$.

Figure VIII.6 gives the values of $S_5(n)$, $S_{10}(n)$, $S_{20}(n)$, $S_{40}(n)$, and $n^{-1} \cdot \varphi(n)$ [in this order] in the range $1 \le n \le 120$.

Hopefully, these diagrams give an impression of the convergence of partial sums of the RAMANUJAN expansion to the function $n^{-1} \cdot \varphi(n)$.

VIII.3. MEAN-VALUE FORMULAE FOR MULTIPLICATIVE FUNCTIONS

RAMANUJAN coefficients are closely connected with mean-values on residue-classes; in order to calculate these coefficients, we need some mean-value formulae, particularly for multiplicative functions. For the sake of completeness, we repeat some results from Chapter VII.

Lemma 3.1. *Assume that* f *is a multiplicative arithmetical function with finite semi-norm* $\|f\|_q$ *for some* $q > 1$. *Then*

(a) $|f(n)| \le C\cdot n^{1/q}$ *for some constant* $C > 0$, *and* $f(n) = o(n^{1/q})$, *as* $n \to \infty$.

(b) $\sum_p p^{-2}\cdot|f(p)|^2 < \infty$.

(c) $\sum_p \sum_{k\ge2} p^{-k}\cdot|f(p^k)| < \infty$.

(d) $\sum_p \sum_{k\ge2} p^{-k}\cdot|f(p^k)|^r < \infty$ *for every* r *in* $1 \le r < q$.

For the **proof** see VII, Lemma 1.4.

Proposition 3.2. *Let* f *be a multiplicative arithmetical function, with a mean-value* $M(f) \ne 0$. *Assume, further, that the series*

$$\sum_{k=0}^{\infty} p^{-k}\cdot|f(p^k)| \tag{3.1}$$

is convergent for every prime p. *Then*

$$M(f) = \lim_{\sigma\to1+} \zeta^{-1}(\sigma)\cdot\sum_{n=1}^{\infty}\frac{f(n)}{n^\sigma} = \lim_{\sigma\to1+}\prod_p\left(1+\frac{f(p)}{p^\sigma}+\dots\right)\cdot\left(1-\frac{1}{p^\sigma}\right)$$

$$= \lim_{\sigma\to1+}\prod_p\left(1+\frac{f(p)-1}{p^\sigma}+\frac{f(p^2)-f(p)}{p^{2\sigma}}+\dots\right).$$

Therefore, for every prime p,

$$1+\frac{f(p)-1}{p}+\frac{f(p^2)-f(p)}{p^2}+\dots \ne 0.$$

Proof. The first assertion is the continuity theorem for DIRICHLET series, which is a simple application of the formula for partial summation. In particular, the DIRICHLET series $\sum_{n=1}^{\infty} \frac{f(n)}{n^\sigma}$ is convergent for $\sigma > 1$. The second assertion is obvious from the first and the assumption $M(f) \neq 0$. □

Remark. Assumption (3.1), for the convergence of $\sum p^{-k} \cdot |f(p^k)|$, is fulfilled, for example, if

$f \in \mathcal{A}^q$, where $q > 1$, and $M(f) \neq 0$, or if f is in $\mathcal{G}$ (see Chapter III), or if $\sum_n |(\mu * f)(n)| < \infty$, or if $f \in \mathcal{A}^1$, and $M(f) \neq 0$ (VII, Thm. 5.1).

Theorem 3.3 (Formulae for mean-values). *Assume that* f *is a multiplicative function in* $\mathcal{E}_q$, *where* $q \geq 1$, *with mean-value* $M(f) \neq 0$. *Then*

$$M(f) = \prod_p \left(1 + \frac{f(p)-1}{p} + \frac{f(p^2)-f(p)}{p^2} + \dots \right).$$

In particular:

(a) *If is completely multiplicative, then*

$$M(f) = \prod_p \left(1 + \left(\frac{f(p)-1}{p}\right) \cdot \left(1 - \frac{f(p)}{p}\right)^{-1} \right).$$

(b) *If is strongly multiplicative, then*

$$M(f) = \prod_p \left(1 + \frac{f(p)-1}{p} \right).$$

(c) *If is 2-multiplicative (this means that* $f(p^k) = 0$ *for every prime* p *and every exponent* $k \geq 2$*), then*

$$M(f) = \prod_p \left(1 - \frac{1}{p}\right) \cdot \left(1 + \frac{f(p)}{p}\right).$$

Proof. We use the formula of Proposition 3.2. Well-known results on infinite products (see the Appendix, Theorem A.7.1) guarantee that the main assertion of Theorem 3.3 is true as soon as the convergence of the three series

$$\sum_p p^{-1} \cdot (f(p) - 1), \ \sum_p p^{-2} \cdot |f(p) - 1|^2, \text{ and } \sum_p \sum_{k \geq 2} p^{-k} \cdot |f(p^k) - f(p^{k-1})|$$

is proved. But this is obvious from the assumption $f \in \mathcal{E}_q$. □

Remark. The assumptions $f \in \mathcal{A}^q$, $q \geq 1$, f multiplicative, and $M(f) \neq 0$ imply that $f \in \mathcal{E}_q$ (see Chapter VII, Theorem 5.1).

A simpler result is the following theorem.

Theorem 3.4. *Assume that* f *is strongly multiplicative,* M(f) $\neq$ 0 *exists, and* $\|f\|_q < \infty$ *for some* $q > 1$. *Then*

$$M(f) = \prod_p \left(1 + \frac{f(p)-1}{p} \right).$$

Proof. We start with $M(f) = \lim_{\sigma \to 1+} \prod_p \left(1 + p^{-\sigma} \cdot (f(p) - 1) \right)$. The product is equal to

$$\exp \Big(\sum_p \big\{ p^{-\sigma} \cdot (f(p) - 1) + \mathcal{O}(p^{-2\sigma} \cdot |f(p) - 1|^2) \big\} \Big).$$

The series $\sum p^{-2} \cdot |f(p) - 1|^2$ is convergent. Therefore,

$$\lim_{\sigma \to 1+} \sum p^{-\sigma} \cdot (f(p) - 1)$$

exists. In the same manner as used in the proof of the DABOUSSI-DELANGE Theorem in Chapter VII, this implies the convergence of $\sum_p p^{-1} \cdot (f(p) - 1)$, and the desired result follows (see the Appendix). □

The following theorem deals with the ERATOSTHENES-MÖBIUS transform $f' = \mu * f$ of an arithmetical function f.

Theorem 3.5. *Assume that* $f \in \mathcal{E}_q$ *is multiplicative, and* $q > 1$. *Then:*

(a) *The mean-values of* f' *and of* |f'| *are zero.*

(b) *The series* $\sum_{n=1}^{\infty} n^{-1} \cdot f'(n)$ *converges, with limit* M(f).

Proof. (a) It suffices to show that $M(|f'|) = 0$. We use Theorem 3.1 from Chapter II to estimate $\sum_{n \le x} |f'(n)|$. The assumption $\sum_{p^k \le y} |f'(p^k)| \cdot \log p^k \le c_1 \cdot y$ is satisfied:

$$\sum_{p^k \le y} |f'(p^k)| \cdot \log p^k = \sum_{p^k \le y} |f(p^k) - f(p^{k-1})| \cdot \log p^k$$

$$\le \sum_{p \le y} \log p + 3 \cdot \sum_{p^k \le y} |f(p^k)| \cdot \log p^k$$

$$\le \mathcal{O}(y) + 3 \Big(\sum_{p^k \le y} |f(p^k)|^q \Big)^{1/q} \cdot \Big(\sum_{p^k \le y} (\log p^k)^{q'} \Big)^{1/q'}$$

$$= \mathcal{O}\Big(y + y \cdot (\log y)^{1 - 1/q'} \Big) = \mathcal{O}\Big(y \cdot (\log y)^{1/q} \Big).$$

Theorem II, 3.1 (3.2) gives

$$x^{-1} \cdot \sum_{n \le x} |f'(n)|$$
$$\le c_2 \cdot (\log x)^{-1+1/q} \cdot \exp\Big\{ \sum_{p \le x} p^{-1} \cdot |f'(p)| + \sum_{p \le x} \sum_{k \ge 2} p^{-k} \cdot |f'(p^k)| \Big\}.$$

The sum

$$\sum_{p \le x} \sum_{k \ge 2} p^{-k} \cdot |f'(p^k)| \le \sum_{p \le x} \sum_{k \ge 2} p^{-k} \cdot \left(|f(p^k)| + |f(p^{k-1})| \right)$$

$$\le 2 \sum_{p \le x} \sum_{k \ge 2} p^{-k} \cdot |f(p^k)| + \sum_{p \le x} p^{-2} \cdot |f(p)| = \mathcal{O}\left(S_{3,q} + S''_{2,q} + 1 \right)$$

is bounded in x, and

$$\sum_{p \le x} p^{-1} \cdot |f'(p)| \le \Big\{ \sum_{p \le x, |f(p)| \le 5/4} p^{-1} \cdot |f(p) - 1|^2 \Big\}^{\frac{1}{2}} \cdot \Big\{ \sum_{p \le x} p^{-1} \Big\}^{\frac{1}{2}} + \mathcal{O}(S''_{2,q})$$

$$= \mathcal{O}\left(\sqrt{\log \log x} \right).$$

Therefore $M(|f'|) = 0$.

(b) $$x^{-1} \cdot \sum_{n \le x} f(n) = x^{-1} \cdot \sum_{n \le x} \sum_{d | n} f'(d) = x^{-1} \cdot \sum_{d \le x} f'(d) \cdot \left(\frac{x}{d} + \mathcal{O}(1) \right)$$

$$= \sum_{d \le x} \frac{f'(d)}{d} + \mathcal{O}\Big(x^{-1} \cdot \sum_{d \le x} | f'(d)| \Big).$$

The existence of $M(f)$ and $M(|f'|) = 0$ now imply the convergence of $\sum d^{-1} \cdot f'(d)$ with limit $M(f)$. □

Remark. If f is strongly multiplicative, we can also prove the convergence of the series

$$\sum_{n \equiv 0 \bmod r} n^{-1} \cdot f'(n), \text{ for } r = 1, 2, \dots .$$

Define multiplicative functions by

$$\chi_r(p^k) = \begin{cases} 1, & \text{if } p \nmid r, \\ 0, & \text{if } p | r, \end{cases} \qquad F(p^k) = \begin{cases} f(p^k), & \text{if } p \nmid r, \\ 1, & \text{if } p | r. \end{cases}$$

Then the convolution relation $F = 1 * (f' \cdot \chi_r)$ gives

$$x^{-1} \cdot \sum_{n \le x} F(n) = \sum_{d \le x} d^{-1} \cdot f'(d) \chi_r(d) + \mathcal{O}\Big(x^{-1} \cdot \sum_{d \le x} |f'(d) \chi_r(d)| \Big).$$

F is in $\mathcal{E}_q$, the mean-value $M(F)$ exists, and $M(|f' \cdot \chi_r|) \le M(|f'|) = 0$; so we obtain the convergence of $\sum_{\gcd(d,r)=1} d^{-1} \cdot f'(d)$. But

$$\sum_{n\equiv 0 \bmod r} n^{-1}\cdot f'(n) = (r^{-1}\cdot f'(r))\cdot \sum_{\gcd(d,r)=1} d^{-1}\cdot f'(d),$$

because f is strongly multiplicative (and f' is 2-multiplicative). Thus the assertion is proved. □

VIII.4. FORMULAE FOR RAMANUJAN COEFFICIENTS

There is a close connection between RAMANUJAN coefficients $a_r(f) = \dfrac{M(f\cdot c_r)}{\varphi(r)}$ and mean-values on residue-classes,

$$M_d(f) = \lim_{x\to\infty} x^{-1}\cdot \sum_{n\le x, n\equiv 0 \bmod d} f(n).$$

Proposition 4.1. *For every arithmetical function* f,

(1) *the existence of all the mean-values* $M_d(f)$ *for* $d = 1, 2, \ldots$ *implies the existence of all* RAMANUJAN *coefficients* $a_r(f)$, $r = 1, 2, \ldots$, *and these coefficients are given by*

$$a_r(f) = \{\varphi(r)\}^{-1}\cdot \sum_{d\mid r} d\cdot\mu(r/d)\cdot M_d(f),\ r = 1, 2, \ldots,$$

(2) *and, conversely, the existence of all* $a_r(f)$, $r = 1, 2, \ldots$, *implies the existence of all* $M_d(f)$, *and*

$$M_d(f) = d^{-1}\cdot \sum_{r\mid d} \varphi(r)\cdot a_r(f).$$

Proof. (1) is obvious: $M(c_r f) = \sum_{d\mid r} d\cdot\mu(r/d)\cdot M_d(f)$.
(2) For every $x > 0$,

$$\sum_{r\mid d}\sum_{n\le x} f(n)\cdot c_r(n) = \sum_{n\le x} f(n)\sum_{r\mid d}\sum_{t\mid (n,r)} t\cdot\mu(r/t)$$

$$= \sum_{t\mid d} t \sum_{\substack{n\le x\\ n\equiv 0 \bmod t}} f(n) \sum_{s\mid (d/t)} \mu(s) = d\cdot \sum_{\substack{n\le x\\ n\equiv 0 \bmod d}} f(n).$$

So $M_d(f)$ exists and the formula given is valid. □

So, for the calculation of RAMANUJAN coefficients, it is crucial to obtain the mean-values $M_d(f)$. In order to be able to calculate these mean-values, we begin with the definition

$$M_{(d)}(f) = \lim_{x \to \infty} x^{-1} \cdot \sum_{n \le x,\ \gcd(n,d) = 1} f(n),$$

if this limit exists.

Proposition 4.2. *Assume that* f *is a multiplicative function, for which*

a) *the mean-value* $M(f)$ *exists and is non-zero,*

b) *for any prime* p *the series* $\sum_{k\ge 0} p^{-k} \cdot |f(p^k)|$ *is convergent,*

c) *and all the mean-values* $M_{(d)}(f)$ *exist.*

Then, for every prime p,

$$1 + p^{-1} \cdot f(p) + p^{-2} \cdot f(p^2) + \dots \neq 0,$$

and for every integer d,

$$M_{(d)}(f) = M(f) \cdot \prod_{p|d} \left(1 + p^{-1} \cdot f(p) + p^{-2} \cdot f(p^2) + \dots \right)^{-1}. \tag{4.1}$$

Proof. According to Proposition 3.2,

$$M(f) = \lim_{\sigma \to 1+} \zeta^{-1}(\sigma) \cdot \sum_{n=1}^{\infty} n^{-\sigma} \cdot f(n),$$

and similarly for $M_{(d)}(f) = M(f \cdot \chi_r)$, where $\chi_d(n) = 1$ if $\gcd(n,d) = 1$, and $= 0$ otherwise. Using the multiplicativity of f and of $f \cdot \chi_r$, and noting

$$\lim_{\sigma \to 1+} \left(1 + p^{-\sigma} \cdot f(p) + p^{-2\sigma} \cdot f(p^2) + \dots \right) = \left(1 + p^{-1} \cdot f(p) + p^{-2} \cdot f(p^2) + \dots \right),$$

we obtain

$$M_{(d)}(f) \cdot \prod_{p|d} \left(1 + p^{-1} \cdot f(p) + p^{-2} \cdot f(p^2) + \dots \right) = M(f).$$

Thus $M_{(d)}(f) \neq 0$ and $\left(1 + p^{-1} \cdot f(p) + p^{-2} \cdot f(p^2) + \dots \right) \neq 0$, and (4.1) is true.

We remark that for a function f in $\mathcal{D}^1$ the mean-values $M(f)$, $M_d(f)$ and $M_{(d)}(f)$ do exist (see Chapter VI, Section 1). Moreover, if $f \in \mathcal{A}^1$ and $M(f) \neq 0$, in the proof of VII, Theorem 5.1, the convergence of $S_{3,1}(f)$ has been shown; hence all the series $\sum_{k\ge 0} p^{-k} \cdot |f(p^k)|$ are convergent. If $\|f\|_q < \infty$ for some $q > 1$, then $|f(p^k)| \le C \cdot p^{k/q}$, therefore the series $1 + p^{-1} \cdot f(p) + p^{-2} \cdot f(p^2) + \ldots$ is absolutely convergent. Thus we obtain the following proposition.

Proposition 4.3. *If* $f \in \mathcal{A}^1$ *is multiplicative, with mean-value* $M(f) \neq 0$, *then all the mean-values* $M_{(d)}(f)$ *exist and formula* (4.1) *holds.*

Theorem 4.4. *Assume that* $f \in \mathcal{A}^1$ *is multiplicative, and that* $M(f) \neq 0$. *Then the mean-values* $M_d(f)$ *and the* RAMANUJAN *coefficients* $a_r(f)$ *exist, and the maps*

$$d \mapsto M_d(f) / M(f), \text{ and } r \mapsto a_r(f) / M(f)$$

are multiplicative. There are product representations for $M_d(f)$ *and* $a_r(f)$ *as follows:*

$$M_d(f) = M(f) \cdot \prod_{p^\delta \| d} \left(\frac{f(p^\delta)}{p^\delta} + \frac{f(p^{\delta+1})}{p^{\delta+1}} + \ldots \right) \cdot \left(1 + \frac{f(p)}{p} + \frac{f(p^2)}{p^2} + \ldots \right)^{-1},$$

and

$$a_r(f) = \frac{M(f)}{\varphi(r)} \cdot \prod_{p^\delta \| r} \left(\sum_{k=0}^{\infty} p^{-k} \cdot \left(f(p^{k+\delta}) - f(p^{k+\delta-1}) \right) \right) \cdot \left(1 + \frac{f(p)}{p} + \frac{f(p^2)}{p^2} + \ldots \right)^{-1}$$

$$= M(f) \cdot \prod_{p^\delta \| r} \left(\sum_{k \geq \delta} p^{-k} \cdot \left(f(p^k) - f(p^{\delta-1}) \right) \right) \cdot \left(\sum_{k \geq 0} p^{-k} \cdot f(p^k) \right)^{-1}.$$

Proof. For a fixed integer $d = p_1^{\delta_1} \cdot \ldots \cdot p_r^{\delta_r}$, write $m = p_1^{\mu_1} \cdot \ldots \cdot p_r^{\mu_r} \cdot \ell$, where $\gcd(\ell,d) = 1$. Then

$$M_d(f) = \lim_{\sigma \to 1+} \zeta^{-1}(\sigma) \cdot \sum_{n \equiv 0 \bmod d} n^{-\sigma} \cdot f(n)$$

$$= d^{-1} \cdot \lim_{\sigma \to 1+} \zeta^{-1}(\sigma) \cdot \sum_m m^{-\sigma} \cdot f(m \cdot d)$$

$$= \lim_{\sigma \to 1+} \zeta^{-1}(\sigma) \cdot \sum_{\substack{\ell \\ (\ell,d)=1}} \frac{f(\ell)}{\ell^\sigma} \cdot \sum_{\mu_1 \geq 0} \ldots \sum_{\mu_r \geq 0} \frac{f(p_1^{\mu_1+\delta_1}) \cdot \ldots \cdot f(p_r^{\mu_r+\delta_r})}{p_1^{\mu_1\sigma+\delta_1} \cdot \ldots \cdot p_r^{\mu_r\sigma+\delta_r}}.$$

By VII, Theorem 5.1, the multi-series $\sum_{\mu_1 \geq 0} \ldots \sum_{\mu_r \geq 0} \ldots$ is absolutely convergent in $\sigma \geq 1$. So we obtain (from (4.1))

$$M_d(f) = \sum_{\mu_1 \geq 0} \ldots \sum_{\mu_r \geq 0} \frac{f(p_1^{\mu_1+\delta_1}) \cdot \ldots \cdot f(p_r^{\mu_r+\delta_r})}{p_1^{\mu_1+\delta} \cdot \ldots \cdot p_r^{\mu_r+\delta_r}} \cdot M_{(d)}(f)$$

$$= M(f) \cdot \prod_{p^\delta \| d} \left(\frac{f(p^\delta)}{p^\delta} + \frac{f(p^{\delta+1})}{p^{\delta+1}} + \dots \right) \cdot \left(1 + \frac{f(p)}{p} + \frac{f(p^2)}{p^2} + \dots \right)^{-1}.$$

This proves the formula for $M_d(f)$ and the multiplicativity of the map $d \mapsto M_d(f)/M(f)$. Proposition 4.1 yields the fact that the function

$$r \mapsto a_r(f)/M(f) = \{\varphi(r)\}^{-1} \cdot \sum_{d|r} d \cdot \frac{M_d(f)}{M(f)} \cdot \mu\left(\frac{r}{d}\right)$$

is multiplicative, and

$$\frac{a_{p^\delta}(f)}{M(f)} = \frac{1}{\varphi(p^\delta)} \cdot \left(\sum_{k=0}^{\infty} p^{-k} \cdot \left(f(p^{k+\delta}) - f(p^{k+\delta-1}) \right) \right) \cdot \left(1 + \frac{f(p)}{p} + \frac{f(p^2)}{p^2} + \dots \right)^{-1}$$

$$= \sum_{k \ge \delta} p^{-k} \cdot \left(f(p^k) - f(p^{\delta-1}) \right) \cdot \left(\sum_{k \ge 0} p^{-k} \cdot f(p^k) \right)^{-1}. \qquad \square$$

Corollary 4.5. *Assume that* $f \in \mathcal{A}^1$ *is* *strongly* *multiplicative, and* $M(f) \neq 0$. *Then*

$$\frac{M_d(f)}{M(f)} = \frac{f(d)}{d} \cdot \prod_{p|d} \left(1 + \frac{f(p)-1}{p} \right)^{-1},$$

$$a_r(f) = M(f) \cdot \frac{\mu(r)}{\varphi(r)} \cdot \prod_{p|r} \left(1 - \frac{f(p)}{1 + \frac{f(p)-1}{p}} \right).$$

Proof. $M(f) = \lim_{\sigma \to 1+} \prod_p \left(1 + p^{-\sigma} \cdot (f(p)-1) \right)$, therefore $\left(1 + \frac{f(p)-1}{p}\right)$ is non-zero for every prime p. Theorem 4.4 implies the above formulae. $\square$

Corollary 4.6. *If* $f \in \mathcal{A}^1$ *is a* *completely* *multiplicative arithmetical function, for which the mean-value* $M(f)$ *is non-zero, then*

$$M_d(f) = M(f) \cdot d^{-1} \cdot f(d),$$

$$a_r(f) = \{\varphi(r)\}^{-1} \cdot M(f) \cdot \sum_{d|r} f(d) \cdot \mu(r/d) = M(f) \cdot \{\varphi(r)\}^{-1} \cdot (\mu * f)(r).$$

Corollary 4.7. *If* $f \in \mathcal{A}^1$ *is* *2-multiplicative* *and* $M(f) \neq 0$, *then*

$$M_{p^\delta}(f) = M(f) \cdot f(p) \cdot \{ p + f(p) \}^{-1} \textit{ if } \delta = 1, \textit{ and } M_{p^\delta}(f) = 0 \textit{ if } \delta \ge 2,$$

$$a_{p^\delta}(f) = \frac{M(f)}{\varphi(p)} \cdot \left(-1 + f(p)/(1 + p^{-1}f(p)) \right) \textit{ if } \delta = 1, \quad a_{p^\delta}(f) = 0 \textit{ if } \delta \ge 3,$$

$$a_{p^\delta}(f) = -M(f) \cdot \{\varphi(p^2)\}^{-1} \cdot \left(f(p)/(1 + p^{-1}f(p)) \right) \textit{ if } \delta = 2.$$

Finally, we give some formulae containing the ERATOSTHENES transform. The WINTNER condition (2.1) implies $\sum_{n\le x} |f'(n)| = o(x)$. Then

$$\sum_{n\le x, n\equiv 0 \bmod d} f(n) = \sum_{n\le x, n\equiv 0 \bmod d} \sum_{t|n} f'(t)$$

$$= \sum_{t\le x} t^{-1}\cdot f'(t)\cdot\left((x/d)\cdot \gcd(d,t) + O(1)\right)$$

$$= x\cdot d^{-1}\cdot \sum_{t\le x} t^{-1}\cdot f'(t)\cdot \gcd(d,t) + o(x);$$

so $M_d(f) = \frac{1}{d}\cdot\sum_t t^{-1}\cdot f'(t)\cdot\gcd(d,t)$ exists and the following result holds.

Theorem 4.8. *If the series* (2.1) *is absolutely convergent, then the mean-values* $M_d(f)$ *and the* RAMANUJAN *coefficients* $a_d(f)$ *exist for* $d = 1, 2, \ldots$. *If* f *is multiplicative in addition, then the mean-values* $M(f)$ *and* $M_d(f)$ *are given by*

$$M(f) = \prod_p \left(1 + p^{-1}\cdot f'(p) + p^{-2}\cdot f'(p^2) + \ldots\right),$$

$$M_d(f) = \prod_{p\nmid d}\left(1 + \frac{f'(p)}{p} + \frac{f'(p^2)}{p^2} + \ldots\right)$$

$$\times \prod_{p^k\| d}\left(1 + f'(p) + \ldots + f'(p^k) + \frac{f'(p^{k+1})}{p} + \frac{f'(p^{k+2})}{p^2} + \ldots\right).$$

If f *is multiplicative and* $M(f) \neq 0$, then

$$\frac{M_d(f)}{M(f)} = \prod_{p^k\| d}\left(1 + f'(p) + \ldots + f'(p^k) + \frac{f'(p^{k+1})}{p} + \ldots\right)\left(1 + \frac{f'(p)}{p} + \frac{f'(p^2)}{p^2} + \ldots\right)^{-1},$$

$$\frac{a_d(f)}{M(f)} = \prod_{p^k\| d}\frac{\left(1 + f'(p) + \ldots + f'(p^{k-1}) + \left(1+\frac{1}{p}\right)\left(f'(p^k) + \frac{f'(p^{k+1})}{p} + \ldots\right)\right)}{1 + \frac{f'(p)}{p} + \frac{f'(p^2)}{p^2} + \ldots},$$

and the maps $d \mapsto M_d(f)/M(f)$, $r \mapsto a_r(f)/M(f)$ *are multiplicative.*

VIII.5. POINTWISE CONVERGENCE OF RAMANUJAN EXPANSIONS

A large class of arithmetical functions f, for which the RAMANUJAN expansion (1.1) is pointwise convergent, is the set of multiplicative functions in $\mathcal{A}^2$. This is a consequence of the main Theorem 5.1 in Chapter VII.

Theorem 5.1. *Assume that* f *is a multiplicative function in* $\mathcal{A}^2$ *with mean-value* $M(f) \neq 0$. *Then its* RAMANUJAN *expansion is pointwise convergent and*

$$\sum_{r=1}^{\infty} a_r(f) \cdot c_r(n) = f(n) \text{ for any } n \in \mathbb{N}.$$

Remark 1. In general, convergence is neither absolute nor uniform in n.

Lemma 5.2. *If* f *is a multiplicative function in* $\mathcal{A}^2$ *with mean-value* $M(f) \neq 0$, *then - denoting by* $a_r^*(f) = \{M(f)\}^{-1} \cdot a_r(f)$ *the normed* RAMANUJAN *coefficients - the following two series are convergent:*

$$\sum_p a_p^*, \tag{5.1}$$

$$\sum_p p \cdot |a_p^*|^2. \tag{5.2}$$

Remark 2. The same proof (with a slight modification in (2)) works, if $f \in \mathcal{A}^q$ for some $q > 1$ is assumed.

Proof. (1) VII, Theorem 5.1 yields $f \in \mathcal{E}_2$, and so, in particular, the series

$$e_p = (1 - p^{-1}) \cdot \sum_{k=0}^{\infty} p^{-k} \cdot f(p^k)$$

is convergent for any prime p. $\|f\|_q < \infty$ implies $|f(n)| \le c \cdot n^{\frac{1}{2}}$, and so there exists a prime p_1 with the property $|\sum_{k\ge1} p^{-k} \cdot f(p^k)| \le \frac{1}{2}$ for all primes $p \ge p_1$. Therefore, $|e_p| \ge \frac{1}{4}$. According to Theorem 4.4, we obtain (for every $p \ge p_1$)

$$-a_p^* = \{p\, e_p\}^{-1} \cdot \sum_{k=0} p^{-k} \cdot \left(f(p^k) - f(p^{k-1}) \right)$$

$$= \frac{1-f(p)}{p} + \frac{1}{e_p}\left(\frac{1-f(p)}{p}\right)^2 + \frac{1}{e_p}\left(\frac{1-f(p)}{p} + 1\right) \cdot \sum_{k\ge1} \frac{f(p^k)-f(p^{k-1})}{p^{k+1}}.$$

Summed over $p \ge p_1$, the three series on the right-hand side are convergent:

$$\sum_p \frac{1-f(p)}{p} = S_1(f), \quad \sum_p \left| \frac{1}{e_p}\left(\frac{1-f(p)}{p}\right)^2 \right| = \mathcal{O}(1) + \mathcal{O}\left(\sum_p \left|\frac{f(p)}{p}\right|^2\right),$$

$$\sum_p \left| \frac{1}{e_p}\left(\frac{1-f(p)}{p} + 1\right) \cdot \sum_{k\ge1} \frac{f(p^k)-f(p^{k-1})}{p^{k+1}} \right| = \mathcal{O}\left(S_2(f) + S_{3,2}(f)\right) + \mathcal{O}(1).$$

This proves the convergence of (5.1).

(2) Choose a prime p_2 such that $p^{-1}\cdot|f(p)| \le \frac{1}{2}$ for every $p \ge p_2$. Then $|\eta_p|^{-1} \le 4$ for $p \ge \max\{p_1, p_2\}$, where

$$\eta_p = \left(1 - p^{-1}\cdot f(p)\right)\cdot \sum_{k\ge 0} p^{-k}\cdot f(p^k). \tag{5.3}$$

Thus

$$\begin{aligned} a_p^* &= \Big(\sum_{k\ge 0} p^{-k}\cdot f(p^k)\Big)^{-1}\cdot\Big(\sum_{k\ge 1} p^{-k}\cdot f(p^k) - \frac{1}{p-1}\Big) \\ &= \frac{1}{\eta_p}\cdot\left(\frac{f(p)-1}{p-1} + \frac{f(p^2)-f^2(p)}{p^2} + \sum_{k\ge 3}\frac{f(p^k)-f(p^{k-1})f(p)}{p^k}\right) \\ &= \mathcal{O}\left(\left|\frac{f(p)-1}{p-1}\right| + \left|\frac{f(p^2)-f^2(p)}{p^2}\right| + p^{-3/2}\right). \end{aligned}$$

This gives

$$p\cdot|a_p{}^*|^2 = \mathcal{O}\left(p^{-1}\cdot|f(p)-1|^2 + p^{-3}\cdot|f(p^2)|^2 + p^{-3}\cdot|f(p)|^4 + p^{-2}\right), \tag{5.4}$$

and so $\sum_p p\cdot|a_p{}^*|^2 < \infty$, estimating the sums over the terms in (5.4) by $\mathcal{O}(S_2(f))$, $\mathcal{O}(S_{3,2}(f))$, $\mathcal{O}(S_2(f))$ and $\mathcal{O}(1)$ respectively. □

Proof of Theorem 5.1. **(1)** We first prove the convergence of the RAMANUJAN expansion at the point n:

$$\sum_{r\le x} a_r\cdot c_r(n) = \sum_{r\le x} a_r\cdot \sum_{d|\gcd(r,n)} d\cdot\mu(r/d) = \sum_{d|n} d\cdot \sum_{r\le x,\, r\equiv 0 \bmod d} a_r\cdot\mu(r/d).$$

Thus, in order to prove the convergence of $\sum_r a_r\cdot c_r(n)$, it is sufficient to show the convergence of the series $\sum_r a_{rd}^*\cdot\mu(r)$ for every d. Write $d = \prod p^\delta = t\cdot D$, where $t = \prod_{p^\delta \| d,\, a_{p^\delta}^* = 0} p^\delta$, and denote the squarefree kernel of t by $\alpha(t) = \prod_{p|t} p$. Then

$$\begin{aligned} \sum_{r\le x} a_{rd}^*\cdot\mu(r) &= \mu(\alpha(t))\, a_{t\alpha(t)}^*\cdot \sum_{r\le x/\alpha(t),\, \gcd(r,t)=1} \mu(r)\cdot a_{rD}^* \\ &= a_D^*\cdot\sum_{r\le x/\alpha(t)} r^{-1}\cdot\upsilon(r), \end{aligned}$$

where

$$\upsilon(r) = \begin{cases} \mu(r)\cdot r\cdot a_{rD}^*\big/a_D^*, & \text{if } \gcd(r,t) = 1, \\ 0, & \text{otherwise.} \end{cases}$$

The convergence of $\sum r^{-1} \cdot \upsilon(r)$ remains to be proved. Since

$$x^{-1} \cdot \sum_{n \le x} (1 * \upsilon)(n) = \sum_{r \le x} r^{-1} \cdot \upsilon(r) + \mathcal{O}\Big(x^{-1} \cdot \sum_{r \le x} |\upsilon(r)| \Big),$$

it suffices to show that

(a) $M(1*\upsilon)$ exists,

(b) $M(|\upsilon|) = 0$.

Proof of (a). The 2-multiplicative function $\Upsilon = 1 * \upsilon$ belongs to $\mathcal{E}_2$. Since

$$p^{-1} \cdot (1 - \Upsilon(p)) = - p^{-1} \cdot \upsilon(p) = a^*_{pD} \big/ a^*_D = a^*_p$$

for every $p \nmid d$, the series $S_1(\Upsilon)$ and $S_2(\Upsilon)$ are convergent (see Lemma 5.2), and

$$S_{3,2}(\Upsilon) = \sum_p \sum_{k \ge 2} p^{-k} \cdot |\Upsilon(p^k)|^2$$

$$= \sum_p \{p(p-1)\}^{-1} \cdot |\Upsilon(p)|^2 = \mathcal{O}\Big(\sum_p |a^*_p|^2\Big) = \mathcal{O}(1).$$

VII, Theorem 5.1 implies $\Upsilon \in \mathcal{A}^2$, and so $M(\Upsilon)$ exists.

Proof of (b). By partial summation the estimate

$$\sum_{r \le x} r^{-1} \cdot |\upsilon(r)|^2 \le \prod_{p \le x} \Big(1 + \frac{|\upsilon(p)|^2}{p}\Big) \le \exp\Big\{ \sum_{p \le x} \frac{|\upsilon(p)|^2}{p} \Big\}$$

$$= \mathcal{O}\Big(\exp \sum_p \frac{|1 - f(p)|^2}{p}\Big) = \mathcal{O}\big(\exp S_2(f)\big) = \mathcal{O}(1)$$

implies $x^{-1} \cdot \sum_{r \le x} |\upsilon(r)|^2 = o(1)$, and (b) and the first part are proved.

(2) For any fixed n, the DIRICHLET series

$$A(\sigma) = \sum_{r \ge 1} r^{-\sigma} \cdot a_r c_r(n)$$

is convergent for any $\sigma > 0$. In fact, it is absolutely convergent since

$$\sum_{r \le x} | r^{-\sigma} \cdot a^*_r c_r(n)| \le \prod_{p \le x} \sum_{k \ge 0} p^{-k\sigma} \cdot |a^*_{p^k} c_{p^k}(n)|,$$

and, because of $c_{p^k}(n) = -1$ if $k = 1$ and $p \nmid n$, and $c_{p^k}(n) = 0$ if $k \ge 2$ and $p \nmid n$, the product is absolutely convergent:

$$\Big(\sum_p p^{-\sigma} \cdot |a^*_p| \Big)^2 \le \sum_p p \cdot |a^*_p|^2 \cdot \sum_p p^{-(1+2\sigma)} < \infty$$

(by Lemma 5.2). In $\sigma > 0$ the DIRICHLET series $A(\sigma)$ has the product representation

$$A(\sigma) = M(f)\cdot \sum_{r\ge 1} r^{-\sigma}\cdot a_r^* c_r(n) = \prod_p b_p(\sigma),$$

with factors

$$b_p(\sigma) = e_p \cdot \sum_{k\ge 0} p^{-k\sigma}\cdot a_{p^k}^* \, c_{p^k}(n).$$

According to the continuity theorem for DIRICHLET series, it suffices to show that:

(c) $$\text{if } p^\delta \| n, \text{ then } e_p\cdot \sum_{k\ge 0} a_{p^k}^* \, c_{p^k}(n) = f(p^\delta),$$

(d) $$\lim_{\sigma\to 0+} \prod_{p>n} b_p(\sigma) = 1.$$

Proof of (c). It is easy to show (see Exercise 3) that

$$\Big(\sum_{0\le k\le\delta} c_{p^k} - p^{-1}\sum_{0\le k\le\delta+1} c_{p^k}\Big)(n) = \begin{cases} p^\delta, & \text{if } p^\delta\|n,\\ 0 & \text{otherwise.}\end{cases}$$

Therefore,

$$\Big(1-\frac{1}{p}\Big)\cdot\sum_{0\le k\le\delta} M(f\cdot c_{p^k}) - \frac{1}{p}M(f\cdot c_{p^{\delta+1}}) = f(p^\delta)\cdot(M(f) - M_p(f)).$$

So we obtain

$$\begin{aligned} b_p(0) &= e_p\cdot\Big(\sum_{0\le k\le\delta} a_{p^k}^*\cdot\varphi(p^k) - a_{p^{\delta+1}}^*\cdot p^\delta\Big)\\ &= \sum_{k\ge 0}\frac{f(p^k)}{p^k}\cdot\frac{1}{M(f)}\cdot\Big(\Big(1-\frac{1}{p}\Big)\cdot\sum_{0\le k\le\delta} M\big(f\cdot c_{p^k}\big) - \frac{1}{p}M\big(f\cdot c_{p^{\delta+1}}\big)\Big)\\ &= \sum_{k\ge 0}\frac{f(p^k)}{p^k}\cdot f(p^\delta)\cdot\Big(1-\frac{M_p(f)}{M(f)}\Big) = f\big(p^\delta\big), \end{aligned}$$

by Theorem 4.4.

Proof of (d). $c_{p^k}(n) = -1$ [resp. 0]. if $k = 1$ [resp. $k \ge 2$] and $p > n$. Therefore,

$$b_p(\sigma) = e_p\cdot(1 - p^{-\sigma}\cdot a_p^*) = 1 + \delta_p(\sigma),$$

where $\delta_p(\sigma) = (1-p^{-\sigma})(e_p - 1)$. The relation

$$e_p - 1 = \sum_{k\ge 1} p^{-k}\cdot\big(f(p^k) - f(p^{k-1})\big) = \frac{1}{p}\big(f(p)-1\big) + p^{-2}f(p^2) + \mathcal{O}(p^{-3/2})$$

shows that the following series are uniformly convergent in $0 \le \sigma \le 1$:

$$\sum_p \delta_p(\sigma) = \sum_p \left(1 - p^{-\sigma}\right) \cdot \frac{1}{p} \left(f(p) - 1 \right) + \mathcal{O}\Big(\sum_p \left(p^{-2} |f(p^2)| + p^{-3/2} \right) \Big),$$

$$\sum_p |\delta_p(\sigma)|^2 = \mathcal{O}\Big(\sum_p \Big|\frac{1}{p} \left(f(p) - 1 \right)\Big|^2 + 1 \Big) = \mathcal{O}(1).$$

Therefore, $\prod_{p>n} b_p(\sigma)$ is uniformly convergent in $0 \le \sigma \le 1$, and

$$\lim_{\sigma \to 0+} \prod_{p>n} b_p(\sigma) = \prod_{p>n} b_p(0) = 1.$$

This concludes the proof of Theorem 5.1. □

VIII.6. STILL ANOTHER PROOF FOR PARSEVAL'S EQUATION

In Chapter VI two proofs for PARSEVAL's equation were presented for functions f in $\mathcal{B}^2$. In this section, in the special case where f is multiplicative in addition (and $M(f) \neq 0$), a third proof is given.

Theorem 6.1. *Assume that* $f \in \mathcal{A}^2$ *is multiplicative, with mean-value* $M(f) \neq 0$. *Then* PARSEVAL*'s equation*

$$\sum_{1 \le r < \infty} |a_r(f)|^2 \cdot \varphi(r) = \|f\|_2^2 = M(|f|^2)$$

holds.

Proof. $f \in \mathcal{A}^2$ implies the existence of $M(|f|^2)$, and this mean-value is $\neq 0$ for $M(|f|^2) \ge |M(f)|^2 > 0$. Moreover, BESSEL's inequality yields the convergence of the series $\sum_{1 \le r < \infty} |a_r(f)|^2 \cdot \varphi(r)$, and so, using multiplicativity of the normed RAMANUJAN coefficients, it has the product representation

$$\sum_{1 \le r < \infty} |a_r(f)|^2 \cdot \varphi(r) = |M(f)|^2 \cdot \sum_{k \ge 0} |a^*_{p^k}|^2 \cdot \varphi(p^k).$$

Theorem 3.3 gives $M(f) = \prod_p e_p$, $M(|f|^2) = \prod_p \left(1 - \frac{1}{p}\right) \cdot \sum_{k \ge 0} p^{-k} \cdot |f(p^k)|^2$.

So it is sufficient to prove

(6.1) $$|e_p|^2 \cdot \sum_{k\geq 0} |a^*_{p^k}|^2 \cdot \varphi(p^k) = \left(1-\frac{1}{p}\right)\cdot \Sigma_{k\geq 0} p^{-k}\cdot |f(p^k)|^2$$

for every prime p. Consider the function

(6.2) $$k_p(n) = e_p \cdot \sum_{k\geq 0} a^*_{p^k} \cdot c_{p^k}(n).$$

This series is finite, since $c_{p^k}(n) = 0$ as soon as $p^{k-1} > n$, and k_p and $|k_p|^2$ are even functions. Calculate the mean-value of $|k_p|^2$ in two different ways.

$$x^{-1}\sum_{n\leq x} |k_p(n)|^2 = |e_p|^2 \sum_{0\leq k,\ell\leq 1+\log x/\log p} a^*_{p^k}\, \overline{a^*_{p^\ell}}\; x^{-1}\sum_{n\leq x} c_{p^k}(n)\, c_{p^\ell}(n).$$

The last sum $\sum_{n\leq x} c_{p^k}(n)\, c_{p^\ell}(n)$ equals $x\cdot\varphi(p^k) + O(p^{2k})$, if $k = \ell$, and it is $O(p^{k+\ell})$, if $k \neq \ell$. And

$$a^*_{p^k} = e_p^{-1} \cdot \sum_{m\geq k} \frac{f(p^m) - f(p^{m-1})}{p^m} = o(p^{-\frac{1}{2}k}).$$

Therefore,

$$x^{-1}\sum_{n\leq x} |k_p(n)|^2 = |e_p|^2 \cdot \sum_{0\leq k\leq 1+\log x/\log p} |a^*_{p^k}|^2 \cdot \varphi(p^k)$$
$$+ |e_p|^2\, x^{-1} \cdot \sum_{0\leq k,\ell\leq 1+\log x/\log p} O(1).$$

This equation implies

$$M(|k_p|^2) = |e_p|^2 \cdot \sum_{k\geq 0} |a^*_{p^k}|^2 \cdot \varphi(p^k).$$

On the other hand, $k_p(n) = f(p^k)$ if $p^k \| n$ (see (c), p.288) and so

$$x^{-1}\sum_{n\leq x} |k_p(n)|^2 = \sum_{0\leq k\leq 1+\log x/\log p} p^{-k}\cdot |f(p^k)|^2 \cdot \left(\frac{x}{p^k}\right)^{-1} \cdot \sum_{m\leq x/p^k,\, p\nmid m} 1;$$

hence,

$$M(|k_p|^2) = \left(1 - \frac{1}{p}\right) \cdot \sum_{k\geq 0} p^{-k} \cdot |f(p^k)|^2.$$

Comparing both representations of $M(|k_p|^2)$, (6.1) is proved. □

VIII.7. ADDITIVE FUNCTIONS

A. HILDEBRAND and the second author of this book [1980] proved the following result for additive functions. We do not prove this here, but refer instead to the paper quoted in the bibliography. Another proof for this result, in sharpened form, was given independently by K.-H. INDLEKOFER.

Theorem 7.1. *Assume that* g *is an additive arithmetical function. If* $q \geq 1$, *then the following three conditions are equivalent:*

(i) $g \in \mathcal{B}^q$.

(ii) *The mean-value* $M(g)$ *exists and* $\|g\|_q < \infty$.

(iii) *The following three series are convergent:*

$$\sum_{|g(p)|\leq 1} p^{-1}\cdot g(p), \quad \sum_{|g(p)|\leq 1} p^{-1}\cdot |g(p)|^2, \quad \textit{and} \quad \sum_{p,\ k\geq 1,\ |g(p^k)|>1} p^{-k}\cdot |g(p^k)|^q.$$

VIII.8. EXERCISES

1) Let $f' = \mu * f$ be the ERATOSTHENES transform of the arithmetical function f. If

$$\sum_{m\geq 1} \sum_{n\geq 1} \frac{|f'(m)\cdot f'(n)|}{\mathrm{lcm}[m,n]} < \infty,$$

then $f \in \mathcal{B}^2$.

2) For any positive integers d and r, prove

$$\lim_{x\to\infty} x^{-1}\cdot \sum_{n\leq x,\, n\equiv 0 \bmod d} c_r(n) = \begin{cases} \frac{\varphi(r)}{d}, & \text{if } r\mid d, \\ 0, & \text{if } r\nmid d. \end{cases}$$

3) (a) Prove that $\sum_{r|d} c_r(n) = d$, if $d|n$, and $\sum_{r|d} c_r(n) = 0$, id $d \nmid n$.

(b) For all integers $n \geq 1$ and $\delta \geq 0$, prove that

$$\Big(\sum_{0\leq k\leq\delta} c_{p^k} - p^{-1} \sum_{0\leq k\leq\delta+1} c_{p^k}\Big)(n) = \begin{cases} p^{\delta}, & \text{if } p^{\delta}\| n, \\ 0 & \text{otherwise.} \end{cases}$$

4) Verify the calculation of the RAMANUJAN coefficients and the pointwise convergence of the RAMANUJAN expansions for the arithmetical functions $f = \sigma/\mathrm{id}$, and $f = \varphi/\mathrm{id}$, given in VIII.2, p.274.

5) Let f be a multiplicative arithmetical function; denote the ERATOSTHENES transform by $f' = f*\mu$. Prove that $\|f'\|_2 < \infty$, whenever $\|f\|_2 < \infty$ is true.

6) Assume that $f \in \mathcal{A}^1$ is multiplicative, and $M(f) \neq 0$. Prove that for all primes p, for which $|f(p)| < p$, the formula

$$\frac{1}{M(f)} \cdot a_{p^\ell} = \frac{\frac{f(p)-1}{p-1} \cdot \frac{f(p^{\ell-1})}{p^{\ell-1}} + \sum_{k\geq\ell} p^{-k} \cdot \big(f(p^k) - f(p^{k-1})\cdot f(p)\big)}{\Big\{1 - \frac{f(p)}{p}\Big\} \cdot \sum_{k\geq 0} p^{-k} \cdot f(p^k)}$$

holds. Hint: use Theorem 4.4.

7) If ν is the function used in the proof of Theorem 5.1, show that

$$M(1*\nu) = \prod_{p\nmid t} \Big(1 + \frac{\nu(p)}{p}\Big).$$

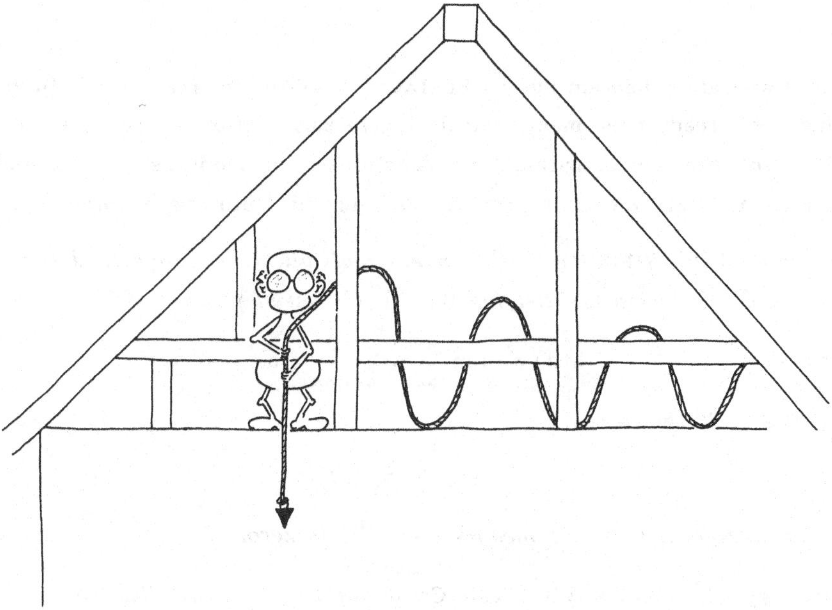

Chapter IX

Mean-Value Theorems and Multiplicative Functions, II

Abstract. *This chapter is a continuation of Chapter II. We are going to give proofs for two, deep mean-value theorems for multiplicative functions, namely one due to E. WIRSING [1967], with a proof by A. HILDEBRAND [1986], and the other due to G. HALÁSZ [1968], with an elementary proof given by H. DABOUSSI and K.-H. INDLEKOFER [1992]. This proof uses ideas from DABOUSSI's elementary proof of the prime number theorem. HILDEBRAND's proof uses a version of the prime number theorem with a [weak] error term, and thus, while HILDEBRAND's proof does not give a new elementary proof of the prime number theorem, the DABOUSSI-INDLEKOFER proof does.*

IX.1. ON WIRSING'S MEAN-VALUE THEOREM

The mean-value theorem due to EDUARD WIRSING for real-valued functions has already been mentioned in II.5. In this section we restrict ourselves to real-valued arithmetical functions f of modulus $|f| \le 1$, and we give A. HILDEBRAND's proof [1986] for the following theorem.

Theorem 1.1 (E. WIRSING, 1967). *For any multiplicative, real-valued arithmetical function* f *satisfying* $|f| \le 1$, *the mean-value*

$$M(f) = \lim_{x \to \infty} x^{-1} \cdot \sum_{n \le x} f(n) \tag{1.1}$$

exists. If the series

$$\sum_p p^{-1} \cdot \big(1 - f(p)\big) \tag{1.2}$$

is divergent, then the mean-value M(f) *is zero.*

Corollary 1.2 (ERDÖS-WINTNER Conjecture). *Any multiplicative arithmetical function assuming only values from the set* {-1, 0, 1} *has a mean-value.*

Corollary 1.3 (Prime Number Theorem). *The MÖBIUS function* $n \mapsto \mu(n)$ *has a mean-value.*

Remark. In fact, the Prime Number Theorem

$$\pi(x) \sim \frac{x}{\log x}, \text{ as } x \to \infty \tag{1.3}$$

follows from Corollary 1.3. However, the proof of Theorem 1.1 (in the stronger version of Theorem 1.4) and of its corollary uses a stronger version of the Prime Number Theorem, and so this result cannot be considered to give a new proof of the Prime Number Theorem.

Corollary 1.2 is obviously a special case of Theorem 1.1, and the assertion of Corollary 1.3 for the MÖBIUS function is contained in Corollary 1.2. The divergence of $\sum_p p^{-1} \cdot (1 - \mu(p)) = \sum_p 2 \cdot p^{-1}$ implies $M(\mu) = 0$. The deduction of the Prime Number Theorem (1.3) in the equivalent form

$$\psi(x) = \sum_{n \le x} \Lambda(n) \sim x \tag{1.4}$$

is possible by elementary (though somewhat tricky) arguments, as shown by E. LANDAU. We start with the arithmetical function

$$h = \log - \tau + 2\mathscr{C}, \tag{1.5}$$

where $\mathscr{C}$ is EULER's constant, and $\tau = 1 * 1$ is the divisor function; ε denotes the unit of the ring of arithmetical functions with convolution, and from Chapter I we know the convolution relations

$$\Lambda = \mu * \log,\ 1 = \mu * \tau,\ \varepsilon = \mu * 1.$$

Therefore,

$$\left\{ \begin{aligned} \sum_{n\le x}\left\{\Lambda(n) - 1 + 2\,\mathscr{C}\cdot\varepsilon(n)\right\} &= \sum_{n\le x}\sum_{d\mid n}\mu(d)\cdot\left\{\log\tfrac{n}{d} - \tau\left(\tfrac{n}{d}\right) + 2\mathscr{C}\right\} \\ &= \sum\sum_{d\cdot m\le x}\mu(d)\cdot h(m). \end{aligned}\right. \tag{1.6}$$

On the other hand,

$$\sum_{n\le x}\left\{\Lambda(n) - 1 + 2\,\mathscr{C}\,\varepsilon(n)\right\} = \psi(x) - [x] + 2\,\mathscr{C},$$

and so the Prime Number Theorem (1.4) is proved as soon as

$$\sum\sum_{d\cdot m\le x}\mu(d)\cdot h(m) = o(x)\ (\text{as } x\to\infty)$$

is proved. A DIRICHLET summation (see I, 2), using the summatory functions

$$M(x) = \sum_{n\le x}\mu(n), \text{ and } H(x) = \sum_{n\le x}h(n),$$

gives, with some parameter $B = B(x)$,

$$\begin{aligned} \sum\sum_{d\cdot m\le x}\mu(d)\cdot h(m) &= \sum_{d\le x/B}\mu(d)\cdot\sum_{m\le x/d}h(m) + \sum_{m\le B}h(m)\cdot\sum_{x/B<d\le x/m}\mu(d) \\ &= \sum_{d\le x/B}\mu(d)\cdot H\left(\tfrac{x}{d}\right) + \sum_{m\le B}h(m)\cdot\left(M\left(\tfrac{x}{m}\right) - M\left(\tfrac{x}{B}\right)\right). \end{aligned}$$

DIRICHLET's result on the divisor function (I, 2) and an application of EULER's summation formula to $\sum_{n\le x}\log(n)$ give the asymptotic relation

$$H(x) = O(\sqrt{x}),$$

and the trivial estimate $|\mu|\le 1$ leads to

$$\left|\sum_{d\le x/B}\mu(d)\cdot H\left(\tfrac{x}{d}\right)\right| \le \gamma_1\sum_{d\le x/B}\left(\tfrac{x}{d}\right)^{\frac12} \le \gamma_2\, x\cdot B^{-\frac12},$$

with some constants γ_1, γ_2, and also to

$$\sum_{m\le B} h(m)\cdot M\left(\frac{x}{B}\right) \le \gamma_1 \cdot \sqrt{B}\cdot \frac{x}{B} \le \gamma_1 \cdot x \cdot B^{-\frac{1}{2}}.$$

Our assumption $M(\mu) = 0$ is only required for the estimation of the last sum $\sum_{m\le B} h(m)\cdot M\left(\frac{x}{m}\right)$. Given $\varepsilon > 0$, fix the parameter B such that

(i) $$\max\left\{\gamma_1 \cdot B^{-\frac{1}{2}},\ \gamma_2\cdot B^{-\frac{1}{2}}\right\} \le \tfrac{1}{3}\,\varepsilon,$$

and

(ii) $$\sum_{n\le B}\left(\log n + \tau(n) + 2\mathcal{C}\right) \le \tfrac{1}{3}\,\varepsilon\cdot B^2.$$

(This is possible, the sum in question being $\le \gamma\cdot B\cdot\log B$.) Then $M(\mu) = 0$ implies that there is an $x_0(B)$ with the property

(iii) $$\max_{m\le B}\left|M\left(\frac{x}{m}\right)\right| \le \frac{x}{B^2}, \text{ if } x \ge x_0(B).$$

Therefore, using (iii) and (ii),

$$\left|\sum_{m\le B} h(m)\cdot M\left(\frac{x}{m}\right)\right| \le \frac{x}{B^2}\cdot\sum_{m\le B}\left\{\log m + \tau(m) + 2\mathcal{C}\right\} \le \tfrac{1}{3}\,\varepsilon\cdot x,$$

if $x \ge x_0(B)$. Therefore,

$$\left|\sum_{d\cdot m\le x}\sum \mu(d)\cdot h(m)\right| \le \varepsilon\cdot x, \text{ if } x \ge x_0(B),$$

and the Prime Number Theorem $\psi(x) \sim x$ is proved. □

The implication $\left(\psi(x) \sim x\right) \Rightarrow \left(M(\mu) = 0\right)$ is also true; an elementary proof of this implication is requested in Exercise 4.

For the proof of Theorem 1.1 we remark that in the case where $\sum_p p^{-1}\cdot\left(1 - f(p)\right)$ is convergent, it is absolutely convergent, and so the function f is related to 1, and the Relationship Theorem (Chapter III, Theorem 4.1) (or, more directly, an application of II. Corollary 2.3) easily gives the existence of M(f), and the value of the mean-value is

(1.7) $$M(f) = \prod_p \left(1 + p^{-1}\cdot(f(p)-1) + p^{-2}\cdot(f(p^2)-f(p)) + \dots\right).$$

Therefore, we may assume that the series

(1.8) $$\sum_p p^{-1}\cdot\left(1 - f(p)\right) = \infty$$

is divergent. In this case the mean-value of f is $M(f) = 0$. More precisely, we prove the following theorem.

Theorem 1.4 (E. WIRSING, A. HILDEBRAND). *If* f *is any multiplicative arithmetical function, assuming only real values from the interval* [-1, 1], *then the divergence of the series* (1.8) *implies the existence of a [universal] constant* γ *[independent from* f*], so that in* $x \ge 2$ *the estimate*

$$(1.9)\qquad \Big| x^{-1} \cdot \sum_{n\le x} f(n) \Big| \le \gamma \cdot \Big(1 + \sum_{p\le x} p^{-1} \cdot (1 - f(p)) \Big)^{-\frac{1}{2}}$$

holds. In particular, the mean-value M(f) *of* f *exists and is zero.*

The proof of this theorem and of the crucial Lemma 2.1 will be postponed until section 2. We collect three auxiliary results, which have been dealt with in former chapters.

Lemma 1.5. *Given any complex numbers* w_n*, then for* $x \ge 2$*, the inequality*

$$\sum_{p\le x} p^{-1} \cdot \Big| \frac{p}{x} \cdot \sum_{n\le x, p|n} w_n - \frac{1}{x} \cdot \sum_{n\le x} w_n \Big|^2 \le c \cdot \frac{1}{x} \cdot \sum_{n\le x} |w_n|^2$$

holds.

See I, Exercise 16.

Lemma 1.6. *Assume that* f *is a non-negative multiplicative arithmetical function, satisfying the two conditions*

$$\sum_{p^k\le x} f(p^k) \cdot \log p^k \le \gamma_1 \cdot x,$$

$$\sum_{p\le x} \sum_{k\ge 2} p^{-k} \cdot f(p^k) \le \gamma_2$$

with some constants $\gamma_1 > 0$, $\gamma_2 > 0$. *Then, with some constant* γ, *depending only on* γ_1 *and* γ_2, *the estimate*

$$x^{-1} \cdot \sum_{n\le x} f(n) \le \gamma \cdot \exp\Big(\sum_{p\le x} p^{-1} \cdot (f(p)-1) \Big)$$

holds.

Proof. This result may be deduced from II, Theorem 3.1 (3.3). □

Examples. Lemma 1.6 can be applied to multiplicative functions f satisfying $0 \le f \le 1$, or $0 \le f(p^k) \le k+1$ for all prime-powers p^k, or

$0 \le f(p^k) \le \lambda_1 \cdot \lambda_2^k$, for all prime powers p^k, with some constants $\lambda_1 > 0,\ 2 > \lambda_2 > 0$.

Lemma 1.7. *Uniformly in* $x \ge 2$, *and for all real-valued multiplicative arithmetical functions* f *satisfying* $-1 \le f \le +1$, *the estimate*

$$\Big| \sum_{n \le x} n^{-1} \cdot f(n) \Big| \le \gamma_3 \cdot \Big(\exp\Big\{ \sum_{p \le x} p^{-1} \cdot f(p) \Big\} + 1 \Big)$$

holds.

Proof. The function $g = 1 * f$ is multiplicative and satisfies $|g(p)| = |1 + f(p)| = 1 + f(p)$, and $|g(p^k)| \le 1 + |f(p)| + \ldots + |f(p^k)| \le k+1$. Therefore Lemma 1.6 implies

$$x^{-1} \cdot \Big| \sum_{n \le x} g(n) \Big| \le x^{-1} \cdot \sum_{n \le x} |g(n)|$$

$$\le \gamma_3 \cdot \exp\Big(\sum_{p \le x} p^{-1} \cdot (|g(p)| - 1) \Big) = \gamma_3 \cdot \exp\Big(\sum_{p \le x} p^{-1} \cdot f(p) \Big).$$

This estimate, together with

$$\sum_{n \le x} n^{-1} \cdot f(n) = \frac{1}{x} \sum_{n \le x} f(n) \cdot \Big[\frac{x}{n} \Big] + \Theta = \frac{1}{x} \cdot \sum_{n \le x} g(n) + \mathcal{O}(1),$$

where $|\Theta| \le 1$, gives the assertion of Lemma 1.7. □

IX.2. PROOF OF THEOREM 1.4.

In Chapter II, the summatory function of f was denoted by $\mathbf{M}(f,x)$ $\big(= \sum_{n \le x} f(n)\big)$. Let us define the function $\mathcal{M}(f,x)$ by

$$\mathcal{M}(f,x) = \frac{1}{x} \cdot \mathbf{M}(f,x) = \frac{1}{x} \cdot \sum_{n \le x} f(n). \tag{2.1}$$

Next, we use the notation

$$S(x) = \sum_{p \le x} p^{-1} \cdot \big(1 - f(p)\big), \tag{2.2}$$

and

$$\delta(x) = \min\Big(\tfrac{1}{4},\ e^{-\frac{1}{3} S(x)} \Big). \tag{2.3}$$

The proof of Theorem 1.4 depends on an oscillation property of $\mathcal{M}(f,x)$, stated in the following lemma.

Lemma 2.1. *With an absolute constant* C, *for all multiplicative functions* f, *satisfying* $-1 \le f \le 1$, *with divergent series* (1.8), *the "oscillation condition"*

(2.4) $$|\,\mathcal{M}(f,y) - \mathcal{M}(f,x)\,| \le C \cdot \left(\log \frac{\log x}{\log(2y/x)} \right)^{-\frac{1}{2}}$$

holds in $3 \le x \le y \le x^{5/4}$.

Proof of Theorem 1.4. We use the notation $S(x)$ and $\delta(x)$, introduced at the beginning of this section. First, with positive [absolute] constants γ_i, $1 \le i \le 4$,

(2.5) $$\delta(x) \ge \gamma_1 \cdot \exp\left(-\tfrac{2}{3} \sum_{p \le x} p^{-1} \right) \ge \gamma_2 \cdot (\log x)^{-\frac{2}{3}},$$

and, if $3 \le x \le y \le x^{1+\delta(x)}$,

(2.6) $$\gamma_3 \cdot \left(\log \frac{\log x}{\log(2y/x)} \right)^{-\frac{1}{2}} \le (\log \delta^{-1}(x))^{-\frac{1}{2}} \le \gamma_4 \cdot (S(x) + 1)^{-\frac{1}{2}}.$$

Applying Lemma 2.1 and inequality (2.6) we obtain, in $3 \le x \le y \le x^{1+\delta(x)}$,

$$\mathcal{M}(f,x) = \mathcal{M}(f,y) + \mathcal{O}\left((S(x)+1)^{-\frac{1}{2}}\right).$$

An integration of this equation gives a representation for $\mathcal{M}(f,x)$ "in the mean",

$$\mathcal{M}(f,x) = \left(\delta(x) \cdot \log x \right)^{-1} \cdot \int_x^{x^{1+\delta(x)}} y^{-1} \cdot \mathcal{M}(f,y)\, dy + \mathcal{O}\left((S(x)+1) \right)^{-\frac{1}{2}}.$$

The integral is

$$\int_x^{x^{1+\delta(x)}} y^{-2} \cdot \sum_{n \le y} f(n)\, dy = \sum_{n \le x^{1+\delta(x)}} f(n) \cdot \int_{\max(x,n)}^{x^{1+\delta(x)}} y^{-2}\, dy$$

$$= \sum_{n \le x^{1+\delta(x)}} f(n) \cdot \left(\{\max(x,n)\}^{-1} - x^{-1-\delta(x)} \right)$$

$$= \sum_{x < n \le x^{1+\delta(x)}} \frac{f(n)}{n} + \mathcal{M}(f,x) - \mathcal{M}(f, x^{1+\delta(x)})$$

$$= \sum_{x < n \le x^{1+\delta(x)}} \frac{f(n)}{n} + \mathcal{O}(1),$$

and we obtain

$$\mathcal{M}(f,x) = \left(\delta(x) \cdot \log x\right)^{-1} \cdot \sum_{x < n \le x^{1+\delta(x)}} \frac{f(n)}{n} + \mathcal{O}\left((S(x)+1) \right)^{-\frac{1}{2}} + \mathcal{O}\left(\delta(x) \cdot \log x \right)^{-1}.$$

The estimation of $\sum_{n\le x^{1+\delta(x)}} \frac{f(n)}{n}$ and $\sum_{n\le x} \frac{f(n)}{n}$ is achieved using Lemma 1.7. The sum $\left\{\sum_{p\le x} p^{-1}\cdot f(p)\right\}$ equals $\sum_{p\le x} p^{-1}\cdot(f(p)-1) + \log\log x + \mathcal{O}(1)$, and the sum up to $x^{1+\delta(x)}$ is treated similarly; the difference

$$\Big|\sum_{x<p\le x^{1+\delta(x)}} p^{-1}\cdot f(p)\Big| = \mathcal{O}(1)$$

is bounded trivially. So we obtain

$$\Big|\sum_{x<n\le x^{1+\delta(x)}} \frac{f(n)}{n}\Big| \le \Big|\sum_{n\le x} \frac{f(n)}{n}\Big| + \Big|\sum_{n\le x^{1+\delta(x)}} \frac{f(n)}{n}\Big|$$

$$\le \gamma_5\cdot \log x\cdot \exp\{-S(x)\}.$$

Finally, a simple calculation, using (2.3) and (2.5), shows that $\delta^{-1}(x)\cdot\exp\{-S(x)\}$, as well as $(\delta(x)\cdot\log x)^{-1}$, may be estimated by $\mathcal{O}\left(1/\sqrt{S(x)+1}\right)$, and Theorem 1.4 is proved. □

Thus it remains to **prove** Lemma 2.1, which states: with an absolute constant C, for all multiplicative functions f, satisfying $-1 \le f \le 1$, with divergent series (1.8), the "oscillation condition"

$$(2.4) \qquad |\mathcal{M}(f,y) - \mathcal{M}(f,x)| \le C\cdot\left(\log\frac{\log x}{\log(2y/x)}\right)^{-\frac{1}{2}}$$

holds in $3 \le x \le y \le x^{5/4}$.

First,

$$(2.7) \qquad |\mathcal{M}(f,y) - \mathcal{M}(f,x)| \le 2x^{-1}\cdot(y - x + 1)$$

is trivial. Given $3 \le x < y$, we define $\delta(x,y)$ by $y = x^{1+\delta(x,y)}$ [therefore $0 < \delta(x,y) = \left(\log y/\log x - 1\right) \le \frac{1}{4}$], and $R(x,y)$ by

$$R(x,y) = \sqrt{\log\delta^{-1}(x,y)} = \sqrt{\log\frac{\log x}{\log(y/x)}}\,.$$

Equation (2.7) gives the assertion of the lemma if $y - x$ is small. Therefore, we may assume that $y - x \gg x\cdot(\log x)^{-\frac{1}{2}}$, and

$$(2.8) \qquad \delta(x,y) \ge (\log x)^{-3/2}.$$

Then $R(x,y)$ is of the order

$$1 \ll \sqrt{\log\frac{\log x}{\log(2y/x)}} \ll R(x,y) \ll \sqrt{\log\frac{\log x}{\log(2y/x)}}$$

and the right-hand side of assertion (2.4) is of the order $1/R(x,y)$.

Next we *remark* that it suffices to prove the assertion of the lemma for the function $|\mathcal{M}(f,x)|$. Thus we have to prove that

(2.9) $$\Big|\, |\mathcal{M}(f,y)| - |\mathcal{M}(f,x)| \,\Big| \ll R^{-1}(x,y).$$

This *remark* is obvious if $\mathcal{M}(f,y)$ and $\mathcal{M}(f,x)$ are both positive or both negative. If not, then there is - due to the oscillation condition (2.7) - a point z lying between x and y, for which $\mathcal{M}(f,z)$ is very small. Application of (2.9) to the two intervals $[x,z]$ and $[z,y]$ proves $\Big|\, \mathcal{M}(f,y) - \mathcal{M}(f,x) \,\Big| \ll R^{-1}(x,y)$.

Without loss of generality, we finally assume

(2.10) $$\sum_{p\le x} p^{-1} \cdot (1 - |f(p)|) \le R(x,y).$$

Otherwise, applied to $|f|$, Lemma 1.6 gives

$$|\mathcal{M}(|f|,y)| \ll e^{-R(x,y)}, \quad |\mathcal{M}(|f|,x)| \ll e^{-R(x,y)},$$

and (2.9) is obvious, since $e^{-R} \le R^{-1}$ in $R > 0$.

Now we are going to prove (2.9), using (2.8) and (2.10). For an application of Lemma 1.5 with $w_n = f(n)$ we transform $\sum_{n\le x, p|n} f(n)$, using the multiplicativity of f, as follows:

$$\frac{p}{x} \sum_{n\le x, p|n} f(n) = f(p)\cdot \mathcal{M}(f,\frac{x}{p}) + E_p$$

with an error term

$$E_p = - f(p) \cdot \frac{p}{x} \cdot \sum_{n\le x/p, p|n} f(n) + \frac{p}{x} \cdot \sum_{n\le x,\, p^2|n} f(n) = \mathcal{O}\left(p^{-1}\right).$$

Using Lemma 1.5, the triangle inequality and the CAUCHY-SCHWARZ inequality, we obtain, for every $u < x$,

$$\sum_{u<p\le x} p^{-1} \cdot \Big|\, f(p)\cdot \mathcal{M}\Big(f,\frac{x}{p}\Big) - \mathcal{M}(f,x) \,\Big|$$

$$\le \sum_{u<p\le x} p^{-1} \cdot \Big|\, \frac{p}{x} \sum_{n\le x, p|n} f(n) - \mathcal{M}(f,x) \,\Big| + \sum_{u<p\le x} p^{-1} \cdot |E_p|$$

$$\le \Big(\sum_{p\le x} p^{-1} \cdot \Big| \frac{p}{x} \sum_{n\le x, p|n} f(n) - \mathcal{M}(f,x)\Big|^2 \Big)^{\frac12} \cdot \Big(\sum_{u<p\le x} p^{-1} \Big)^{\frac12} + \mathcal{O}\Big(\sum_{u<p\le x} p^{-2} \Big)$$

$$\leq \left(c \cdot x^{-1} \cdot \sum_{n \leq x} |f(n)|^2 \right)^{\frac{1}{2}} \cdot \left(\sum_{u<p\leq x} p^{-1} \right)^{\frac{1}{2}} + \mathcal{O}(1) = \mathcal{O}\big(R(x,y)\big),$$

where $u = x^{\sqrt{\delta(x,y)}} = \exp\{\log x \cdot \sqrt{\delta(x,y)}\}$ is chosen, and (see I.6)

$$\sum_{u<p\leq x} p^{-1} = \log \frac{\log x}{\log u} = \tfrac{1}{2} R^2 + \mathcal{O}(1)$$

is used. Relation (2.10) permits, to replace $p^{-1} \cdot f(p)$ by p^{-1} "in the mean", with a small error, and so we obtain, with the given choice of u,

$$\text{(2.11)} \qquad |\mathcal{M}(f,x)| \cdot \Big(\sum_{u<p\leq x} p^{-1} \Big) = \sum_{u<p\leq x} p^{-1} \cdot |\mathcal{M}(f,x/p)| + \mathcal{O}(R(x,y)).$$

Similarly, with $v = x^{\delta(x,y)+\sqrt{\delta(x,y)}}$, we obtain

$$\text{(2.12)} \qquad |\mathcal{M}(f,y)| \cdot \Big(\sum_{v<p\leq y} p^{-1} \Big) = \sum_{v<p\leq y} p^{-1} \cdot |\mathcal{M}(f,y/p)| + \mathcal{O}(R(x,y))$$

and

$$\sum_{v<p\leq y} p^{-1} = \log \frac{(1+\delta(x,y))}{(\delta(x,y)+\sqrt{\delta(x,y)})} + \mathcal{O}(1) = \tfrac{1}{2} R^2 + \mathcal{O}(1).$$

The difference between (2.11) and (2.12) is, in absolute value,

$$\text{(2.13)} \quad \begin{cases} \tfrac{1}{2} \cdot R^2(x,y) \,\Big|\, |\mathcal{M}(f,x)| - |\mathcal{M}(f,y)| \,\Big| + \mathcal{O}(1) \\[2ex] \quad = \Big| \sum_{u<p\leq x} p^{-1} \cdot |\mathcal{M}(f,x/p)| - \sum_{v<p\leq y} p^{-1} \cdot |\mathcal{M}(f,y/p)| \Big| + \mathcal{O}\big(R(x,y)\big). \end{cases}$$

We have to show that the second line of this equation is $\ll R(x,y)$. We begin by replacing the sums over the primes by integrals, using partial summation and (a sharper version of) the prime number theorem: $\sum_{p\leq x} p^{-1} = \log\log x + \gamma + \mathcal{O}(\delta^2(x,y)/\log x)$. For brevity, write $\delta = \delta(x,y)$, $R = R(x,y)$. Put $u = x^{\sqrt{\delta}}$ as before, and $U = x^{\exp(-R)}$. Remember that $|\mathcal{M}(f,x)| \leq 1$.

Then $\sum_{U<p\leq x} p^{-1} \cdot |\mathcal{M}(f,x/p)| \ll R(x,y)$, and so we restrict ourselves to the range $u < z \leq U$.

In $z \leq \upsilon \leq z + \delta z$, by the trivial estimate (2.7), we obtain

$$\Big|\, |\mathcal{M}(f,x/\upsilon)| - |\mathcal{M}(f,x/z)| \,\Big| \ll z^{-1} \cdot (\upsilon - z) + \upsilon/x \ll \delta + (z/x) \ll \delta,$$

for $z/x \leq (x^{1-1/e^R})^{-1} \ll (\log x)^{-3/2} \ll \delta$. Therefore,

$$\sum_{z<p\leq z(1+\delta)} p^{-1} \cdot |\mathcal{M}(f,x/p)| =$$

$$= |\mathcal{M}(f,x/z)| \cdot \Sigma_{z<p\le z(1+\delta)} p^{-1} + \mathcal{O}\Big(\delta \cdot \Sigma_{z<p\le z(1+\delta)} p^{-1} \Big)$$

$$= |\mathcal{M}(f,x/z)| \cdot \int_z^{z(1+\delta)} \frac{dv}{v \log v} + \mathcal{O}\Big(\frac{\delta^2}{\log z}\Big) + \mathcal{O}\Big(\delta \int_z^{z(1+\delta)} \frac{dv}{v \log v}\Big)$$

$$= \int_z^{z(1+\delta)} |\mathcal{M}(f,x/v)| \frac{dv}{v \log v} + \mathcal{O}\Big(\delta \int_z^{z(1+\delta)} \frac{dv}{v \log v}\Big).$$

Splitting up the range of summation into intervals of the type $]z, (1+\delta)\cdot z]$ (with a possible incomplete interval at the end), and taking into account our former estimate for the interval $]U, x]$, we obtain

$$\sum_{u<p\le x} p^{-1} \cdot |\mathcal{M}(f,x/p)| + \mathcal{O}(R)$$

$$= \int_u^x (\upsilon \log \upsilon)^{-1} \cdot |\mathcal{M}(f,x/\upsilon)| \, d\upsilon + \mathcal{O}\Big(\delta \int_u^x (\upsilon \log \upsilon)^{-1} d\upsilon \Big)$$

$$= \int_1^{x/u} |\mathcal{M}(f,w)| \cdot (w \cdot \log(x/w))^{-1} dw + \mathcal{O}(R).$$

And in the same way, the relation

$$\sum_{v<p\le y} p^{-1} \cdot |\mathcal{M}(f,y/p)| = \int_1^{x/u} |\mathcal{M}(f,w)| \cdot (w \cdot \log(y/w))^{-1} dw + \mathcal{O}(R)$$

is proved (note that $y/v = x/u$). Therefore, the desired estimate of the difference on the right-hand side of (2.13) is

$$\ll \int_1^{x/u} |\mathcal{M}(f,w)| \cdot \Big((w \cdot \log(x/w))^{-1} - (w \cdot \log(y/w))^{-1}\Big) dw + R$$

$$\ll \log(y/x) \cdot \int_1^{x/u} (\log(x/w))^{-2} w^{-1} dw + R \ll \sqrt{\delta} + R \ll R,$$

and Lemma 2.1 (and hence Theorem 1.4 also) is proved. □

IX.3. THE MEAN-VALUE THEOREM OF G. HALÁSZ

As outlined in the Introduction, the ERDÖS-WINTNER conjecture states that any multiplicative function f assuming only the values 0, 1, and -1, has a mean-value M(f). This conjecture includes the prime number theorem because the assertion $M(\mu) = 0$, where μ denotes the MÖBIUS function, is [elementarily] equivalent to the prime number theorem, as was shown at the beginning of this century by E. LANDAU. The ERDÖS-WINTNER conjecture was not proved until 1967, by EDUARD WIRSING, as already mentioned; in particular, real-valued multiplicative functions f of absolute value $|f| \le 1$ do have a mean-value. In 1968 G. HALÁSZ dealt with multiplicative complex-valued functions f of modulus $|f| \le 1$, making use of the technique of complex integration in a very skilful manner. His proof contains the following result.

Theorem 3.1 (HALÁSZ's Theorem). *Let f be a complex-valued multiplicative arithmetical function, satisfying $|f| \le 1$.*

(1) *If there exists a real number a_0, for which the series*

$$\sum_p p^{-1} \cdot \left(1 - \operatorname{Re}\left\{f(p)\cdot p^{-ia_0}\right\}\right) \tag{3.1}$$

*is **convergent**, then the asymptotic relation*

$$\sum_{n\le x} f(n) = x \cdot \frac{x^{ia_0}}{1+ia_0} \cdot \prod_{p\le x} \left(\left(1-p^{-1}\right)\cdot\left(1 + \sum_{r=1}^{\infty} \frac{f(p^r)}{p^{r(1+ia_0)}}\right) \right) + o(x) \tag{3.2}$$

holds.

(2) *If the series $\sum_p p^{-1} \cdot \left(1 - \operatorname{Re}\left\{f(p)\cdot p^{-it}\right\}\right)$ is **divergent** for every real number t, then*

$$\lim_{x\to\infty} x^{-1} \cdot \sum_{n\le x} f(n) = 0. \tag{3.3}$$

(3) *In both cases there are constants c, a_0 and a slowly oscillating function L of modulus $|L| = 1$, so that the asymptotic formula*

$$\sum_{n\le x} f(n) = c \cdot x^{1+ia_0} \cdot L(\log x) + o(x) \tag{3.4}$$

is true.

Remark 1. Simple examples (see Exercises 1 and 2) show that complex-valued bounded arithmetical functions need not have a mean-value; of course, this is also obvious from (3.2).

The result (3.2), where (3.1) is convergent, can be obtained comparatively easily, for example, it follows from the DELANGE Theorem. Therefore, we concentrate our efforts on case (2) only, where the series

(3.5) $\sum_p p^{-1} \cdot (1 - \operatorname{Re}\{ f(p)\cdot p^{-it}\})$ *is divergent for every* $t \in \mathbb{R}$.

Put

(3.6) $$\alpha(w) = e^{-w} \cdot \sum_{n \le \exp(w)} f(n),$$

and

(3.7) $$\alpha = \limsup_{w \to \infty} |\alpha(w)| \le 1.$$

"Without loss of generality" (using the results of Chapter III - the proof is reduced to the simplest possible case), we may *assume* that the values $f(p^k)$ are zero if $k = 2, 3, \dots$ (see Exercise 3). First we are going to prove the following proposition.

Proposition 3.2. *If* f *is multiplicative,* $|f| \le 1$, *and if the values* $f(p^k)$, $k \ge 2$, *at higher prime-powers, are zero, then for every* $\delta \ge 1$

(3.8) $$\lim_{x \to \infty} x^{-1} \cdot \int_0^x |\alpha(w)|^\delta \, dw = \alpha^\delta.$$

Remark 2. For a fixed $\delta \ge 1$, this result is equivalent to

(3.9) $$\int_0^\infty |\alpha(w)|^\delta \cdot e^{-cw(\sigma-1)} \, dw = c^{-1} \cdot (\sigma-1)^{-1} \cdot \alpha^\delta + o(\,(\sigma-1)^{-1}\,), \text{ as } \sigma \to 1+,$$

for every $c > 0$.

The implication (3.9) ⇒ (3.8), which is not needed in the proof, follows from the HARDY-LITTLEWOOD-KARAMATA TAUBERIAN Theorem (see the Appendix, A.4).

The other implication (3.8) ⇒ (3.9) is achieved by partial integration, giving

$$\int_0^\infty |\alpha(w)|^\delta \cdot e^{-w/x} \, dw = x^{-1} \int_0^\infty e^{-w/x} \cdot \Big(\int_0^w |\alpha(u)|^\delta \, du \Big) \, dw = \alpha^\delta \cdot x + o(x).$$

Using $x = (c(\sigma-1))^{-1}$, this is (3.9).

Proposition 3.3. (DABOUSSI-INDLEKOFER). *Assume that* $f\colon \mathbb{N} \to \mathbb{C}$ *is multiplicative,* $|f| \le 1$, *and the series* (3.5) *is divergent for every real* t. *Define* α *by* (3.6) *and* (3.7). *Assume, furthermore, for simplicity, that* $f(p^k) = 0$ *for every prime* p *and every* $k \ge 2$. *Then*

(3.10) $$\lim_{x \to \infty} x^{-1} \cdot \int_0^x |\alpha(w)| \, dw = \alpha.$$

The proof of this result, which uses ideas from DABOUSSI's elementary proof of the prime number theorem, will be postponed until section 4.

Proof of (3.8), Proposition 3.2. In $1 < \delta < \infty$, HÖLDER's inequality gives

$$\int_0^x |\alpha(w)| \, dw \le \left(\int_0^x |\alpha(w)|^\delta dw \right)^{1/\delta} \cdot \left(\int_0^x dw \right)^{1-(1/\delta)}$$

$$= x \cdot \left(x^{-1} \int_0^x |\alpha(w)|^\delta \, dw \right)^{1/\delta},$$

and, using (3.10), we obtain

$$\liminf_{x \to \infty} x^{-1} \cdot \int_0^x |\alpha(w)|^\delta \, dw \ge \alpha^\delta.$$

The estimate

$$\limsup_{x \to \infty} x^{-1} \cdot \int_0^x |\alpha(w)|^\delta \, dw \le \alpha^\delta$$

is trivial on behalf of the definition of α: $\alpha = \limsup_{x \to \infty} |\alpha(x)|$. This proves Proposition 3.2, (3.8) [and thus (3.9) also]. □

By partial summation (see I.1), for a bounded arithmetical function f, in Re s > 1, the generating DIRICHLET series $\sum_{n=1}^\infty n^{-s} \cdot f(n)$ is represented by a LAPLACE integral,

$$(3.11) \quad \begin{cases} F(s) = \sum_{n=1}^{\infty} n^{-s} \cdot f(n) \\ \quad = s \cdot \int_1^\infty \sum_{n \le u} f(n) \cdot u^{-(s+1)} du = s \cdot \int_0^\infty \alpha(w) \cdot e^{-w(\sigma-1)-iwt} \, dw, \end{cases}$$

and PARSEVAL's formula gives

$$(3.12) \qquad \int_{-\infty}^{+\infty} \left| \frac{F(s)}{s} \right|^2 dt = 2\pi \cdot \int_0^\infty |\alpha(w)|^2 \cdot e^{-2w(\sigma-1)} \, dw.$$

The following lemma will be useful.

Lemma 3.4. *Assume* g *is a multiplicative arithmetical function, uniformly bounded at the prime-powers,* $|g(p^k)| \le \gamma$ *for* $k = 1, 2, \ldots,$ *and every prime* p. *Then*

$$(3.13) \qquad \sum_{n \le x} |g(n)| \ll x \cdot \exp\left(\sum_{p \le x} p^{-1} \cdot (|g(p)|-1) \right),$$

and, for the generating DIRICHLET series $G(s) = \sum_{n=1}^\infty n^{-s} \cdot g(n)$,

$$(3.14) \quad \int_{-\infty}^{\infty} \left| \frac{G(s)}{s} \right|^2 dt \ll \int_0^\infty \exp\Big(2 \sum_{p \le \exp(w)} p^{-1} \cdot (|g(p)|-1) \Big) \cdot e^{-2w(\sigma-1)} dw,$$

as $\sigma \to 1+$.

The **proof** of (3.13) is given in II, Theorem 3.1. The proof of (3.14) is immediate from the PARSEVAL equation (3.12), where the function α has to be replaced by $\alpha_g(w) = e^{-w} \cdot \sum_{n\le\exp(w)} |g(n)|$. □

Proof of G. HALÁSZ's Theorem. We assume that the series (3.5) diverges for every real t, and that $\alpha \neq 0$; we have to come to a contradiction. The inequality $\sigma > 1$ is always assumed. For every real t, as $\sigma \to 1+$,

$$\sum_p p^{-\sigma}\cdot\left(1 - \operatorname{Re}\left\{f(p)\cdot p^{-it}\right\}\right) \tag{3.15}$$

converges monotonically increasing to infinity. The divergence of (3.15) is uniform on every intervall $[-K,K]$. This can be seen from a variant of DINI's theorem [see P. D. T. A. ELLIOTT [1979], Lemma 6.7, p.241, and the remark on p.242)] or directly in the following way. Define

$$G_\sigma(t) = \sum p^{-\sigma}\cdot\left(1-\operatorname{Re}\left(\frac{f(p)}{p^{it}}\right)\right),\ \sigma > 1,$$

and assume that divergence of (3.15) is not uniform. Then there exists a constant $c > 0$, a sequence $\sigma_1 > \sigma_2 > \ldots > 1$, $\sigma_n \to 1$, and points $t_n \in [-K,K]$ such that $G_{\sigma_n}(t_n) \le c^{-1}$ for every positive integer n. Taking a suitable subsequence we may assume $t_n \to t_0$. Fix $\sigma \in\ ]1,\sigma_0[$. There exists an integer n_0 such that $\sigma_n \le \sigma$ for every $n \ge n_0$, and so

$$G_\sigma(t_n) \le G_{\sigma_n}(t_n) \le c^{-1}.$$

Hence $G_\sigma(t_0) \le c^{-1}$ in $\sigma \in\ ]1,\sigma_0[$. This contradicts the divergence of (3.15) at $t = t_0$, if $\sigma \to 1+$.

Using the PARSEVAL equation (3.12), and then (3.9) (with $\delta = 2$, $c = 2$), we obtain

$$\int_{-\infty}^{\infty} \left|\frac{F(s)}{s}\right|^2 dt = \pi\cdot(\sigma-1)^{-1}\cdot\alpha^2 + o\left((\sigma-1)^{-1}\right),\ \text{as } \sigma \to 1+. \tag{3.16}$$

Fix some large constant K. As mentioned above, the divergence of (3.15) to infinity is uniform on $|t| \le K$. This implies that

$$\lim_{\sigma\to 1+} |F(s)|\cdot\zeta^{-1}(\sigma) = 0,\ \text{uniformly in } |t| \le K. \tag{3.17}$$

The proof of (3.17) is straightforward; the arguments were used, for example, in the proof of WIRSING's Theorem in Chapter II. First there is the product representation

$$F(s)\cdot\zeta^{-1}(\sigma) = \prod_p \left(1 + \frac{f(p)}{p^{\sigma+it}} - \frac{1}{p^\sigma} - \frac{f(p)}{p^{\sigma+s}}\right).$$

The product

$$\prod_p \left(1+\frac{f(p)}{p^{\sigma+it}}-\frac{1}{p^\sigma}-\frac{f(p)}{p^{\sigma+s}}\right)\cdot\left(1+\frac{f(p)}{p^{\sigma+it}}-\frac{1}{p^\sigma}\right)^{-1}$$

being convergent in Re $s \geq \frac{3}{4}$, it is sufficient to examine the product

$$P(\sigma) = \left|\prod_p \left(1+\frac{f(p)}{p^{\sigma+it}}-\frac{1}{p^\sigma}\right)\right| = \prod_p \left|1-\frac{1}{p^\sigma}\cdot\left(1-\frac{f(p)}{p^{it}}\right)\right|.$$

The maximum of the holomorphic function $(1+z)\cdot e^{-z}$, $|z| \leq r$, is taken at the boundary, and an easy determination of the maximum gives

$$|(1+z)\cdot e^{-z}| \leq \exp(|z|^2).$$

Therefore,

$$P(\sigma) \ll \exp\left\{-\sum_p \operatorname{Re}\left(\frac{1}{p^\sigma}\cdot\left(1-\frac{f(p)}{p^{it}}\right)\right)\right\} \to 0, \text{ as } \sigma\to 1+,$$

uniformly in $|t| \leq K$ because of the divergence of the series (3.5). So (3.17) is proved.

Splitting the range of integration $]-\infty,\infty[$ into $[-K,K]$ and two infinite intervals, and using (3.17) in $|t| \leq K$, we obtain, as $\sigma \to 1+$,

$$\int_{-\infty}^{\infty}\left|\frac{F(s)}{s}\right|^2 dt = \int_{-K}^{K}\left|\frac{F(s)}{s}\right|^{3/2}\left|\frac{F(s)}{\zeta(\sigma)}\right|^{\frac12}\left|\frac{\zeta(\sigma)}{s}\right|^{\frac12} dt + \int_{|t|\geq K}\left|\frac{F(s)}{s}\right|^2 dt$$

$$(3.18)\qquad = o\left((\sigma-1)^{-\frac12}\cdot\int_{-K}^{K}\left|\frac{F(s)}{s}\right|^{3/2}|s|^{-\frac12}dt\right) + \int_{|t|\geq K}\left|\frac{F(s)}{s}\right|^2 dt.$$

Our aim is to obtain the estimate $\int_{-\infty}^{\infty}\left|\frac{F(s)}{s}\right|^2 dt = o(\sigma-1)^{-1}$, as $\sigma \to 1+$. This result contradicts (3.16) [remember our assumption that $\alpha \neq 0$]. To achieve this result, the two integrals in the last line of equation (3.18) must be estimated sufficiently well. First, splitting the range of integration into intervals of length ≤ 2,

$$\int_{|t|\geq K}\left|\frac{F(s)}{s}\right|^2 dt \leq 4\cdot\sum_{|m|\geq K} m^{-2}\cdot\int_{|t-m|\leq 1}|F(s)|^2\, dt$$

$$\ll \sum_{|m|\geq K} m^{-2}\cdot\int_{-1}^{1}\left|\sum_{n=1}^{\infty} f(n)\cdot n^{-s}\cdot n^{-im}\right|^2 dt.$$

We assume that $1 < \sigma < 2$. Then we insert a factor $|s|^{-2} \geq \frac{1}{5}$ into the integral $\int_{-1}^{1} \ldots dt$, extending the range of integration to the whole interval $]-\infty, \infty[$, and use Lemma 3.4 to obtain

$$\int_{|t|\geq K}\left|\frac{F(s)}{s}\right|^2 dt \ll \sum_{|m|\geq K} m^{-2}\cdot\int_{-\infty}^{\infty}\left|s^{-1}\cdot\sum_{n=1}^{\infty} n^{-s}f(n)\cdot n^{-im}\right|^2 dt$$

$$<< \sum_{|m|\geq K} m^{-2} \cdot \int_0^\infty e^{-2w(\sigma-1)}\, dw << K^{-1} \cdot \tfrac{1}{2}(\sigma-1)^{-1}.$$

In order to estimate the integral $\int_{-K}^{K} \left| \frac{F(s)}{s} \right|^{3/2} \cdot |s|^{-\frac{1}{2}}\, dt$, we estimate $|F(s)|^{\frac{3}{4}}$ in the half-plane $\sigma \geq 1$. Similarly, as before, using the product representation of the generating DIRICHLET series F(s), we obtain

$$|F(s)|^{\frac{3}{4}} << \left|\exp\Big(\sum_p \tfrac{3}{4}\, f(p) \cdot p^{-s}\Big)\right| << \left| \sum_{n=1}^{\infty} (\tfrac{3}{4})^{\omega(n)}\, f(n) \cdot n^{-s} \right| = |H(s)|,$$

say. Therefore, using PARSEVAL's equation,

$$\int_{-K}^{K} \left| \frac{F(s)}{s} \right|^{3/2} |s|^{-\frac{1}{2}}\, dt << \int_{-\infty}^{\infty} |H(s)|^2 \cdot |s|^{-2}\, dt$$

$$= 2\pi \int_0^\infty \Big| e^{-w} \cdot \sum_{n\leq \exp(w)} (\tfrac{3}{4})^{\omega(n)}\, f(n)\Big|^2 \cdot e^{-2w(\sigma-1)} dw.$$

Using Lemma 3.4, we obtain

$$\int_{-K}^{K} \left| \frac{F(s)}{s} \right|^{3/2} |s|^{-\frac{1}{2}}\, dt << \int_0^\infty \exp\Big(- 2 \sum_{p\leq \exp(w)} \tfrac{1}{4}\, p^{-1}\Big) \cdot e^{-2w(\sigma-1)} dw$$

$$<< \int_{\log 2}^{\infty} w^{-\frac{1}{2}} \cdot e^{-2w(\sigma-1)}\, dw << (\sigma-1)^{-\frac{1}{2}},$$

and so

$$\int_{-\infty}^{\infty} \left| \frac{F(s)}{s} \right|^2 dt = o\, (\sigma-1)^{-1}, \text{ as } \sigma \to 1+,$$

in contradiction with (3.16). Thus, in case of the divergence of the series (3.5), the mean-value of f is zero. □

In our proof, f was assumed to have the property $f(p^k) = 0$, if $k \geq 2$. The deduction of the general case from this special result forms the content of Exercise 3.

IX.4. PROOF OF PROPOSITION 3.3.

Denoting by $p(n) = p_{max}(n)$ the greatest prime factor of n, we define

$$\mathcal{M}_y(f,t) = t^{-1} \cdot \sum_{n\leq t,\, p(n)\leq y} f(n). \tag{4.1}$$

In this chapter, the notation $\mathcal{M}(f,t) = t^{-1} \cdot \sum_{n\leq t} f(n)$ has been used already; so $\mathcal{M}(f,t) = \mathcal{M}_{\infty}(f,t)$. First, we estimate the constant $\alpha = \limsup_{x\to\infty} x^{-1} \cdot |\sum_{n\le x} f(n)|$ (see 3.7) from above.

Lemma 4.1. *For every* $y > 0$,

$$\alpha \le \prod_{p\le y} \left(1 - p^{-1}\right) \cdot \int_1^{\infty} |\mathcal{M}_y(f,t)|\, t^{-1}\, dt. \tag{4.2}$$

The **Proof** of this lemma is nearly the same as that given in II.9, where DABOUSSI's elementary proof of the prime number theorem was presented. The only difference is that the functions $M(x) = \sum_{n\le x} \mu(n)$ and $M_y(x)$ have to be replaced by summatory functions, where the MÖBIUS function μ is replaced by the function f. □

Lemma 4.2. *For any* $\beta > \alpha$, *as* $y \to \infty$, *the estimate*

$$\int_y^{\infty} |\mathcal{M}_y(f,t)|\, t^{-1}\, dt \le \beta \cdot (C-1) \cdot \log y + o(\log y) \tag{4.3}$$

is true, where C is the limit

$$C = \lim_{y\to\infty} \log^{-1} y \cdot \prod_{p\le y} \left(1-p^{-1}\right)^{-1}. \tag{4.4}$$

The limit (4.4) exists by elementary results from prime number theory (see I.6); note that the prime number theorem is not needed to prove the existence of this limit.

Proof. The function

$$\Lambda_f = f^{-1(*)} * (f \cdot \log)$$

has values $\Lambda_f(n) = 0$ if n is not a power of a prime, and

$$\Lambda_f(p^k) = (-1)^{k-1} \cdot (f(p))^k \cdot \log p$$

otherwise; obviously $|\Lambda_f(n)| \le \Lambda(n)$. Again, we have to apply the techniques used in DABOUSSI's proof, given in II.9. The definition and the results concerning the function h (see II.9.) will be used, and we proceed as in the proof of (II.9.4ii). The only difference is, again, that in sums using the MÖBIUS function this has to be replaced by the arithmetical function f. □

Thus, an elementary proof of the HALÁSZ Theorem is concluded; in contrast to HILDEBRAND's proof in section 1, 2, this time it is not necessary to use the prime number theorem.

IX.5. EXERCISES

1) Prove: if $t \neq 0$ is real, then the arithmetical function $n \mapsto n^{it}$ does not have a mean-value.

2) (a) Show that $\sum_{n \le x} n^{-1} \cdot \varphi(n) = 6\pi^{-2} \cdot x + o(x)$, as $x \to \infty$. [It is easy to obtain a sharper result].

(b) Use (a) and partial summation to show that the arithmetical function $n \mapsto n^{-1} \cdot \varphi(n) \cdot n^{it}$ does not have a mean-value if $t \neq 0$ is real.

3) (a) Assume that f is multiplicative, $|f| \le 1$, and in addition, $f(2^k) = 0$, for $k = 1, 2, 3, \ldots$. Use the relationship theorem of Chapter III, to show that $M(f) = 0$, if the series (3.5) is divergent for every real t.

(b) If f is any multiplicative function, for which $|f| \le 1$, and for which the series (3.5) is divergent for every real t. Then deduce a convolution representation $f = h * f_0$, where $\sum n^{-1} \cdot |h(n)| < \infty$, and where f_0 is a function satisfying the assumptions of part (a) of this exercise.

Use this representation to prove the HALÁSZ theorem for f.

4) The Prime Number Theorem $\psi(x) \sim x$ implies $M(\mu) = 0$.

Hints: use (and prove, if necessary)

$$\left| \sum_{n \le x} n^{-1} \cdot \mu(n) \right| \le 2 \text{ (see Chapter I, Corollary 2.5)},$$

an asymptotic evaluation for $\sum_{n \le x} \log(x/n)$, and

$$\sum_{n \le x} \mu(n) \cdot \log x = - \sum_{n \le x} M(n) \cdot \psi(x/n) + O(x).$$

Replace $\psi(x/n)$ by (x/n) + error, and get the desired result by a careful estimate of the resulting sums.

5) Let x_0, c_0 be positive constants, and let f be a real-valued arithmetical function with the property

$$|\mathcal{M}(f,y) - \mathcal{M}(f,x)| \le c_0 \cdot x^{-1} \cdot (y - x + 1), \text{ for all } x_0 \le x \le y.$$

Prove: If $\lim_{x \to \infty} |\mathcal{M}(f,x)|$ exists, then the limit $\lim_{x \to \infty} \mathcal{M}(f,x)$ exists.

N. G. DE BRUIJN

A. HILDEBRAND

K.-H. INDLEKOFER

G. TENENBAUM

J.-L. MAUCLAIRE

P. D. T. A. ELLIOTT

J. KNOPFMACHER

E. FOUVRY, G. TENENBAUM

H. DABOUSSI

P. BATEMAN,
J.-L. MAUCLAIRE

H. DELANGE

A. PERELLI, A. IVIĆ,
P. ERDÖS

J. KUBILIUS

L. LUCHT

Appendix

A.1. THE STONE-WEIERSTRASS THEOREM, TIETZE'S THEOREM

WEIERSTRASS's Approximation Theorem states that real-valued, continuous function F, defined on [-1, +1], can be uniformly approximated by polynomials. We state several standard extensions of this theorem to compact spaces. The results referred to may be found, for example, in HEWITT-STROMBERG [1965]. We use the following *notation*:

$\mathscr{C}^r(X)$ [resp. $\mathscr{C}(X)$] is the vector-space of *real-valued* [resp. complex-valued] continuous functions on the topological space X. Its topology is induced by the supremum norm $\|F\|_u = \sup_{x \in X} |F(x)|$.

A subset $\mathscr{S} \subset \mathscr{C}(X)$ *"separates the points of X"* if, for any given points $x, y \in X$, $x \neq y$, there is a function $F \in \mathscr{S}$ such that $F(x) \neq F(y)$.

Theorem A.1.1. [STONE-WEIERSTRASS Theorem] *Let* $X \neq \emptyset$ *be a compact HAUSDORFF space. Assume that* $\mathscr{S} \subset \mathscr{C}^r(X)$ *separates the points of* X, *and contains the constant function* $1 \colon x \mapsto 1$. *Then the* $\mathbb{R}$*-algebra of polynomials* p *with real coefficients in the functions of* $\mathscr{S}$,

$$p(x) = \sum_{i_1} \dots \sum_{i_k} a_{i_1, \dots, i_k} \cdot f_1^{i_1}(x) \cdot \dots \cdot f_k^{i_k}(x),$$

is $\|\,.\,\|_u$*-dense in* $\mathscr{C}^r(X)$.

A classical **example** is: $X = [a, b]$, $\mathscr{S} = \{1, x\}$.

Theorem A.1.2. *Let* $X \neq \emptyset$ *be a compact* HAUSDORFF *space, as before, and let* $\mathcal{S}$ *be a subset of* $\mathcal{C}(X)$ *separating the points of* X, *which contains the constant function* 1 **and** *which has the property that* $F \in \mathcal{S}$ *implies that the complex conjugate function* $\overline{F}$ *is in* $\mathcal{S}$, *too. Then the* $\mathbb{C}$-*algebra of polynomials* $\mathfrak{p}$ *with complex coefficients in the functions of* $\mathcal{S}$, *is* $\|.\|_u$-*dense in* $\mathcal{C}(X)$.

Corollary. (1) *Polynomials with real coefficients are* $\|.\|_u$-*dense in the algebra of all continuous real-valued functions on the compact interval* $[a,b] \subset \mathbb{R}$.

(2) *Polynomials with complex coefficients are* $\|.\|_u$-*dense in the algebra of continuous complex-valued functions on the torus* $\mathbb{T} = \{ z \in \mathbb{C}, |z| = 1 \}$.

(2') *Trigonometric polynomials*

$$\sum_{-N \le n \le N} \alpha_n \cdot \exp(2\pi i \cdot n \cdot x), \quad \alpha_n \in \mathbb{C},$$

are dense in the space of complex-valued, 1- *periodic functions on* $\mathbb{R}$.

Theorem A.1.3. [TIETZE's *Extension Theorem*]. *Let* $Y \neq \emptyset$ *be a compact subset of the locally compact* HAUSDORFF *space* X. *Let* U *be an open set "between"* Y *and* X, $Y \subset U \subset X$. *Given a continuous function* $f: Y \to \mathbb{C}$, *then there is a continuous extension*

$$F: X \to \mathbb{C} \text{ with compact support, } F|_Y = f,$$

vanishing outside U.

A.2. ELEMENTARY THEORY OF HILBERT SPACE

Let X be a HILBERT-space with inner product $\langle \ , \ \rangle$ and norm $\|e\| = \sqrt{\langle e,e \rangle}$. $E \subset X$ is called an *orthogonal* [resp. *orthonormal*] set if $\langle e,e' \rangle = 0$, if $e \neq e'$ both are in E [and, in case of orthonormality,

$\|e\| = 1$ for each $e \in E$]. The *FOURIER coefficient* of $x \in X$ with respect to $e \in E$ is denoted by

$$\hat{x}(e) = \langle x,e \rangle.$$

An orthonormal set E is *"complete"* if $\langle x,e \rangle = 0$ for every $e \in E$ implies $x = 0$. For example, the set of functions $x \mapsto \exp(2\pi i \cdot nx)$, $n = \dots, -2, -1, 0, 1, 2, \dots$, is complete in $L^2([0,1], \lambda)$, where λ is the LEBESGUE measure.

The GRAM-SCHMIDT orthonormalization process permits us to construct from any at most countable set $E^* = \{ e_1^*, e_2^*, \dots \}$ of linearly independent elements of X an orthonormal set $E = \{ e_1, e_2, \dots \}$ with the additional property

$$\mathrm{Lin}_{\mathbb{C}}\{e_1^*, \dots, e_n^*\} = \mathrm{Lin}_{\mathbb{C}}\{e_1, \dots, e_n\} \text{ for } n = 1, 2, \dots .$$

If $X \supsetneq \{0\}$ contains a *dense*, countable, linearly independent set, then the GRAM-SCHMIDT orthonormalization process leads to a *complete* orthonormal set E in X.

BESSEL's Inequality. *Given an orthonormal set* $E \neq \emptyset$ *in an inner-product space* X. *Then*

$$\sum_{e \in E} | \langle x,e \rangle |^2 \le \|x\|^2,$$

and so the set $\{ e \in E, \hat{x}(e) \neq 0 \}$ *is countable.*

Theorem A.2 1. *Assume that* X *is a HILBERT space, and* $E \subset X$ *an orthonormal set. Then the following properties are equivalent:*

(i) E *is complete.*

(ii) *The smallest linear subspace of* X *containing* E *is dense in* X.

(iii) *For every* $x \in X$, *PARSEVAL's equation is true:*

$$\|x\|^2 = \sum_{e \in E} | \langle x,e \rangle |^2.$$

The series contains at most countably many non-zero summands.

(iv) *For every* x, y *the [generalized] PARSEVAL equation holds:*

$$\langle x,y \rangle = \sum_{e \in E} \langle x,e \rangle \cdot \overline{\langle y,e \rangle}.$$

(v) *Every* $x \in X$ *has a* FOURIER *series*

$$x \sim \sum_{e \in E} \langle x,e \rangle \cdot e$$

with at most countably many non-zero coefficients $\langle x,e \rangle$ *[the corresponding* e*'s are denoted by* $e_1, e_2, \ldots$ *], and*

$$\lim_{N \to \infty} \| x - \sum_{n \le N} \langle x,e_n \rangle \cdot e_n \| = 0.$$

Theorem A.2.2. *Let* F *be a bounded linear functional* $F: X \to \mathbb{C}$, *defined on the* HILBERT *space* X. *Then there exists a unique element* $y \in X$ *"representing"* F:

$$F(x) = \langle x,y \rangle \text{ for every } x \in X.$$

Moreover, the operator norm $\|F\| = \sup_{\|x\| = 1} |F(x)|$ *equals* $\|y\|$.

Theorem A.2.3. *Assume that* X *is a* HILBERT *space. The set* $\mathcal{B}(X)$ *of all bounded linear operators* $T : X \to X$ *is [with composition as multiplication] a* BANACH *algebra with unit element. For every* $T \in \mathcal{B}(X)$ *there is a unique "adjoint"* $T^* \in \mathcal{B}(X)$, *defined by*

$$\langle T(x),y \rangle = \langle x,T^*(y) \rangle \text{ for every } x,y \in X.$$

Moreover, $T^{**} = T$, *and [the operator norm]* $\|T^*\| = \|T\|$.

Some properties of the adjoint operator are listed below:

(i) $(T_1 + T_2)^* = T_1^* + T_2^*$

(ii) $(\alpha \cdot T)^* = \overline{\alpha} \cdot T^*$

(iii) $(T_1 \circ T_2)^* = T_2^* \circ T_1^*$

(iv) $\|T^* \circ T\| = \|T\|^2$

(v) If $\langle T(x),y \rangle = 0$ for all $x,y \in X$, then $T = 0$

(vi) If $\langle T(x),x \rangle = 0$ for every $x \in X$, then $T = 0$.

A. 3. INTEGRATION

Our standard reference are the books by W. RUDIN [1966] and HEWITT-STROMBERG [1965].

Given a *measure space* $(X, \mathcal{A}, \mu)$, where $\mathcal{A}$ is a σ-algebra of subsets of X, and μ is a non-negative measure, a function $f: X \to \mathbb{C}$ is termed *measurable* if $f^{-1}(\mathcal{O})$ is in $\mathcal{A}$ for every open set $\mathcal{O}$ in $\mathbb{C}$. A [measurable] simple function is a finite linear combination of characteristic functions of [measurable] sets. The integral of a measurable simple function [over a set E in $\mathcal{A}$] can be defined in an obvious way as a finite sum. If $f: X \to [0, \infty]$ is measurable, its integral [over $E \in \mathcal{A}$] is defined as

$$\int_E f \, d\mu = \sup_{0 \le s \le f,\ s \text{ simple, measurable}} \int_E s \, d\mu .$$

The extension of this definition to functions $f: X \to \mathbb{C}$ is done via linearity.
A property is said to "hold almost- everywhere" if the set of points, where it does not hold, has measure zero.

Concerning countable limit operations, there are important convergence theorems:

Lebesgue's Monotone Convergence Theorem. *Assume that* $\{f_n\}$ *is a sequence of measurable functions* $f_n : X \to \mathbb{R}$, *satisfying* $0 \le f_1(x) \le f_2(x) \le \ldots \le \infty$ *and* $\lim_{n \to \infty} f_n(x) = f(x)$ *for almost every* $x \in X$. *Then* f *is measurable, and*

$$\lim_{n \to \infty} \int_X f_n \, d\mu = \int_X f \, d\mu .$$

Fatou's Lemma. *If* $f_n : X \to [0, \infty]$ *is a sequence of measurable functions, then*

$$\int_X \liminf_{n \to \infty} f_n \, d\mu \le \liminf_{n \to \infty} \int_X f_n \, d\mu .$$

Lebesgue's Dominated Convergence Theorem. *Let* $\{f_n\}$ *be a sequence of complex-valued measurable functions defined a.e. on* X *with the property*

$$\sum_{n=1}^{\infty} \int_X |f_n| \, d\mu < \infty.$$

(This is equivalent to $\int_X \sum_{n=1}^{\infty} |f_n| \, d\mu < \infty$.) *Then the series*

$$\sum_{n=1}^{\infty} f_n(x) = f(x)$$

is convergent for almost all $x \in X$ *to a measurable function* f, *and*

$$\int_X \sum_{n=1}^{\infty} f_n \, d\mu = \sum_{n=1}^{\infty} \int_X f_n \, d\mu.$$

A useful discrete version of the Dominated Convergence Theorem is the following result.

Corollary. *Let a sequence* $f_N\colon \mathbb{N} \to \mathbb{C}$, $N = 1, 2, \ldots,$ *of arithmetical functions be given, and let* $F\colon \mathbb{N} \to [0, \infty]$ *be a function satisfying* $\sum_{n=1}^{\infty} F(n) < \infty$. *If*

$$|f_N| \le F \text{ for every } N \in \mathbb{N},$$

and if the pointwise limits $\lim_{N\to\infty} f_N(n)$ *exist for every integer* n, *then the limit* $\lim_{N\to\infty} \sum_{n=1}^{\infty} f_N(n)$ *exists, and*

$$\lim_{N\to\infty} \sum_{n=1}^{\infty} f_N(n) = \sum_{n=1}^{\infty} \lim_{N\to\infty} f_N(n).$$

An important tool for the construction of integrals is the

Theorem of F. Riesz. *Let* Λ *be a monotonic* $(x \le y \Rightarrow \Lambda(x) \le \Lambda(y))$ *linear functional on the vector-space of continuous functions on the locally compact* Hausdorff *space* X *with compact support. Then there is*

- *a* σ*-algebra* $\mathcal{A}$ *of subsets of* X, *containing all the* Borel *sets of* X, *and*
- *a [positive] measure* μ *on* $\mathcal{A}$

such that

$$\Lambda(f) = \int_X f \, d\mu$$

for every continuous function f *on* X *with compact support.*

In addition, the measure has the following properties:

(a) $\mu(K) < \infty$ *for compact sets* K.

(b) μ *is outer regular, which means* $\mu(E) = \inf_{V \supset E,\ V \text{ open}} \mu(V)$ *for any* E *in* $\mathcal{A}$.

(c) μ *is inner regular, which means* $\mu(E) = \sup_{K \subset E,\ K \text{ compact}} \mu(K)$ for E in $\mathcal{A}$

if E *is open* ***or if*** E *has finite measure* ***or if*** X *is* σ*-compact* (X *is a countable union of compact sets*) *and* HAUSDORFF.

(d) μ *is complete: a subset of a set* E *in* $\mathcal{A}$ *with measure zero is in* $\mathcal{A}$ *again and has measure zero.*

Theorem of FUBINI. *Let* $(X, \mathcal{A}, \mu)$ *and* $(Y, \mathcal{B}, \nu)$ *be* σ*-finite* (*a countable union of finite measurable sets*) *measure spaces. If* f *is a complex-valued* $(\mathcal{A} \times \mathcal{B})$*-measurable function, defined (a.e.) on* $X \times Y$, *which satisfies*

$$\int_{X\times Y} |f| \, d(\mu\times\nu) < \infty \ \textbf{\textit{or}} \ \int_X \Big(\int_Y |f| \, d\nu\Big) d\mu < \infty \ \textbf{\textit{or}} \ \int_Y \Big(\int_X |f| \, d\mu\Big) d\nu < \infty,$$

then

$$\int_{X\times Y} f \, d(\mu\times\nu) = \int_X \Big(\int_Y f \, d\nu\Big) d\mu = \int_Y \Big(\int_X f \, d\mu\Big) d\nu \ .$$

The product-σ-algebra $\mathcal{A} \times \mathcal{B}$ is generated by the measurable rectangles $E \times F$, where $E \in \mathcal{A}$, $F \in \mathcal{B}$.

A.4. TAUBERIAN THEOREMS
(HARDY-LITTLEWOOD-KARAMATA, LANDAU-IKEHARA)

A good reference for the topics dealt with in this section is WIDDER [1946]. The HARDY-LITTLEWOOD-KARAMATA Tauberian Theorem will be formulated in a version for LAPLACE-Integrals, from which a version for DIRICHLET series, as well as for power series, will be deduced. We need the notion of a *slowly oscillating* function

$$L\colon [\, a,\infty [\ \to \mathbb{R}.$$

L *is called slowly oscillating if* L *is continuous, positive, and satisfies*

$$\lim_{x\to\infty} L(cx) \,\big/\, L(x) = 1$$

for every c *in* $0 < c < \infty$.

For example, the functions $x \mapsto (\log x)^k$, $x \mapsto \log\log x$, $x \mapsto \exp\left(\sqrt{\log x}\right)$ are slowly oscillating.

Theorem A.4.1 (HARDY-LITTLEWOOD-KARAMATA). *Assume that* A(.) *is a real-valued, non-decreasing function defined on the interval* $[0,\infty[$, *and* $A(0) = 0$. *Let the* LAPLACE-*Integral*

$$\mathcal{L}(\sigma) = \sigma \cdot \int_0^\infty A(u) \cdot e^{-\sigma u}\, du$$

be convergent for any $\sigma > 0$, *and suppose that for some slowly oscillating function* L *and some* $\tau \geq 0$ *the relation*

$$\lim_{\sigma \to 0+} \mathcal{L}(\sigma) \cdot \left\{ \sigma^{-\tau} \cdot L(\sigma^{-1}) \right\}^{-1} = \gamma$$

holds (where $L(\sigma^{-1}) \nearrow \infty$ *in case that* $\tau = 0$*). Then, as* $x \to \infty$,

$$A(x) \sim \frac{\gamma}{\Gamma(\tau+1)} \cdot x^\tau \cdot L(x).$$

For a proof, see, for example, HARDY [1949], p.166, or SCHWARZ [1969].

Partial integration gives

$$\sum_{n=1}^{\infty} a_n \cdot n^{-\sigma} = -\int_1^\infty \sum_{n<u} a_n \cdot (-\sigma) \cdot u^{-\sigma-1}\, du = \sigma \cdot \int_0^\infty \Big(\sum_{n<\exp(v)} a_n \Big) \cdot e^{-\sigma v} dv,$$

and so we obtain the following Corollary.

Corollary A.4.2. *Assume that* $\tau \geq 0$ *and that the* DIRICHLET *series*

$$\mathcal{D}(\sigma) = \sum_{n=1}^{\infty} a_n \cdot n^{-\sigma}$$

with non-negative coefficients a_n *is convergent for* $\sigma > 0$, *and assume that*

$$\lim_{\sigma \to 0+} \mathcal{D}(\sigma) \cdot \left\{ \sigma^{-\tau} \cdot L(\sigma^{-1}) \right\}^{-1} = \gamma$$

(where $L(\sigma^{-1}) \nearrow \infty$ *in case that* $\tau = 0$*). Then*

$$\sum_{n \leq x} a_n \sim \frac{\gamma}{\Gamma(\tau+1)} \cdot (\log x)^\tau \cdot L(\log x), \text{ as } x \to \infty.$$

Theorem A.4.3 (LANDAU-IKEHARA). *Let* A(.) *be a real-valued, monotonically non-decreasing function defined on the interval* $[0,\infty[$, *and let* $A(0) = 0$. *Assume that the* LAPLACE-*Integral*

$$\mathcal{L}(s) = s \cdot \int_0^\infty A(u) \cdot e^{-su}\, du$$

is convergent in the half-plane Re s > 1. *Suppose that for every* $\lambda > 0$ *the expression*

$$(\sigma+it)^{-1} \cdot \mathcal{L}(\sigma+it) - a \cdot (\sigma+it-1)^{-1}$$

converges to a [continuous] function h(t), *uniformly in* $|t| \le 2\lambda$. *If* $a \neq 0$, *then*

$$\lim_{x \to \infty} e^{-x} \cdot A(x) = a.$$

A far-reaching generalization of this theorem may be found, for example, in H. DELANGE [1954]. For TAUBERIAN Theorems with remainder term, see, for example, GANELIUS [1971].

A.5. THE CONTINUITY THEOREM FOR CHARACTERISTIC FUNCTIONS

A standard reference for this section is LUKACS [1970], *Characteristic Functions*, Section 2,3. A function $F: \mathbb{R} \to \mathbb{R}$ is a *"distribution function"*, if it is monotonically non-decreasing, continuous from the right, and satisfies $F(-\infty) = 0$, $F(+\infty) = 1$. Its FOURIER-STIELTJES transform

$$f(t) = \int_{-\infty}^{+\infty} e^{itx}\, dF(x)$$

is called the *characteristic function* of the distribution function F. Simple properties are given in the following theorem.

Theorem A.5.1. (i) *If* F *is a distribution function with characteristic function* f, *then* $f(0) = 1$, $|f| \le 1$, *and* $f(-t) = \overline{f(t)}$.

(ii) *Any characteristic function is uniformly continuous on* $\mathbb{R}$.

(iii) *A linear combination of characteristic functions* $f_1, \ldots, f_n$ *with coefficients* $a_1, \ldots, a_n$, *satisfying* $0 \le a_\nu \le 1$, $\sum_{1 \le \nu \le n} a_\nu = 1$, *is a characteristic function again. In particular,* $\mathrm{Re}\, f = \frac{1}{2} f + \frac{1}{2} \overline{f}$ *is a characteristic function if* f *is.*

(iv) *The distribution functions* F_1 *and* F_2 *are identical if and only if their characteristic functions* f_1 *and* f_2 *are equal.*

(v) *The distribution function* F *is the convolution of* F_1 *and* F_2,

$$F(x) = \int_{-\infty}^{\infty} F_1(x-\xi) \cdot dF_2(\xi)$$

if and only if $f = f_1 \cdot f_2$ *for the corresponding characteristic functions.*

So: if f_1, f_2 *are characteristic functions, then* $f_1 \cdot f_2$ *and* $|f_1|^2$ *are characteristic functions also.*

(vi) *A distribution function* F *is purely discrete [this means, that* $F(x) = \sum_k \alpha_k \cdot \delta(x-\xi_k)$ *with non-negative coefficients* α_k, *satisfying* $\sum \alpha_k = 1$; δ *is the function, defined by* $\delta(x) = 0$ if $x < 0$, $\delta(x) = 1$ *for* $x \geq 0$] *if and only if its characteristic function is almost periodic on* $\mathbb{R}$.

The determination of the distribution function F from a given characteristic function f is possible via the following theorem.

Theorem A.5.2 [Inversion Theorem]. *If* f *is the characteristic function of the distribution function* F, *and if* x+h *and* x *are points of continuity for* F, *then*

$$F(x+h) - F(x) = \lim_{T \to \infty} \frac{1}{2\pi} \cdot \int_{-T}^{T} \frac{1-e^{-ith}}{it} \cdot e^{-itx} \cdot f(t)\, dt.$$

For applications to arithmetical functions, the following Continuity Theorem is of importance. A sequence of functions F_n is said to *converge weakly* to F if

$$\lim_{n \to \infty} F_n(x) = F(x)$$

for every point of continuity of F. Note that the set of discontinuity points of a distribution function is at most denumerable.

Theorem A.5.3 [Continuity theorem]. *Let* $\{F_n\}$ *be a sequence of distribution functions and denote by* $\{f_n\}$ *the corresponding sequence of characteristic functions. Then* F_n *converges weakly to a distribution function* F *if and only if the sequence* $f_n(t)$ *converges pointwise to a function* f(t) *which is continuous at* t = 0. *The limit function* $f(t) = \lim_{n \to \infty} f_n(t)$ *is then the characteristic function of* F.

Remark. If the sequence $\{f_n(t)\}$ of characteristic functions converges [pointwise] to a characteristic function f(t), then the convergence is uniform in every finite interval $-T \le t \le T$.

A.6. GELFAND'S THEORY OF COMMUTATIVE BANACH ALGEBRAS

A commutative BANACH-Algebra $\mathcal{A}$ is a vector-space over $\mathbb{C}$,

- with a multiplication $(x, y) \mapsto x \cdot y$, which is associative, commutative and distributive with addition (and is compatible with scalar multiplication, $\alpha\cdot(x\cdot y) = x\cdot(\alpha y) = (\alpha x)\cdot y$),
- with a norm $\|.\|: \mathcal{A} \to [0,\infty[$ making $\mathcal{A}$ into a normed vector-space and which satisfies the inequality $\|x \cdot y\| \le \|x\| \cdot \|y\|$,
- and which is complete with respect to the norm $\|.\|$.

We always assume in this section that $\mathcal{A}$ has a unit element e, and that $\|e\| = 1$. An element x is invertible if there exists an $x^{-1} \in \mathcal{A}$ with the property $x\cdot x^{-1} \left[= x^{-1}\cdot x \right] = e$. The set $\mathcal{G}$ of invertible elements of $\mathcal{A}$ is open, and the map $x \mapsto x^{-1}$ is an homeomorphism. The **spectrum** of an element x in $\mathcal{A}$,

$$\operatorname{spec}(x) = \left\{ \lambda \in \mathbb{C} \ ; \ x-\lambda e \text{ is **not** invertible} \right\},$$

is a non-empty, compact set in $\mathbb{C}$, $|\lambda| \le \|x\|$ for all $\lambda \in \operatorname{spec}(x)$. The spectral radius is

$$\rho(x) := \sup_{\lambda \in \operatorname{spec}(x)} |\lambda|,$$

and the *spectral radius formula* states

$$\rho(x) = \lim_{n\to\infty} \|x^n\|^{1/n}.$$

We denote the *set of algebra-homomorphisms* defined on $\mathcal{A}$ by

$$\Delta \ [= \Delta_{\mathcal{A}}] = \{ h: \mathcal{A} \to \mathbb{C}, \text{ h is an algebra-homomorphism} \}.$$

Then the following assertions are true (see, for example, RUDIN [1966], 18.17):

(i) *If* m *is a maximal ideal of* $\mathcal{A}$, *then there is some* h *in* Δ *such that*

$$m = \text{kernel}(h).$$

(ii) λ *is in* spec(x) *if and only if there is some* $h \in \Delta$ *for which* $h(x) = \lambda$.

(iii) x *is invertible in* $\mathcal{A}$ *if and only if* $h(x) \neq 0$ *for every* h *in* Δ.

(iv) *If* $x \in \mathcal{A}$, $h \in \Delta$, *then* $h(x) \in \text{spec}(x)$.

(v) $|h(x)| \leq \rho(x) \leq \|x\|$.

(vi) *Every* h *in* Δ *is continuous.*

According to these results the set Δ of algebra-homomorphisms is in 1-1-correspondence with the set of maximal ideals, and so Δ is often referred to as *"the maximal ideal space"* of $\mathcal{A}$.

The **radical** of $\mathcal{A}$ is the intersection of all its maximal ideals, and $\mathcal{A}$ is termed **semi-simple** if the radical of $\mathcal{A}$ is the null-ideal, $\text{rad}(\mathcal{A}) = \{0\}$.

The **GELFAND transform** of an element $x \in \mathcal{A}$ is the map

$$\hat{x} : \Delta \to \mathbb{C}, \text{ defined by } \hat{x}(h) = h(x);$$

so the GELFAND transform ^ is a map

$$\hat{} : \mathcal{A} \to \hat{\mathcal{A}} = \{ \hat{x} : \Delta \to \mathbb{C},\ x \in \mathcal{A} \}.$$

The GELFAND **topology** on Δ is the weakest topology making all the functions $\hat{x}$ continuous. With this topology, Δ becomes a **compact** HAUSDORFF space.

A $\mathbf{B^*}$**-algebra** is a [commutative] BANACH algebra (with unit element e) with an involution *: $\mathcal{A} \to \mathcal{A}$, $x \mapsto x^*$, satisfying

$$\| x \cdot x^* \| = \| x \|^2$$

in addition to the usual conditions for an involution

$$(x+y)^* = x^* + y^*,\ (\lambda \cdot x)^* = \bar{\lambda} \cdot x^*,\ (x \cdot y)^* = y^* \cdot x^*,\ (x^*)^* = x.$$

If $\mathcal{A}$ is commutative and semi-simple, then every involution is continuous.

Theorem of GELFAND and NAIMARK. *If* $\mathcal{A}$ *is a commutative* B^**-algebra with maximal ideal space* Δ, *then the* GELFAND *transform ^ is an isometric isomorphism from* $\mathcal{A}$ *onto the space of continuous functions on* Δ,

$$\hat{}: \mathcal{A} \to \mathcal{C}(\Delta).$$

In addition, for $x \in \mathcal{A}$, $h \in \Delta$, *the relation*

$$h(x^*) = \overline{h(x)}, \text{ equivalent with } (x^*)^\wedge = (\hat{x})^-,$$

holds.

A little more detailed (see RUDIN [1973], 11.12): *assume that* $\mathcal{A}$ *is a commutative* BANACH *algebra. Then*

(a) *The map* $x \mapsto \hat{x}$ *is an isometry if and only if* $\|x^2\| = \|x\|^2$ *for every* x *in* $\mathcal{A}$.

(b) $\mathcal{A}$ *is semi-simple and* $\mathcal{A}^\wedge$, *the image* $\mathcal{A}$ *of under the* GELFAND *transform, is closed in* $\mathcal{C}(\Delta_{\mathcal{A}})$ *if and only if there is some constant* $K < \infty$, *so that* $\|x\|^2 \le K \cdot \|x^2\|$ *for every* x *in* $\mathcal{A}$.

Corollary. *If* $\mathcal{A}$ *is a commutative* B^**-algebra, and if* $x = x^*$, *and if there is no* $\lambda \le 0$ *in* spec(x), *then there exists a square-root of* x:

$$\exists\, y \in \mathcal{A},\ y = y^*, \text{ and } y^2 = x.$$

A.7. INFINITE PRODUCTS

References for this section consist of textbooks of analysis or the theory of functions, for example the books by E. C. TITCHMARSH or J. B. CONWAY. Denote the partial products of the product $P = \prod_{1 \le \nu < \infty} \left(1 + a_\nu\right)$ (with complex a_ν) by $P_n = \prod_{1 \le \nu \le n} \left(1 + a_\nu\right)$. Assume, for simplicity, that all the factors $\left(1 + a_\nu\right)$ of P are non-zero. The product P is said to be convergent if $\lim_{n \to \infty} P_n$ exists **and is $\neq$ 0.** A necessary condition for convergence is $\lim_{n \to \infty} a_n = 0$. P is said to be absolutely convergent if $\prod \left(1 + |a_\nu|\right)$ is convergent; absolute convergence implies convergence, and absolute convergence is equivalent to the absolute convergence of the series $\sum |a_n|$. The factors of an absolutely convergent product may be taken in any order. The convergence of $\sum_n |u_n(z)|$ in some (closed) region of $\mathbb{C}$ is a sufficient condition for the uniform convergence of the product $\prod_n \left(1+u_n(z)\right)$.

By taking logarithms, infinite products are connected with infinite series if some precautions are taken. Denote by log the principal branch of the logarithm function.

Assume that $\mathrm{Re}(1+a_n) > 0$ *for all* n. *Then* $\prod_n \left(1+a_n\right)$ *converges to a non-zero limit if and only if the series* $\sum_n \log\left(1+a_n\right)$ *is convergent.*

If $\mathrm{Re}\, a_n > -1$, *then the series* $\sum_n \log\left(1+a_n\right)$ *converges absolutely if and only if the series* $\sum_n a_n$ *is absolutely convergent.*

If $\mathrm{Re}\, a_n > 0$, *then the product* $\prod_n \left(1+a_n\right)$ *is absolutely convergent if and only if the series* $\sum_n a_n$ *is absolutely convergent.*

Finally, we give a result on infinite products which is useful in number theory.

Theorem A.7.1. *Assume that the two series*

$$\sum_p p^{-1}\cdot a_p, \quad \sum_p p^{-2}\cdot |a_p|^2$$

where p *runs over the primes (in ascending order), and where the* a_p *are complex numbers, are convergent. Let*

$$\sigma \mapsto g(p,\sigma) : [1, 1+\delta] \to \mathbb{C}$$

be a continuous functions satisfying

$$|g(p,\sigma)| \le b_p, \text{ and } \sum_p b_p < \infty.$$

Then

(a) *the product* $\prod_p \left(1+\frac{a_p}{p^\sigma}+g(p,\sigma)\right)$ *is convergent for every* $\sigma \in [1, 1+\delta]$, *and,*

(b) *if it is supposed in addition, that* $A = \lim_{\sigma\to 1+} \prod_p \left(1+\frac{a_p}{p^\sigma}+g(p,\sigma)\right)$ *exists, then*

$$A = \prod_p \left(1+\frac{a_p}{p}+g(p,1)\right).$$

Proof. In $|z| \le \frac{1}{2}$, write $1+z = \exp\left(z+R(z)\right)$, where $R(z) = \log(1+z)-z = \mathcal{O}(|z|^2)$. Choose a prime p_0 so large that $|p^{-1}\cdot a_p| + b_p \le \frac{1}{2}$ for every $p \ge p_0$. Then

$$\left| \frac{a_p}{p^\sigma} + g(p,\sigma) \right| \le \tfrac{1}{2} \text{ for } p \ge p_0, \text{ and } \sigma \in [1, 1+\delta].$$

Then, for $p_1 \ge p_0$ and $\sigma \in [1, 1+\delta]$,

$$\prod_{p_0 \le p \le p_1} \left(1 + \frac{a_p}{p^\sigma} + g(p,\sigma)\right)$$

$$= \exp\left\{ \sum_{p_0 \le p \le p_1} \left(\frac{a_p}{p^\sigma} + g(p,\sigma)\right)\right\} \exp\left\{ \sum_{p_0 \le p \le p_1} R\left(\frac{a_p}{p^\sigma} + g(p,\sigma)\right)\right\}.$$

The convergence of $\sum_p p^{-1} \cdot a_p$ implies uniform convergence of $\sum_p p^{-\sigma} \cdot a_p$ in $\sigma \ge 1$, and $\lim_{\sigma \to 1+} \sum_p p^{-\sigma} \cdot a_p = \sum_p p^{-1} \cdot a_p$. The assumptions of the theorem imply the uniform convergence of $\sum_p g(p,\sigma)$ and $\sum_p R\left(p^{-\sigma} \cdot a_p + g(p,\sigma)\right)$ in $1 \le \sigma \le 1+\delta$, and we obtain the assertion (a) by letting p_1 tend to infinity and σ to $1+$. (b) is then obvious. □

A.8. THE LARGE SIEVE

References for this section are, for example, E. BOMBIERI, *Le grand crible dans la théorie analytique des nombres,* astérisque 18 (1974), H. DAVENPORT [1967], H. HALBERSTAM & K. F. ROTH [1966], M. N. HUXLEY [1972], H. L. MONTGOMERY [1971], H. L. MONTGOMERY & R. C. VAUGHAN, *The Large Sieve,* Mathematika **20**, 119–135 (1973), H.-E. RICHERT, *Sieve Methods,* Bombay 1976, W. SCHWARZ, *Einführung in Siebmethoden der analytischen Zahlentheorie,* Bibl. Inst. (1974).

We only need **one** aspect of the "Large Sieve", namely an estimate of an exponential sum

$$S(x) = \sum_{M < n \le M+N} a_n \cdot \exp(2\pi i \cdot n \cdot x)$$

in the mean, taken over well-spaced points.

Theorem A.8.1. *Let* $x_1, x_2, \dots x_R$, *where* $R \ge 2$, *be real numbers, distinct* modulo one. *Put*

(A.8.1) $$\delta = \min_{r \ne s} \| x_r - x_s \| ,$$

where $|||\,.\,|||$ denotes the distance to the nearest integer. Then

$$\sum_{1\le r\le R} |S(x_r)|^2 \le (\pi N + \delta^{-1}) \cdot \sum_{M<n\le M+N} |a_n|^2.$$

The expression $(\pi N + \delta^{-1})$ on the right-hand side may be replaced by $(N + \delta^{-1})$, as was shown by H. L. MONTGOMERY and R. C. VAUGHAN. A rather simple proof of theorem 8.1 may be obtained using GALLAGHER's Lemma, as follows.

Lemma A.8.2. *If* f *is a continuously differentiable function,* δ *is as above, and* $X + \frac{1}{2}\delta \le x_r \le Y - \frac{1}{2}\delta$*, then*

$$\sum_{1\le r\le R} |f(x_r)|^2 \le \delta^{-1} \cdot \int_X^Y |f(x)|^2\,dx + \Big(\int_X^Y |f(x)|^2\,dx\Big)^{\frac{1}{2}} \cdot \Big(\int_X^Y |f'(x)|^2\,dx\Big)^{\frac{1}{2}}.$$

Theorem A.8.1, specialized to rational numbers a/q, immediately gives the following theorem.

Theorem A.8.3.

$$\sum_{q\le Q}\ \sum_{1\le a\le q,\ \gcd(a,q)=1} \Big|S\Big(\frac{a}{q}\Big)\Big|^2 \le (\pi N + Q^2)\cdot \sum_{M<n\le M+N} |a_n|^2.$$

The factor $(\pi N + Q^2)$ may be replaced by $(N + Q^2)$.

Another kind of "Large Sieve-inequality" *with weights* is the following theorem.

Theorem A.8.4. *Put* $\delta_r = \min_{s \ne r} |||\, x_r - x_s\, |||$. *Then*

$$\sum_{1\le r\le R} \Big(N + \frac{3}{2} \cdot \delta_r^{\,-1}\Big)^{-1} \cdot \big| S(x_r)\big|^2 \le \sum_{M<n\le M+N} |a_n|^2,$$

and

$$\sum_{q\le Q} \Big(N + \frac{3}{2} \cdot qQ\Big)^{-1} \cdot \sum_{1\le a\le q,\ \gcd(a,q)=1} \Big|S\Big(\frac{a}{q}\Big)\Big|^2 \quad \sum_{M<n\le M+N} |a_n|^2.$$

A.9. DIRICHLET SERIES

A convenient reference for this section is E. C. TITCHMARSH, *The Theory of Functions*, Oxford 1932. The well-known ABEL Theorem for power series: *If* $\sum a_n = s$ *is convergent, then* $f(x) = \sum a_n \cdot x^n$ *is uniformly convergent in* $0 \le x \le 1$, *and* $\lim_{x \to 1-} f(x) = s$, which may be formulated for complex x, too (uniform convergence is then true in some angle), has a counterpart for the DIRICHLET series.

Continuity Theorem for DIRICHLET Series.

If the DIRICHLET series $\sum a_n \cdot n^{-s}$ *is convergent in* $s = s_0$, *then it is uniformly convergent in the STOLZ angle* $|\arg(s - s_0)| \le \frac{1}{2}\pi - \delta$, *where* $0 < \delta < \frac{1}{2}\pi$ *is fixed.*

The region of convergence and of absolute convergence of a DIRICHLET series $\sum_{n=1}^{\infty} a_n \cdot n^{-s}$ is a half-plane (these half-planes may differ). The abscissa of absolute convergence is

$$\bar{\sigma} = \limsup_{n \to \infty} (\log n)^{-1} \cdot \log\left(|a_1| + \ldots + |a_n|\right).$$

If all the coefficients a_n are non-negative, then the real point on the vertical boundary of the half-plane of convergence is a singularity of the DIRICHLET series. The product of absolutely convergent DIRICHLET series $\sum_{n=1}^{\infty} a_n \cdot n^{-s}$, $\sum_{n=1}^{\infty} b_n \cdot n^{-s}$, is the DIRICHLET series $\sum_{n=1}^{\infty} c_n \cdot n^{-s}$ with coefficients

$$c_n = \sum_{d|n} a_d \cdot b_{n/d}.$$

The **uniqueness theorem** states that DIRICHLET series $\sum a_n \cdot n^{-s}$, $\sum b_n \cdot n^{-s}$, which are identical on some [small] region of $\mathbb{C}$, have the same coefficients.

The "method of complex integration" is based on PERRON's formula. *If* $c > 0$, x *is not an integer, and* $\sigma > \sigma_0 - c$, *where* σ_0 *is the abscissa of convergence of the DIRICHLET series* $f(s) = \sum_{n=1}^{\infty} a_n \cdot n^{-s}$, *then*

$$\sum_{n \le x} a_n \cdot n^{-s} = \{2\pi i\}^{-1} \cdot \int_{c-i\infty}^{c+i\infty} f(s+w)\, w^{-1} \cdot x^{w}\, dw.$$

Bibliography

APOSTOL, T. M. [1976], *Introduction to Analytic Number Theory,* New York · Heidelberg · Berlin 1976

BABU, G. J. [1972], *On the distribution of additive arithmetical functions of integral polynomials,* Sankhyā A **34**, 323-334 (1972)

BABU, G. J. [1973a,b], *Some results on the distribution of additive arithmetical functions, II,* Acta Arithm. **23**, 315-328, 1973, *III*, Acta Arithm. **25**, 39-49, (1973)

BABU, G. J. [1978], *Probabilistic methods in the theory of arithmetic functions,* New Delhi, vii + 118 pp, 1978

BAKSTYS, A. [1968], *On the asymptotic distribution of multiplicative numbertheoretic functions* [Russian], Litovsk. Mat. Sbornik **8**, 5-20 (1968)

BARBAN, M. B. [1966], *The "large sieve" method and its applications to number theory,* Usp. Mat. Nauk **21**, 51-102 (1966), Russ. Math. Surveys **21** , 49-104, (1966)

BEDIN, J. F. & DEUTSCH, Ch. [1972], *Valeur moyenne de fonctions multiplicatives généralisées,* C. R. Acad. Paris **275**, A 245-247, (1972)

BELLMAN, R. [1950] *Ramanujan sums and the average value of arithmetical functions,* Duke Math. J. **17**, 159-168 (1950)

BESICOVICH, A. S. [1932], *Almost periodic functions,* Cambridge 1932

BIEBERBACH, L. [1955], *Analytische Fortsetzung,* Berlin - Göttingen - Heidelberg 1955

BOMBIERI, E. [1965], *On the large sieve,* Mathematika **12**, 201-225, (1965)

BOMBIERI, E. [1974], *Le grand crible dans la théorie analytique des nombres,* astérisque **18**, 1974

BRINITZER, E. [1977], *Eine asymptotische Formel für Summen über die reziproken Werte additiver Funktionen,* Acta Arithm. **32**, 387-391 (1977)

BRUIJN, N. G. de, [1943] *Bijna periodieke multiplicative functies,* Nieuw Archief voor Wiskunde **32**, 81-95 (1943)

CARMICHAEL, R. [1932] *Expansions of arithmetical functions in infinite series.* Proc. London Math. Soc. (2) **34**, 1-26 (1932)

CHANDRASEKHARAN, K. [1970], *Arithmetical Functions,* Berlin - Heidelberg - New York 1970

CODECÀ, P. [1981], *Sul comportamento quasi periodico del resto di una certa classe di funzioni fortemente moltiplicative.* Ann. Univ. Ferrara, Sez. VII Sci. Mat. 27, 229-244 (1981)

CODECÀ, P. & NAIR, M. [1992], *On Elliott's Theorem on Multiplicative Functions,* Proc. Amalfi Conf. on Analytic Number Theory (1989), 17-34, 1992

COHEN, E. [1952], *Rings of arithmetical functions,* Duke Math. J. **19**, 115-129 (1952)

COHEN, E. [1955], *A class of arithmetical functions,* Proc. Nat. Acad. Sci. USA, **41**, 939-944 (1955)

COHEN, E. [1958], *Representations of even functions (mod r),* Duke Math. J. **25**, 4o1-421 (1958)

COHEN, E. [1961], *Fourier expansions of arithmetical functions,* Bull. Amer. Math. Soc. **67**, 145-147 (1961)

COHEN, E. [1961/62], *Almost even functions of finite abelian groups,* Acta Arithm. 7, 311-323 (1961/62)

CORDUNEANU, C. [1968], *Almost Periodic Functions,* New York · London · Sydney · Toronto 1968

CORPUT, VAN DER, J. G. [1939], *Une inégalité relative au nombre des diviseurs,* Proc. Kon. Nederl. Akad. Wetensch. **42**, 547-553, (1939)

DABOUSSI, H. [1975], *Fonctions multiplicatives presque périodiques B.* D'après un travail commune avec Hubert Delange. J. Arithm. Bordeaux (Conf. 1974), Astérisque **24/25**, 321-324 (1975)

DABOUSSI, H. [1979], *On the density of direct factors of the set of positive integers,* J. London Math. Soc. (2) **19**, 21-24 (1979)

DABOUSSI, H. [1980], *Caractérisation des fonctions multiplicatives p.p. B^λ à spectre non vide.* Ann. Inst. Fourier Grenoble **30**, 141-166 (1980)

DABOUSSI, H. [1981], *Sur les fonctions multiplicatives ayant une valeur moyenne non nulle.* Bull. Soc. Math. France **109**, 183-205 (1981)

DABOUSSI, H. [1984], *Sur le Théorème des Nombres Premiers,* C. R.

Acad. Sc. Paris **298**, 161-164 (1984)

DABOUSSI, H. & DELANGE, H. [1974], *Quelques propriétés des fonctions multiplicatives du module au plus égal à 1.* C. R. Acad. Sci. Paris Sér. A **278**, 657-660 (1974)

DABOUSSI, H. & DELANGE, H. [1976], *On a theorem of P. D. T. A. Elliott on multiplicative functions.* J. London Math. Soc. (2) **14**, 345-356 (1976)

DABOUSSI, H. & DELANGE, H. [1982], *On multiplicative arithmetical functions whose modulus does not exceed one.* J. London Math. Soc. (2) **26**, 245-264 (1982)

DABOUSSI, H. & DELANGE, H. [1985], *On a class of multiplicative functions,* Acta Scient. Math. Szeged **49**, 143-149 (1985)

DABOUSSI, H. & INDLEKOFER, K.-H. [1992], *Two elementary proofs of Halász's theorem,* Math. Z. **209**, 43-52 (1992)

DABOUSSI, H. & MENDES FRANCE, M. [1974], *Spectrum, almost periodicity and equidistribution modulo 1,* Studia Scient. Math. Hung. **9**, 173-180, 1974

DAVENPORT, H. [1967], *Multiplicative Number Theory,* Chicago 1967

DELANGE, H. [1954], *Généralisation du théorème de Ikehara,* Ann. Scient. de l'École Norm. Sup. (3) **71**, 213-242 (1954)

DELANGE, H. [1961a], *Sur les fonctions arithmétiques multiplicatives,* Ann. Sci. de l' École Norm. Sup. **78**, 273-304 (1961)

DELANGE, H. [1961b], *Un théorème sur les fonctions arithmétiques multiplicatives et ses applications,* Ann. Sci. de l'École Norm. Sup. **78**, 1-29 (1961)

DELANGE, H. [1963a], *On a class of multiplicative functions,* Scripta Math. **26**, 121-141 (1963)

DELANGE, H. [1963b], *Application de la méthode du crible à l'étude des valeurs moyennes de certaines fonctions arithmétiques,* Séminaire Delange-Pisot, 3e année 1961/62, Paris 1963

DELANGE, H. [1967], *A theorem on multiplicative arithmetic functions,* Proc. Amer. Math. Soc. **18**, 743-749 (1967)

DELANGE, H. [1969], *Sur certaines fonctions additives à valeurs entières,* Acta Arithm. **16**, 195-206 (1969)

DELANGE, H. [1970a], *A remark on multiplicative functions,* Bull. London Math. Soc. **2**, 183-185 (1970)

DELANGE, H. [1970b], *Sur les fonctions multiplicatives de plusieurs entiers,* L'Énseignement Math. **16**, 273–304 (1970)

DELANGE, H. [1971], *Sur des formules de Atle Selberg,* Acta Arithm. **19**, 105–146 (1971)

DELANGE, H. [1972], *Sur les fonctions multiplicatives de module au plus égal à un,* C. R. Acad. Paris **275**, Sér. A, 781–784 (1972)

DELANGE, H. [1974/75], *On finitely distributed additive functions,* J. London Math. Soc. (2) **9**, 483–489 (1974/75)

DELANGE, H. [1975], *Quelques résultats sur les fonctions multiplicatives.* C. R. Acad. Sci. Paris Sect. A **281**, A 997–A 1000 (1975)

DELANGE, H. [1976], *On Ramanujan expansions of certain arithmetical functions,* Acta Arithm. **31**, 259–270 (1976)

DELANGE, H. [1981], *Sur certaines sous-ensembles de $\mathbb{N}^*$ de fonction caractéristique presque-périodique-B.* Journées de Théorie Analytique et Élémentaire des Nombres. 4 pp. Reims 1981

DELANGE, H. [1984], *Generalization of Daboussi's theorem,* Topics in classical number theory, Vol. I, II. Coll. Math. Soc. János Bolyai **34**, Budapest 1981, 305–318 (1984)

DELANGE, H. [1985], *On a theorem of J.-L. Mauclaire on limit-periodic sequences.* Bull. London Math. Soc. **17**, 518–526 (1985)

DELANGE, H. [1987], *On a formula for almost-even arithmetical functions,* Illinois J. Math. **31**, 24–35 (1987)

DELANGE, H. [1988], *Probabilistic Number Theory,* Ramanujan Revisited, Proc. Cent. Conf. Univ. of Illinois at Urbana-Champaign, 1987, Academic Press 1988, 153–165

DELSARTE, J. M. [1945], *Essai sur l'application de la théorie des fonctions presque-périodiques à l'arithmétique,* Ann. Sci. de l'École Norm. Sup. **62**, 185–204 (1945)

DIAMOND, H. G. [1982], *Elementary methods in the study of the distribution of the prime numbers*, Bull. Amer. Math. Soc. 7, 553–589 (1982)

DIAMOND, H. G. & ERDÖS, P. [1981], *On sharp elementary prime number estimates,* L'Enseignement mathém. **26** (1980), 313–321 (1981)

DUNCAN, R. L. [1971], *Some applications of the Turán-Kubilius inequa-*

lity, Proc. Amer. Math. Soc. **30**, 69-72 (1971)

EDWARDS, D. A. [1957], *On absolutely convergent Dirichlet series*, Proc. Amer. Math. Soc. **8**, 1067-1074 (1957)

ELLIOTT, P. D. T. A. [1967, 1968], *On certain additive functions I, II,* Acta Arithm. **12** , 365-384 (1967), ibid. **14**, 51-64 (1968)

ELLIOTT, P. D. T. A. [1970], *The Turán - Kubilius inequality, and a limitation theorem for the large sieve,* Amer. J. Math. **92**, 293-300 (1970)

ELLIOTT, P. D. T. A. [1971], *On the limiting distribution of additive functions (mod 1),* Pacific J. Math. **38**, 49-59 (1971)

ELLIOTT, P. D. T. A. [1974], *On the position of a theorem of Kubilius in Probabilistic Number Theory,* Math. Ann. **209**, 201-209 (1974)

ELLIOTT, P. D. T. A. [1975], *A mean-value theorem for multiplicative functions,* Proc. London Math. Soc. (3) **31**, 418-438 (1975)

ELLIOTT, P. D. T. A. [1977], *The Turán-Kubilius Inequality*, Proc. Amer. Math. Soc. **65**, 8-10 (1977)

ELLIOTT, P. D. T. A. [1979, 1980a], *Probabilistic Number Theory, Vol. I, II.* New York - Heidelberg- Berlin 1979, 1980

ELLIOTT, P. D. T. A. [1980b], *Mean-value theorems for functions bounded in mean α-power, $\alpha > 1$.* J. Austral. Math. Soc. **29**, 177-205 (1980)

ELLIOTT, P. D. T. A. [1980c], *High-power analogues of the Turán-Kubilius inequality, and an application in number theory.* Canadian Math. J. **32**, 893-907 (1980)

ELLIOTT, P. D. T. A. [1987a], *Functional analysis and additive arithmetical functions.* Bull. Amer. Math. Soc. **16**, 179-223 (1987)

ELLIOTT, P. D. T. A. [1987b], *Applications of elementary functional analysis to the study of arithmetic functions,* Colloquia Mathematica Soc. János Bolyai 51, Number Theory, Budapest, 35-43 (1987)

ELLIOTT, P. D. T. A. [1987c], *A local Turán-Kubilius inequality,* Acta Arithm. **49**, 127-139 (1987)

ERDÖS, P. [1946], *On the distribution function of additive functions,* Annals of Math. (2) **47**, 1-20 (1946)

ERDÖS, P. [1947], *Some asymptotic formulas for multiplicative functions,*

Bull. Amer. Math. Soc. **53**, 536–544 (1947)

ERDÖS, P. [1949a], *On a new method in elementary number theory which leads to an elementary proof of the prime number theorem,* Proc. Nat. Acad. Sci. USA **35**, 374–384 (1949)

ERDÖS, P. [1949b], *On a Tauberian theorem connected with the new proof of the prime number theorem,* J. Indian Math. Soc. **13**, 131–144 (1949)

ERDÖS, P. [1952], *On the sum* $\sum_{k=1}^{x} d(f(k))$, J. London Math. Soc. **27**, 7–15 (1952)

ERDÖS,, P. [1957], *Some unsolved problems,* Mich. Math. J. **4**, 291–300 (1957)

ERDÖS,, P. & HALL, R. R. [1980], *On the Möbius function,* J. Reine Angew. Math. **315**, 121–126 (1980)

ERDÖS, P. & MAXSEIN, T. & SMITH, P. R. [1990], *Primzahlpotenzen in rekurrenten Folgen,* Analysis 10, 71–83 (1990)

ERDÖS, P. & RÉNYI, A. [1965], *On the mean value of nonnegative multiplicative number-theoretical functions,* Mich. Math. J. **12**, 321–338 (1965)

ERDÖS, P. & RUSZA, I. & SÁRKÖZY, A. [1973], *On the number of solutions of* f(n) = a *for additive functions,* Acta Arithm. **24**, 1–9 (1973)

ERDÖS, P. & SAFFARI, B. & VAUGHAN, R. C. [1979], *On the asymptotic density of sets of integers, II,* J. London Math. Soc. (2) **19**, 17–20, (1979)

ERDÖS, P. & WINTNER, A. [1939], *Additive arithmetical functions and statistical independence,* Amer. J. Math. **61**, 713–721 (1939)

ERDÖS, P. & WINTNER, A. [1940], *Additive functions and almost periodicity* (B^2), Amer. J. Math. **62**, 635–645 (1940)

ESTERMANN, T. [1952], *Introduction to Modern Prime Number Theory,* Cambridge 1952

GALAMBOS, J. [1970a], *A probabilistic approach to mean values of multiplicative functions,* J. London Math. Soc. (2) 2, 405–419 (1970)

GALAMBOS, J. [1970b], *Distribution of arithmetical functions. A survey,* Ann. Inst. H. Poincaré, Sect. B **6**, 281–305 (1970)

GALAMBOS, J. [1971a], *Distribution of additive and multiplicative func-*

tions, in: The Theory of Arithmetic Functions, Conf. Western Michigan Univ., Springer Lecture Notes 251, 127-139 (1971)

GALAMBOS, J. [1971b], *On the distribution of strongly multiplicative functions,* Bull. London Math. Soc. **3**, 307-312 (1971)

GALAMBOS, J. [1972], *On the asymptotic distribution of values of arithmetical functions,* Accad. Nazionale dei Lincei, Ser. VIII, Vol. **52**, 84-89 (1972)

GALAMBOS, J. [1973], *Approximation of arithmetical functions by additive ones,* Proc. Amer. Math. Soc. **39**, 19-25 (1973)

GALLAGHER, P. X. [1967], *The large sieve,* Mathematika **14**, 14-20 (1967)

GANELIUS, T. H. [1971], *Tauberian Remainder Theorems,* Berlin · Heidelberg · New York 1971

GÁT, G. [1991], *On almost even arithmetical functions via orthonormal systems on Vilenkin groups,* Acta Arithm. **60**, 105-123 (1991)

GELFAND, I. M. [1941], *Normed Rings,* Mat. Sbornik **9**, 3-24 (1941)

HALÁSZ, G. [1968], *Über die Mittelwerte multiplikativer zahlentheoretischer Funktionen,* Acta Math. Sci. Hung. **19**, 365-403 (1968)

HALÁSZ, G. [1969], *Über die Konvergenz multiplikativer zahlentheoretischer Funktionen,* Studia Scient. Math. Hung. **4**, 171-178 (1969)

HALÁSZ, G. [1971], *On the distribution of additive and the mean-values of multiplicative arithmetic functions,* Studia Scient. Math. Hung. **6**, 211-233 (1971)

HALÁSZ, G. [1972], *Remarks to my paper "On the distribution of additive and the mean-values of multiplicative arithmetic functions",* Acta Math. Acad. Sci. Hung. **23**, 425-432 (1972)

HALÁSZ, G. [1974], *On the distribution of additive arithmetic functions,* Acta Arithm. **27**, 143-152 (1974)

HALBERSTAM, H. & RICHERT, H.-E. [1971], *Mean value theorems for a class of arithmetic functions,* Acta Arithm. **18**, 243-256 (1971)

HALBERSTAM, H. & RICHERT, H.-E. [1974], *Sieve Methods,* London, New York 1974

HALBERSTAM, H. & ROTH, K. [1966], *Sequences,* Oxford 1966

HALL, R. R. & TENENBAUM, G. [1988], *Divisors,* Cambridge 1988

HARDY, G. H. [1921], *Note on Ramanujan's trigonometrical function* $c_q(n)$ *and certain series of arithmetical functions.* Proc. Cambridge Phil. Soc. **20**, 263-271 (1921)

HARDY, G. H. [1949], *Divergent Series,* Oxford 1949

HARDY, G. H. & RAMANUJAN, S. [1917], *The normal number of prime factors of an integer,* Quart. J. Math. (Oxford) **48**, 76-92, (1917)

HARDY, G. H. & WRIGHT, E. M. [1956], *An introduction to the theory of numbers,* 3rd edition, Oxford 1956

HARTMANN, P. & WINTNER, A. [1940], *On the almost-periodicity of additive number-theoretical functions,* Amer. J. Math. **62**, 753-758 (1940)

HASSE, H. [1964], *Vorlesungen über Zahlentheorie,* Berlin · Göttingen · Heidelberg · New York 1964

HEPPNER, E. [1973], *Die maximale Ordnung primzahl-unabhängiger multiplikativer Funktionen,* Archiv Math. **24**, 63-66 (1973)

HEPPNER, E. [1980], *Über benachbarte multiplikative zahlentheoretische Funktionen mehrerer Variabler,* Archiv Math. **35**, 454-460 (1980)

HEPPNER, E. [1981], *Über Mittelwerte multiplikativer zahlentheoretischer Funktionen mehrerer Variabler,* Monatsh. Math. **91**, 1-9 (1981)

HEPPNER, E. [1982], *Über Mittelwerte multiplikativer, zahlentheoretischer Funktionen,* Ann. Univ. Sci. Budapest **25**, 85-96 (1982)

HEPPNER, E. [1984], *On the existence of mean-values of multiplicative functions.* Topics in Classical Number theory, Vol. I, II. Coll. Math. Soc. János Bolyai **34**, Budapest 1981, 717-729 (1984)

HEPPNER, E. & MAXSEIN, T. [1985], *Potenzreihen mit multiplikativen Koeffizienten,* Analysis **5**, 87-95 (1985)

HEPPNER, E. & SCHWARZ, W. [1983], *Benachbarte multiplikative Funktionen.* Studies in Pure Mathematics (To the Memory of Paul Turán). Budapest 1983, 323-336

HEWITT, E. & ROSS, K. A. [1963, 1970], *Abstract harmonic analysis, I, II.* Berlin-Heidelberg-New York 1963, 1970

HEWITT, E. & STROMBERG, K. [1965], *Real and abstract analysis,* Berlin-

Heidelberg-New York 1965

HEWITT, E. & WILLIAMSON, J. H. [1957], *Note on absolutely convergent Dirichlet series,* Proc. Amer. Math. Soc. **8**, 863-868 (1957)

HILDEBRAND, A. [1984], *Über die punktweise Konvergenz von Ramanujan-Entwicklungen zahlentheoretischer Funktionen.* Acta Arithm. **44**, 109-140 (1984)

HILDEBRAND, A. [1986], *On Wirsing's mean value theorem for multiplicative functions,* Bull. London Math. Soc. **18**, 147-152 (1986)

HILDEBRAND, A. [1988], *On the number of prime factors of an integer,* Ramanujan Revisited, Proc. Cent. Conf. Univ. of Illinois at Urbana-Champaign, 1987, Academic Press 1988, 167-185

HILDEBRAND, A. & SCHWARZ, W. & SPILKER, J. [1988], *Still another proof of Parseval's equation for almost-even arithmetical functions,* Aequationes Math. **35**, 132-139 (1988)

HILDEBRAND, A. & SPILKER, J. [1980], *Charakterisierung der additiven, fastgeraden Funktionen,* Manuscripta Math. **32**, 213-230 (1980)

HOOLEY, C. [1976], *Applications of sieve methods to the theory of numbers,* Cambridge · London · New York · Melbourne 1976

HUA, LOO KENG [1982], Introduction to Number Theory, Berlin · Heidelberg · New York 1982

HUXLEY, M. N. [1972], *The Distribution of Prime Numbers,* Oxford 1972

HUXLEY, M. N. [1981], *A note on polynomial congruences,* Recent Progress in Analytic Number Theory (Durham 1979), Vol. 1, 193-196, London · New York et al. 1981

HUXLEY, M. N. & WATT, N. [1987], *The Hardy-Littlewood Method for Exponential Sums,* Colloquia Mathematica Soc. János Bolyai 51, Number Theory, Budapest, 173-191 (1987)

INDLEKOFER, K.-H. [1972], *Multiplikative Funktionen mehrerer Variabler,* J. f. Reine u. Angew. Math. **256**, 180-184 (1972)

INDLEKOFER, K.-H. [1976], *On the distribution of values of additive arithmetical functions,* Coll. Math. Soc. J. Bolyai, Debrecen 1974, 111-128 (1976)

INDLEKOFER, K.-H. [1980], *A mean-value theorem for multiplicative functions*, Math. Z. **172**, 255-271 (198o)

INDLEKOFER, K.-H. [1981a], *Remark on a theorem of G. Halász*, Archiv Math. **36**, 145-151 (1981)

INDLEKOFER, K.-H. [1981b], *Some remarks on almost-even and almost-periodic functions*, Archiv Math. **37**, 353-358 (1981)

INDLEKOFER, K.-H. [1981c], *Limiting distributions and mean-values of arithmetical functions*, J. Reine Angew. Math. **328**, 116-127 (1981)

INDLEKOFER, K.-H. [1984], *On multiplicative arithmetical functions.* Topics in Classical Number Theory, Vol. I, II. Coll. Math. Soc. János Bolyai **34**, Budapest 1981, 731-748 (1984)

INDLEKOFER, K.-H. [1986], *Limiting distributions and mean-values of complex-valued multiplicative functions*, Prob. Theory and Math. Stat. Vol. 1, 547-552, VNU Science Press 1986

INDLEKOFER, K.-H. [1992], *Remarks on an elementary proof of Halász's theorem*, Preprint 1992

INGHAM, A. E. [1962], *On absolutely convergent Dirichlet series*, Studies in Mathematical Analysis and Related Topics. Essays in Honor of G. Pólya, Stanford Calif. p. 156-164 (1962)

IVIC, A. [1985], *The Riemann Zeta-Function*, New York et al. 1985

JURKAT, W. & PEYERIMHOFF, A. [1976], *A constructive approach to Kronecker approximation and its application to the Mertens conjecture*, J. Reine Angew. Math. **286/287**, 322-340 (1976)

KAC, M. & KAMPEN, E. R. van & WINTNER, A. [1940], *Ramanujan sums and almost periodic behaviour*, Amer. J. Math. **62**, 107-114 (1940)

KAMPEN, E. R. van [1940], *On uniformly almost periodic multiplicative and additive functions*, Amer. J. Math. **62**, 627-634 (1940)

KAMPEN, E. R. van & WINTNER, A. [1940], *On the almost-periodic behaviour of multiplicative number-theoretical functions*, Amer. J. Math. **62**, 613-626 (1940)

KANOLD, H. J. [1961], *Über periodische multiplikative zahlentheoretische Funktionen*, Math. Ann. **144**, 135-141 (1961)

KANOLD, H. J. [1962], *Über periodische zahlentheoretische Funktionen*, Math. Ann. **147**, 269-274 (1962)

KÁTAI, I. [1968a,b], *On sets characterizing number-theoretical functions,* Acta Arithm. **13**, 315-320, *II (The set of "prime plus one" is a set of quasi-uniqueness),* Acta Arithm. **16**, 1-4 (1968)

KÁTAI, I. [1969], *On the distribution of arithmetical functions,* Acta Math. Acad. Sci. Hung. **20**, 69-87 (1969)

KLUSCH, D. [1986], *Mellin Transforms and Fourier-Ramanujan Expansions,* Math. Z. **193**, 515-526 (1986)

KNOPFMACHER, J. [1975], *Abstract analytic number theory,* Amsterdam/ Oxford 1975

KNOPFMACHER, J. [1976], *Fourier analysis of arithmetical functions.* Annali Mat. Pura Appl. (IV) **109**, 177-201 (1976)

KONINCK, J. M. de & GALAMBOS, J. [1974], *Sums of reciprocals of additive functions,* Acta Arithm. **25**, 159-164 (1974)

KRYŽIUS, Z. [1985a], *Almost even arithmetic functions on semigroups (Russ.).* Litovsk. Mat. Sbornik **25**, No. 2, 90-101 (1985)

KRYŽIUS, Z. [1985b], *Limit periodic arithmetical functions.* Litovsk. Mat. Sbornik **25**, No. 3, 93-103 (1985),

KUBILIUS, J. [1956], *Probabilistic methods in the theory of numbers, (Russ.),* Uspechi Mat. Nauk **11**, 31-66 (1956)

KUBILIUS, J. [1964], *Probabilistic methods in the theory of numbers,* Amer. Math. Soc. Translations of Math. Monographs, Vol. 11, 1-182 (1964)

KUBILIUS, J. [1968], *On local theorems for additive number-theoretic functions,* Abh. aus Zahlentheorie und Analysis zur Erinnerung an Edmund Landau, VEB Dt. Verl. d. Wiss. Berlin 1968, 175-191

KUBILIUS, J. [1975], *On an inequality for additive arithmetic functions,* Acta Arithm. **27**, 371-383 (1975)

KUIPERS, L. & NIEDERREITER, H. [1974], *Uniform Distribution of Sequences,* New York · London et al. 1974

KUNTH, P. [1988], *Einige funktionalanalytische Aspekte in der Theorie der zahlentheoretischen Funktionen,* Dissertation Frankfurt 1988

KUNTH, P. [1989a], *Einige Ergebnisse aus der nicht-multiplikativen* B^q*-Theorie,* Archiv Math. **53**, 373-383 (1989)

KUNTH, P. [1989b], *Der topologische Dualraum des Raums* B^q *der q-fast-*

geraden zahlentheoretischen Funktionen ($0 < q < \infty$), Archiv Math. **53** , 553-564 (1989)

KUNTH, P. [1989c], *Über eine arithmetische Charakterisierung der multiplikativen, fast-geraden Funktionen,* Manuscripta Math. **65**, 275-279 (1989)

KUNTH, P. [1990], *Über die asymptotische Dichte gewisser Teilmengen der natürlichen Zahlen,* Acta Arithm. **55**, 95-106 (1990)

LEITMANN, D. & WOLKE, D. [1976], *Periodische und multiplikative zahlentheoretische Funktionen,* Monatsh. Math. **81**, 279-289 (1976)

LEVIN, B. V. & FAINLEIB, A. S. [1967], *Anwendung gewisser Integralgleichungen auf Fragen der Zahlentheorie, (Russ.),* Uspechi Mat.Nauk **22**, 122-199 (1967)

LEVIN, B. V. & FAINLEIB, A. S. [1970], *Multiplicative functions and probabilistic number theory,* Russian. Izv. Akad. Nauk SSSR, Ser. Mat. **34**, 1064-1109 (1970)

LEVIN, B. V. & TIMOFEEV, N. M. [1970], *Sums of multiplicative functions,* Dokl. Akad. Nauk SSSR **193**, 992-995 (1970)

LEVIN, B. V. & TIMOFEEV, N. M. [1974/75], *On the distribution of values of additive functions,* Acta Arithm. **26**, 333-364 (1974/75)

LOOMIS, L., [1953], *An introduction to abstract harmonic analysis*, Princeton N. J. 1953

LUCHT, L. [1974], *Asymptotische Eigenschaften multiplikativer Funktionen,* J. f. Reine u. Angew. Math. **266**, 200-220 (1974)

LUCHT, L. [1978], *Über benachbarte multiplikative Funktionen*, Archiv Math. **30**, 40-48 (1978)

LUCHT, L. [1979a], *Mittelwerte zahlentheoretischer Funktionen und lineare Kongruenzsysteme,* J. Reine Angew. Math. **306**, 212-220 (1979)

LUCHT, L. [1979b], *Mittelwerte multiplikativer Funktionen auf Linearformen,* Archiv Math. **32**, 349-355 (1979)

LUCHT, L. [1981], *Power series with multiplicative coefficients,* Math. Z. **177**, 359-374 (1981)

LUCHT, L. [1991], *An application of Banach algebra techniques to multiplicative functions*, Preprint 1991

LUCHT, L. & TUTTAS, F. [1979], *Mean-values of multiplicative functions and natural boundaries of power-series with multiplicative coefficients,* J. London Math. Soc. (2) **19**, 25–34 (1979)

LUKACS, E. [1970], *Characteristic functions.* 2nd ed. 350 pp, London 1970

MANSTAVIČIUS, E. [1974], *Application of the method of Dirichlet generating series in the theory of distribution of values of arithmetic functions, (Russ.)* Litovskii Mat. Sbornik **16**, 99–111 (1974)

MAUCLAIRE, J.-L. [1980a,b], *Suites limite-périodiques et théorie des nombres,* II,III, Proc. Japan Acad. Ser. A Math.Sci. **56**, 223–224, **56**, 294–295 (1980)

MAUCLAIRE, J.-L. [1981], *Fonctions arithmétiques et analyse harmonique,* Analytic Number Theory, Proc. Symp. Tokyo 1980, 83–94 (1981)

MAUCLAIRE, J.-L. [1983a,b, 1984], *Suites limite-périodiques et théorie des nombres,* VII, VIII, Proc. Japan Acad. Ser. A Math. Sci. **59**, 26–28, **59**, 164–166 (1983), **60**, 130–133 (1984)

MAUCLAIRE, J.-L. [1986], *Intégration et Théorie des Nombres,* Paris 1986

MAXSEIN, T. [1985], *Charakterisierung gewisser Klassen zahlentheoretischer Funktionen,* Dissertation, Frankfurt 1985

MAXSEIN, T. [1989a], *Potenzreihen mit fastgeraden Koeffizienten,* Acta Math. Hung. **53**, 263–270 (1989)

MAXSEIN, T. [1989b], *Potenzreihen, die einer Wachstumsbedingung genügen und additive Koeffizienten besitzen,* Acta Math. Hung. **54**, 3–7 (1989)

MAXSEIN, T. [1990], *Potenzreihen mit additiven Koeffizienten,* Analysis **10**, 17–21 (1990)

MAXSEIN, T. & SCHWARZ, W. [1992], *On a theorem concerning functions in different spaces of almost-periodic functions,* Proc. Amalfi Conference on Analytic Number Theory (1989), 315–323, (1992)

MAXSEIN, T. & SCHWARZ, W. & SMITH, P. [1991], *An example for Gelfand's theory of commutative Banach algebras,* Math. Slovaca **41**, 299–310 (1991)

MCCARTHY, Paul J. [1986], *Arithmetical Functions,* New York - Berlin

- Tokyo 1986

MOLITOR-BRAUN, CARINE [1991], *La propriété de Wiener,* Séminaire de mathématique de Luxembourg, Trav. Math. **3**, 33-44 (1991)

MONTGOMERY, H. [1971], *Topics in multiplicative number theory,* Berlin - Heidelberg - New York 1971

MONTGOMERY, H. & VAUGHAN, R. C. [1973], *On the large sieve,* Mathematika **20**, 119-134 (1973)

MONTGOMERY, H. & VAUGHAN, R. C. [1977], *Exponential sums with multiplicative coefficients,* Inventiones Math. **43**, 69-82 (1977)

MOTOHASHI, Y. [1970], *An asymptotic formula in the theory of numbers,* Acta Arithm. **16**, 255-264 (1970)

MOTOHASHI, Y. [1973], *On the distribution of the divisor function in arithmetic progressions,* Acta Arithm. **22**, 175-199 (1973)

NAGEL, T. [1919], *Über höhere Kongruenzen nach einer Primzahlpotenz als Modulus,* Norsk Matem. Tidsskr. **1**, 95-98 (1919)

NAGEL, T. [1923], *Zahlentheoretische Notizen I, Ein Beitrag zur Theorie der höheren Kongruenzen,* Vidensk. Skrifter, Ser. I, Mat. Nat.Kl. No. 13, 3-6 (1923)

NAGELL, T. [1951], *Introduction to Number Theory,* New York · Stockholm 1951, reprinted 1964

NAIR, M. [1982], *A new method in elementary prime number theory,* J. London Math. Soc. (2) **25**, 385-391 (1982)

NAIR, M. [1982], *On Chebychev-type inequalities for primes,* Amer. Math. Monthly **89**, 126-129 (1982)

NEUBAUER, G. [1963], *Eine empirische Untersuchung zur Mertens'schen Funktion,* Num. Math. **5**, 1-13 (1963)

NEWMAN, D. J. [1975], *A simple proof of Wiener's $\frac{1}{f}$ - Theorem,* Proc. Amer. Math. Soc. **48**, 264-265 (1975)

NOVOSELOV, E. V. [1964], *A new method in probabilistic number theory,* [Russian], Izv. Akad. Nauk SSSR ser. Mat. **28**, 307-364 (1964)

NOVOSELOV, E. V. [1982], *Introduction to polyadic analysis* (Russian), Petrozavodsk, Gos. Univ. 112 pp (1982)

ODLYZKO, A. M. & TE RIELE, H. J. J. [1985], *Disproof of the Mertens Conjecture,* J. Reine Angew. Math. **357**, 138-160 (1985)

ORE, O. [1921], *Anzahl der Wurzeln höherer Kongruenzen,* Norsk Mat. Tidsskrift, **3**, 63-66 (1921)

ORE, O. [1922], *Über höhere Kongruenzen,* Norsk mat. Forenings Skrifter **1**, Nr. 7, 15pp (1922)

PAUL, E. M. [1962a,b], *Density in the light of probability theory,* I, II, Sankhyā, The Indian Journ. of Statistics, Ser. A **24**, 103-114, 209-212 (1962)

PAUL, E. M. [1963], *Density in the light of probability theory,* III, Sankhyā, The Indian Journ. of Statistics, Ser. A **25**, 273-280 (1963)

PHILLIPS, R. S. [1951], *Spectral theory for semi-groups of linear operators,* Transactions Amer. Math. Soc. **71**, 393-415 (1951)

PÓLYA, G. & SZEGÖ, G. [1925], *Aufgaben und Lehrsätze aus der Analysis, I, II,* Berlin 1925

PRACHAR, K. [1957], *Primzahlverteilung,* Berlin · Göttingen · Heidelberg 1957

RAMANUJAN, S. [1918], *On certain trigonometrical sums and their application in the theory of numbers,* Transactions Cambr. Phil. Soc. **22**, 259-276 (1918); *Collected papers,* 179-199

RÉNYI, A. [1955], *On the density of certain sequences of integers,* Publ. Inst. Math. Acad. Serbe Sci. **8**, 157-162 (1955)

RÉNYI, A. [1965], *A new proof of a theorem of Delange,* Publ. Math. Debrecen **12**, 323-329 (1965)

RÉNYI, A. [1970], *Probability Theory,* Amsterdam · London 1970

RIBENBOIM, P. [1989], *The Book of Prime Number Records,* 2nd edition, Berlin · Heidelberg · New York 1989

RICHERT, H.-E. [1976], *Lectures on Sieve Methods,* Bombay 1976

RIEGER, G. J. [1960], *Ramanujan'sche Summen in algebraischen Zahlkörpern,* Math. Nachrichten **22**, 371-377 (1960)

RIEGER, G. J. [1965], *Zum Teilerproblem von Atle Selberg,* Math. Nachr. **30**, 181-192 (1965)

ROSSER, J. B. & SCHOENFELD, L. [1962], *Approximate formulas for some functions of prime numbers,* Illinois J. Math. **6**, 64-94 (1962)

ROSSER, J. B. & SCHOENFELD, L. [1975], *Sharper bounds for the Chebychev functions $\vartheta(x)$ and $\psi(x)$,* Math. Comp. **29**, 243-269

(1975)

RUBEL, L. & STOLARSKY, K. [1980], *Subseries of the power series for e^x*, Amer. Math. Monthly **87**, 371-376 (1980)

RUDIN, W. [1962], *Fourier analysis on groups*, New York 1962

RUDIN, W. [1966], *Real and complex analysis*, New York 1966

RUDIN, W. [1973], *Functional analysis*, New York 1973

RYAVEC, C. [1970], *A characterization of finitely distributed additive functions*, J. Number Theory **2**, 393-403 (1970)

SAFFARI, B. [1968], *Sur quelques applications de la méthode de l'hyperbole de Dirichlet à la théorie des nombres premiers*, L'Énseignement Mathématique **14**, 205-224 (1968)

SAFFARI, B. [1976a], *Existence de la densité asymptotique pour les facteurs directs de $\mathbb{N}^*$*, C. R. Acad. Sci. Paris **282**, A 255-258 (1976)

SAFFARI, B. [1976b], *On the asymptotic density of sets of integers*, J. London Math. Soc. (2) **13**, 475-485 (1976)

SATHE, L. G. [1953a,b, 1954a,b], *On a problem of Hardy and Littlewood on the distribution of integers having a given number of prime factors, I, II*, J. Indian Math. Soc. **17**, 63-82, 83-141 (1953), *III, IV*, ibid. **18**, 27-42, 43-81 (1954)

SCHOENFELD, L. [1976], *Sharper bounds for the Chebychev functions $\vartheta(x)$ and $\psi(x)$. II*, Math. Comp. **30**, 337-360 (1976)

SCHWARZ, W. [1969], *Einführung in Methoden und Ergebnisse der Primzahltheorie*, Mannheim 1969

SCHWARZ, W. [1973a], *Ramanujan-Entwicklungen stark multiplikativer zahlentheoretischer Funktionen*, Acta Arithm. **22**, 329-338 (1973)

SCHWARZ, W. [1973b], *Ramanujan-Entwicklungen stark multiplikativer Funktionen*, J. f. Reine u. Angew. Math. **262/263**, 66-73 (1973)

SCHWARZ, W. [1973c], *Eine weitere Bemerkung über multiplikative Funktionen*, Colloquium Math. **28**, 81-89 (1973)

SCHWARZ, W. [1974], *Einführung in Siebmethoden der analytischen Zahlentheorie*, Mannheim/Wien/Zürich 1974

SCHWARZ, W. [1976], *Aus der Theorie der zahlentheoretischen Funktionen*, Jber. Deutsche Math.-Verein. **78**, 147-167 (1976)

SCHWARZ, W. [1979a], *Some applications of Elliott's mean-value theorem,* J. Reine Angew. Math. **3o7/3o8**, 418-423 (1979)

SCHWARZ, W. [1979b], *Periodic, multiplicative number-theoretical functions,* Monatsh. Math. **87**, 65-67 (1979)

SCHWARZ, W. [1981], *Fourier-Ramanujan-Entwicklungen zahlentheoretischer Funktionen mit Anwendungen,* Festschrift Wiss. Ges. Univ. Frankfurt, 399-415, Wiesbaden 1981

SCHWARZ, W. [1985], *Remarks on the theorem of Elliott and Daboussi, and applications,* Elementary and analytic theory of numbers (Warsaw 1982), 463-498, Banach Senter Publ.**17**, PWN, Warsaw 1985

SCHWARZ, W. [1986], *A correction to: "Remarks on Elliott's theorem on mean-values of multiplicative functions" (Durham, 1979) and some remarks on almost-even number-theretical functions*, Analytic and elementary number theory, Marseille 1983, 139-158, Univ.Paris XI, Orsay 1986

SCHWARZ, W. [1987a], *Einführung in die Zahlentheorie,* 2nd edition, Darmstadt 1987

SCHWARZ, W. [1987b], *Almost-even number-theoretical functions,* Probability theory and mathematical statistics, Vol. II (Vilnius 1985), 581-587, Utrecht 1987

SCHWARZ, W. [1988], *Ramanujan expansions of arithmetical functions,* Ramanujan Revisited, Proc. Cent. Conf. Univ. of Illinois at Urbana-Champaign, 1987, Academic Press Boston 1988, 187-214

SCHWARZ, W. & SPILKER, J. [1971], *Eine Anwendung des Approximationssatzes von Weierstraß-Stone auf Ramanujan-Summen.* Nieuw Archief voor Wisk. (3) **19**, 198-209 (1971)

SCHWARZ, W. & SPILKER, J. [1974], *Mean values and Ramanujan expansions of almost even arithmetical functions,* Coll. Math. Soc. J. Bolyai **13**. Topics in Number Theory, 315-357, Debrecen 1974

SCHWARZ, W. & SPILKER, J. [1979], *Wiener-Lévy-Sätze für absolut konvergente Reihen,* Archiv Math. **32**, 267-275 (1979)

SCHWARZ, W. & SPILKER, J. [1981], *Remarks on Elliott' s theorem on mean-values of multiplicative functions.* Recent Progress in Analytic Number Theory, Durham 1979, 325–339, London 1981

SCHWARZ, W. & SPILKER, J. [1983], *Eine Bemerkung zur Charakterisierung der fastperiodischen multiplikativen zahlentheoretischen Funktionen mit von Null verschiedenem Mittelwert.* Analysis **3**, 205–216 (1983)

SCHWARZ, W. & SPILKER, J. [1986], *A variant of proof of Daboussi's theorem on the characterization of multiplicative functions with non-void Fourier-Bohr spectrum.* Analysis **6**, 237–249 (1986)

SELBERG, A. [1949], *An elementary proof of the prime number theorem,* Ann. of Math. (2) **50**, 305–313 (1949)

SELBERG, A. [1950], *An elementary proof of the prime number theorem for arithmetic progressions,* Canad. J. Math. **2**, 66–78 (1950)

SELBERG, A. [1954], *Note on a paper by L. G. Sathe,* J. Indian Math. Soc. **18**, 83–87 (1954)

SIEBERT, H. & WOLKE, D. [1971], *Über einige Analoga zum Bombierischen Primzahlsatz,* Math. Z. **122**, 327–341 (1971)

SIVARAMAKRISHNAN, R. [1989], *Classical Theory of Arithmetic Functions,* 1989

SPILKER, J. [1979], *A simple proof of an analogue of Wiener's 1/f Theorem,* Archiv Math. **32**, 265–266 (1979)

SPILKER, J. [1980], *Ramanujan expansions of bounded arithmetical functions,* Archiv Math. **35**, 451–453 (1980)

SPILKER, J. & SCHWARZ, W. [1979], *Wiener-Levy-Sätze für absolut konvergente Reihen,* Archiv Math. **32**, 267.275 (1979)

TE RIELE, H. J. J. [1985], *Some historical and other notes about the Mertens conjecture, and its disproof,* Nieuw Archief voor Wiskunde (4) **3**, 237–243 (1985)

TITCHMARSH, E. C. [1951], *The Theory of the Riemann Zeta-Function,* Oxford 1951

TULYAGANOV, S. T. [1991], *Comparison of Sums of Multiplicative Functions,* Soviet Math. Dokl. **42**, 829–833 (1991); Russian original 1990

TURÁN, P. [1934], *On a theorem of Hardy and Ramanujan,* J. London Math. Soc. **9**, 274–276, 1934

TURÁN, P [1936], *Über einige Verallgemeinerungen eines Satzes von Hardy und Ramanujan,* J. London Math. Soc. **11**, 125-133 (1936)

TURÁN, P. [1984], *On a New Method of Analysis and its Applications,* New York · Chichester · et al. 1984

TUTTAS, F. [1980], *Über die Entwicklung multiplikativer Funktionen nach Ramanujan-Summen.* Acta Arithm. **36**, 257-270 (1980)

UŠDAVINIS, P. [1967], *An analogue of the theorem of Erdös-Wintner for the sequence of polynomials with integer coefficients (Russ.),* Litovskii Mat. Sbornik 7, 329-338 (1967)

VAUGHAN, R. C. [1977], *On the estimation of trigonometrical sums over primes, and related questions,* Institut Mittag-Leffler, Report No. 9, 1977

WAIBEL, M. [1985], *Charakterisierung (exponentiell) multiplikativer $\mathcal{B}^q$-fastgerader Funktionen mit nichtverschwindendem Mittelwert,* Diplomarbeit Freiburg 1985

WARLIMONT, R. [1983], *Ramanujan expansions of multiplicative functions,* Acta Arithm. **42**, 111-120 (1983)

WEYL, H. [1916], *Über die Gleichverteilung von Zahlen modulo Eins,* Math. Ann. 77, 313-352 (1916)

WIDDER, D. V. [1946], *The Laplace Transform,* Princeton 1946

WINTNER, A. [1942], *On a statistics of the Ramanujan sum,* Amer. J. Math. **64**, 106-114 (1942)

WINTNER, A. [1943], *Eratosthenian Averages,* Baltimore 1943

WINTNER, A. [1944], *The Theory of Measure in Arithmetical Semigroups,* Baltimore 1944

WIRSING, E. [1956], *Über die Zahlen, deren Primteiler einer gegebenen Menge angehören,* Archiv Math. 7, 263-272 (1956)

WIRSING, E. [1961], *Das asymptotische Verhalten von Summen über multiplikative Funktionen,* Math. Annalen **143**, 75-102 (1961)

WIRSING, E. [1967], *Das asymptotische Verhalten von Summen über multiplikative Funktionen, II,* Acta Math. Acad. Sci. Hung. **18**, 414-467 (1967)

WIRSING, E. [1981], *Additive and completely additive functions with restricted growth,* Recent Progress in Analytic Number Theory, Vol. II (Durham 1979), p.231-280 (1981)

WOLKE, D. [1971a], *Über das summatorische Verhalten zahlentheoretischer Funktionen,* Math. Ann. **194**, 147–166 (1971)

WOLKE, D. [1971b], *Multiplikative Funktionen auf schnell wachsenden Folgen,* J. Reine Angew. Math. **251**, 54–67 (1971)

WOLKE, D. [1972], *A new proof of a theorem of van der Corput,* J. London Math. Soc. (2) **5**, 609–612 (1972)

WOLKE, D. [1973], *Über die mittlere Verteilung der Werte zahlentheoretischer Funktionen auf Restklassen, I,* Math. Ann. **202**, 1–25 (1973)

Author Index

Photographs

Subject Index

P. Elliott, R. Vaughan, A. Ivić, J. Mozzochi

E. Wirsing

A. Schinzel, P. Erdös, E. Wirsing

A. Hildebrand

I. Kátai, K.-H. Indlekofer

H. L. Montgomery

A. Schinzel, P. Erdös, Freĭman

P. Elliott, G. Tenenbaum

M. Mendes France, A. Laurinčikas, G. Stépanauskas

Š. Porubský, K. Ramachandra, J. Kubilius

H.-E. Richert, U. Vorhauer

E. Wirsing, J. Pintz, A. Schinzel

Acknowledgements

The photographs reproduced in this book mainly show mathematicians having worked in the fields of Arithmetical Functions, Prime Number Theory or Sieve Theory. In the authors' opinion, photographs of mathematicians ought to be published more often. Unfortunately, in this book, there are several omissions of photographs, but the authors were, for various reasons, unable to obtain photographs of some mathematicians they wanted for publication in this book.

For help with the photographs reproduced here and for permission to publish these the authors are grateful to

- the Master and Fellows and the Librarian of Trinity College, Cambridge, England,[1] and Prof. J. W. S. CASSELS for his kind help,
- The Ferdinand Hamburger Jr. Archives (The John Hopkins University, Baltimore) and Greystone Studies,[2]
- ULRIKE VORHAUER, Ulm,
- the collection of photos at the *Mathematisches Forschungsinstitut Oberwolfach*, and to Prof. K. JACOBS, Erlangen,
- Prof. S. J. PATTERSON, Göttingen, and Dr. H. ROHLFING, Niedersächsische Staats- und Universitätsbibliothek, Göttingen,[3]
- Prof. C. J. MOZZOCHI,
- Dr. U. KOHLENBACH, Frankfurt,
- several mathematicians kindly cooperating by sending photos.

Some photographs were taken by the first author.

WOLFGANG SCHWARZ, Frankfurt am Main

JÜRGEN SPILKER, Freiburg im Breisgau

---

[1]Photographs of DAVENPORT, HARDY, LITTLEWOOD, RAMANUJAN.

[2]Photograph of A. WINTNER.

[3]Photograph of C. L. SIEGEL.

Printed in the United States
By Bookmasters